Teach (Test) yourself...
Introduction to Calculus

Theory and Tests with
Step by Step Solutions

Marcel Sincraian, Ph.D.

ISBN: 978-1-7775022-7-0   Electronic Book

ISBN: 978-1-7775022-6-3   Printed Book

Marcel Sincraian Email: msincraian@yahoo.ca

To:

My dad

Acknowledgement

I would like to thank my daughter Stefany, for proofreading and editing all my books, and for all her valuable suggestions.

Other books by Marcel Sincraian, P.hD.

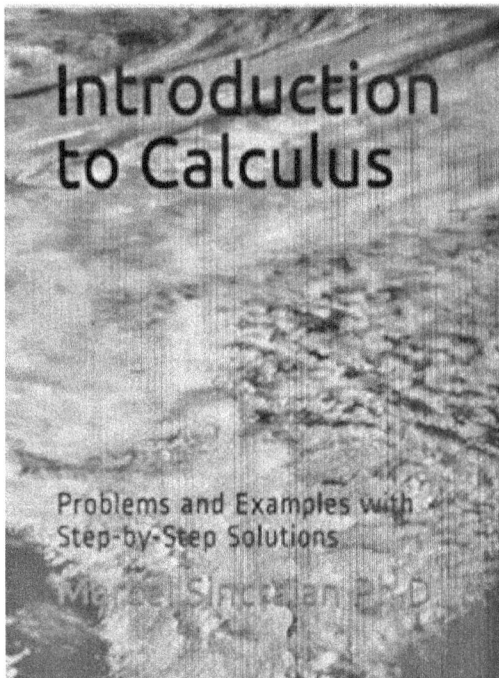

Content

Author's Notes

This book provides students with a tool to improve their knowledge in preparation to learning Calculus. This book is intended to be a good review of the introduction to Calculus. The book follows the main precalculus concepts needed in order to understand Calculus. It starts with the Cartesian System, Geometry and Trigonometric ratios, Algebra. From introduction to powers, and logarithms to polynomials, graphs of linear equations and quadratic equations. Then it follows Calculus I curriculum for high schools. Chapter 5 is a review of Functions. Chapter 6 is about Limits; Chapter 7 deals with Differentiation, and Chapter 8 is all about Integrals.

Before each set of TESTS, a short review of the main theoretical concepts will be presented. There are examples given to help better understand and review the concepts. The TESTS apply the theory and concepts reviewed.

How to use this book?

After the number of the question, for example **27.**, each question starts with a set of numbers and letters: for example:

1B.a. 1). Here;

1 – the chapter number

B – Subchapter

a - section a of subchapter B

1) – question 1 from chapter 1, Subchapter B, section a.

Next the structure of tests, the chapters are taken from and the number of questions on each test is presented.

TEST #	Chapter #	Number of Questions	TEST #	Chapter #	Number of Questions
1	1	10	13	5-8	32
2	1	22	14	6-8	30
3	1-2	30	15	5-8	30
4	1-4	27	16	6-8	34
5	1-4	29	17	6-8	34
6	1-5	29	18	4, 6-8	31
7	3-6	28	19	1-5, 7	30
8	4-7	31	20	1-5, 8	30
9	4-7	30	21	1-4, 8	30

10	5~7	30	22	1-4, 8	30
11	5-7	30	23	2-6, 8	30
12	5-8	32	24	4~7	30

At the end, after all the tests, the FULL SOLUTIONS or step by step solutions are presented for each chapter. Each Chapter has the problems in solved in order. If we want to check the solution for question **17.** 4D.a. 9) in TEST 22, we will go to FULL SOLUTIONS, Chapter 4 Subchapter B, section a and check the solution from question 9.

$3 \times (7 - 4)$ is equal with $3 \times 7 - 3 \times 4$

$3 \times (7 - 4) = 3 \times 3 = 9$

$3 \times 7 - 3 \times 4 = 21 - 12 = 9$

CHAPTER 1

CARTESIAN SYSTEM

1.A Introduction

Remember when you learned about the number line?

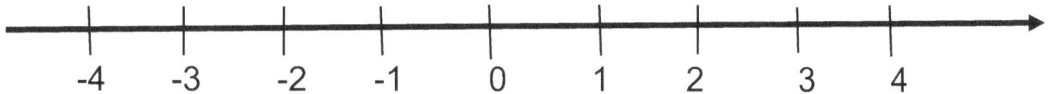

Each number is a certain distance from the origin 0. Either to the left (negative numbers) or to the right (positive numbers).

In the 17th century, Rene Descartes came up with the idea of using two number lines, one horizontal, and the other vertical. From here, each point in the 2-dimensional plane received a pair of ordered coordinates, x for the horizontal line, and y for the vertical line. (x , y)

This is called; the Cartesian System of axes.

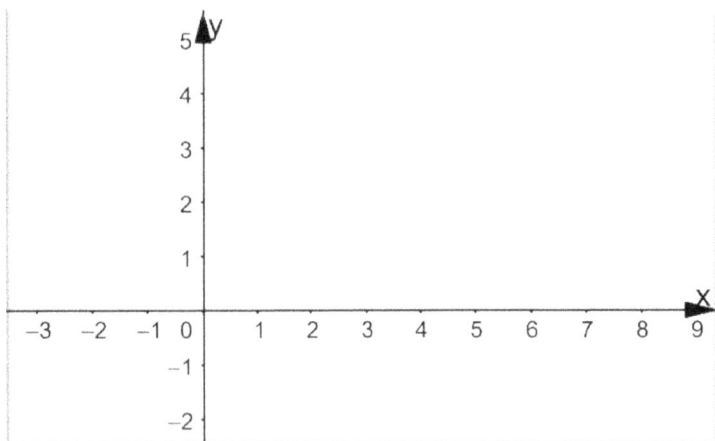

EXAMPLE

If we want to represent any points situated anywhere in the 2-dimensional plane, we will have to give the points the coordinates x and y.

(x, y)

The intersection of the axes is called the <u>system origin</u> or center. The origin has the coordinates zero and zero. (0,0)

NOTE:

<u>Make sure to always write the coordinates in the same order, first the x coordinate then the y coordinate.</u>

The system has four quadrants.

Quadrant 1 has both coordinates positive.

Quadrant 2 has x negative and y positive.

Quadrant 3 has x and y coordinates negative.

Quadrant 4 has x positive and y negative.

Q2	Q1
Q3	Q4

Let's represent few points in the Cartesian system.

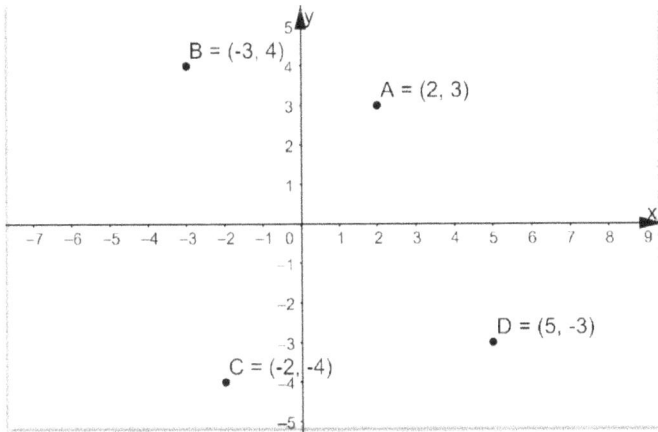

EXAMPLE

Represent the following points.

A(2,3), B(-3,4), C(-2,-4), D(5,-3)

As you can see, point A has an x coordinate of 2, and a y coordinate of 3. Point D, for example, has an x coordinate of 5, and a y coordinate of negative 3.

PRACTICE

Represent the following points on the graph below.

A (1,1)	B (3,5)
C (-2,4)	D (-3,1)
E (-3,-5)	F (-4,0)
G (4,-2)	H (3,-3)

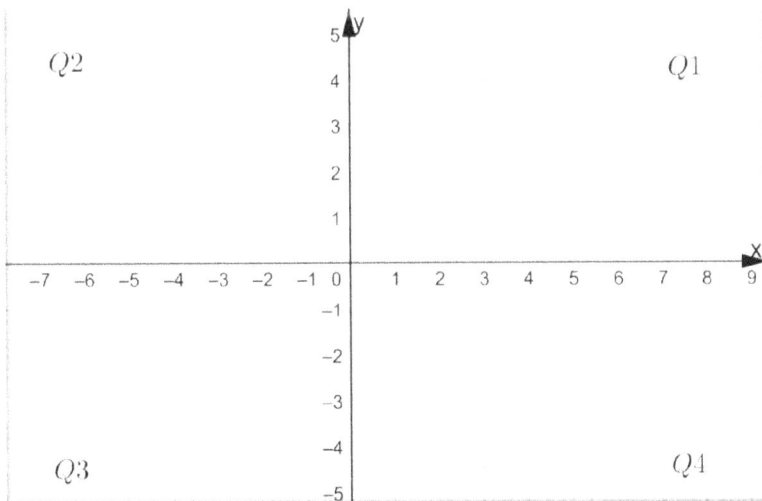

In which quadrants are the points below situated?

A (1,1)	B (3,5)
C (-2,4)	D (-3,1)
E (-3,-5)	F (-4,0)

3

1.B Distance between points

a. Horizontal distance

Let's suppose we have the points A (1,3) and B (5,3) As you can see, the y coordinate is the same. If we represent these points on the cartesian system, we get a horizontal segment that belong to a horizontal line.

The distance between point A and point will be 5 – 1 = 4 units.

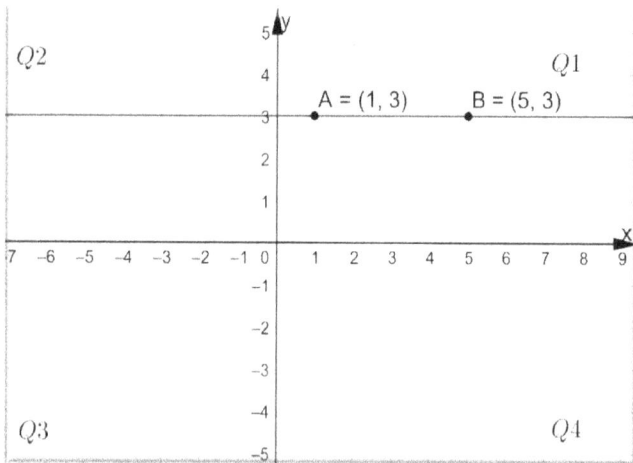

Absolute value

Remember that the <u>absolute value</u> is always a positive number. We can think about the absolute value as the distance between the origin and that particular point, wherever the point is.

EXAMPLE

The absolute value of -7 is 7
The absolute value of -3.56 is 3.56
The notation for showing the absolute value is the number written between two vertical lines like below.
$|-7| = 7$ or $|-3.56| = 3.56$

If we are a bit more general, and consider Point A as point 1 and Point B as point 2, the coordinates of these two points will be written as:
A (x_1, y_1) and B (x_2, y_2)
The horizontal distance between A and B will be $|x_2 - x_1| = |5 - 1| = 4$
$|x_2 - x_1|$ represents the absolute value of the difference between the x coordinates of the points.
What happens when one of the x coordinates is negative?

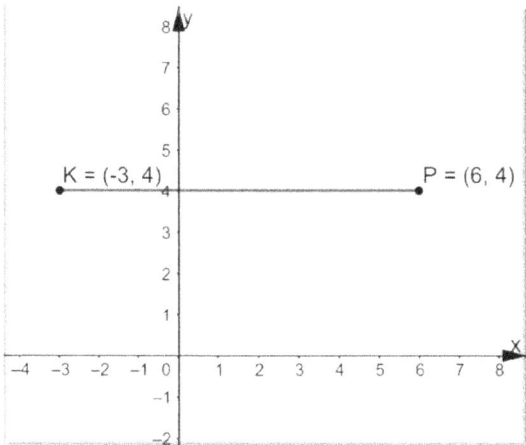

EXAMPLE

Let's calculate the horizontal distance between the points K (-3,4) and P (6, 4)

K (x_1, y_1) and P (x_2, y_2)

The horizontal distance between K and P will be $|x_2 - x_1| = |-3 - 6| = 9$

1.B Distance between points

b. Vertical distance

Let's suppose we have the points A (-3,1) and B (-3,6) As you can see, their x coordinates are the same. If we represent these points on the cartesian system we get a vertical segment that belongs to a vertical line.

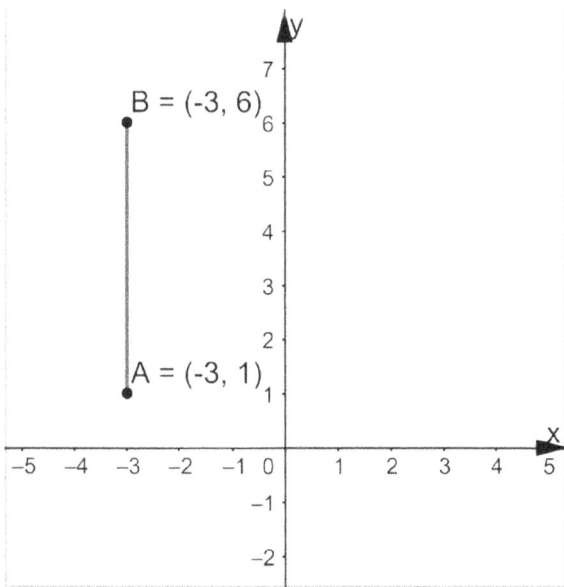

The vertical distance between point A and point B will be 6 – 1 = 5 units.

If we are a bit more general, and consider Point A as point 1 and Point B as point 2, the coordinates of these two points will be written as:

A (x_1, y_1) and B (x_2, y_2)

The vertical distance between A and B will be $|y_2 - y_1| = |6 - 1| = 5$

$|y_2 - y_1|$ represents the absolute value of the difference between the y coordinates of the points.

What happens when one of the y coordinates is negative?

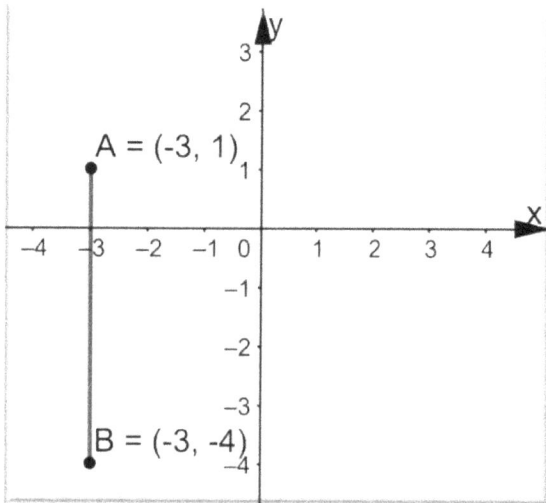

As you can see here, the y coordinates are 1 and -4 respectively. The vertical distance between Point A and B will be:

$$|y_2 - y_1| = |1 - (-4)| = 5$$

1.B Distance between points

c. Non horizontal and Non vertical distance

What happens when the points are in a straight line tilted from the lower left to the upper right? What is the distance in this case?

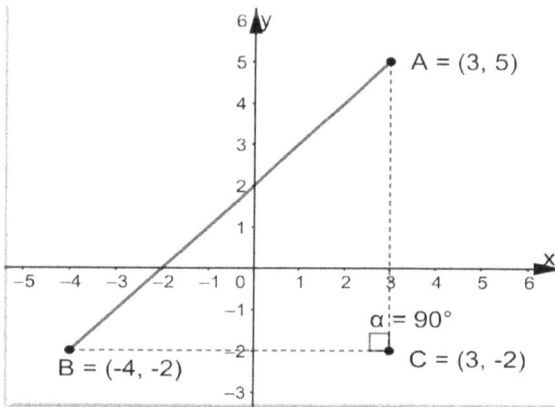

We want to calculate the distance between point A and B. the easiest way is to create a right-angle triangle as follows:

We draw a horizontal line from B to C that will intersect the vertical line from A to C.

The angle in the triangle ABC is a right-angle (90^0).

In this triangle, the distance between A and B represents the hypotenuse.

As we remember, the Pythagorean theorem tells us that in a right-angle triangle, the hypotenuse squared equals the sum of one side squared plus the other side of the triangle squared.

Here we have:

$AB^2 = BC^2 + AC^2$

BC is a horizontal segment; so, distance BC = $|x_2 - x_1| = |-4 - 3| = |-7| = 7$

AC is a vertical segment: so, distance AC = $|y_2 - y_1| = |5 - (-2)| = |7| = 7$

Then AB = $\sqrt{BC^2 + AC^2} = \sqrt{7^2 + 7^2} = \sqrt{98} = 9.89$

1.B Distance between points

d. Midpoint coordinates

If we have two points $A(x_1, y_1)$ *and* $B(x_2, y_2)$, we need to find the coordinates of the point $C(x, y)$ situated at the halfway point between A and B.

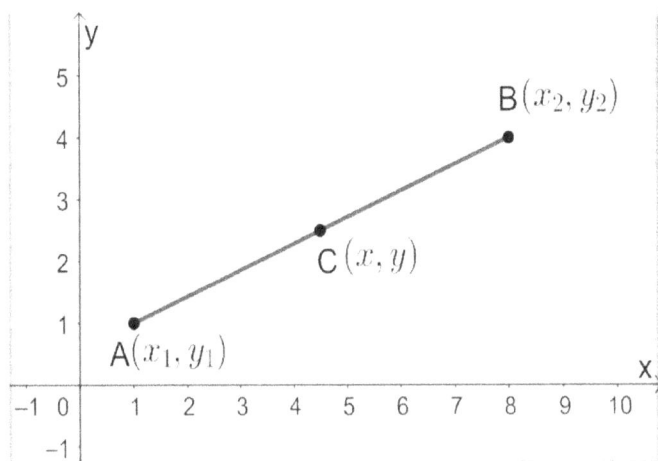

Because the distance between A and C equals the distance between C and B, to //find the x coordinate, we can write:

$x - x_1 = x_2 - x$

$2x = x_2 + x_1$

$x = \frac{x_2 + x_1}{2}$

We do the same calculation on the y axis and have the formula of the y coordinate for the midpoint.

$y - y_1 = y_2 - y$

$2y = y_2 + y_1$

$y = \frac{y_2 + y_1}{2}$

EXAMPLE

Let's suppose the coordinates of A and B are shown in the figure below:

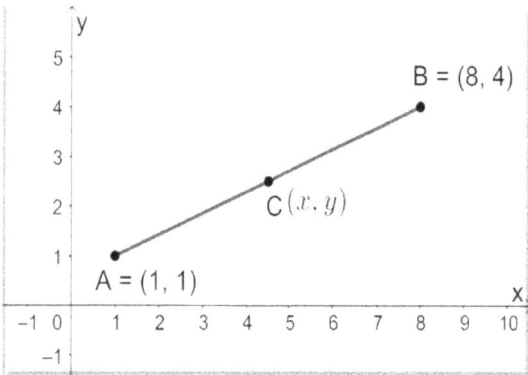

As we saw above, applying the formulas for the x and y coordinates of the midpoint C, we have:

$$x = \frac{x_2+x_1}{2} = \frac{8+1}{2} = \frac{9}{2} = 4.5$$

$$y = \frac{y_2+y_1}{2} = \frac{4+1}{2} = \frac{5}{2} = 2.5$$

So,

The midpoint coordinates are (4.5, 2.5)

1.C slope of a line

A slope of a line is the ratio between the "rise" of the line and the "run" of the line.

The "rise" is the vertical distance or the difference between the y coordinates of any two points situated on the line

The "run" is the horizontal distance or the difference between the x coordinates of any two points situated on the line.

Let's say we want to find the slope of the line that passes through points P (x_1, y_1) and R (x_2, y_2).

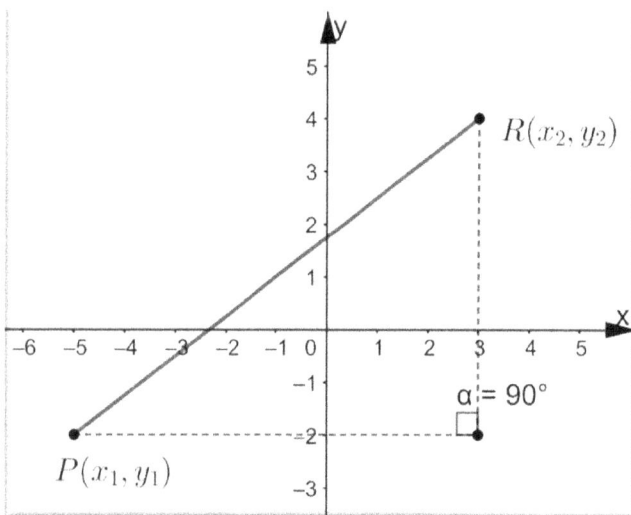

$$Slope = \frac{Rise}{Run} = \frac{Vertical\ distance}{Horizontal\ distance} =$$

$$\frac{y_2-y_1}{x_2-x_1} = m$$

This ratio can be interpreted as showing how fast the vertical distance increases as the horizontal distance increases.

EXAMPLE

We have a line that passes through points A (3,4) and B (-5,-2)

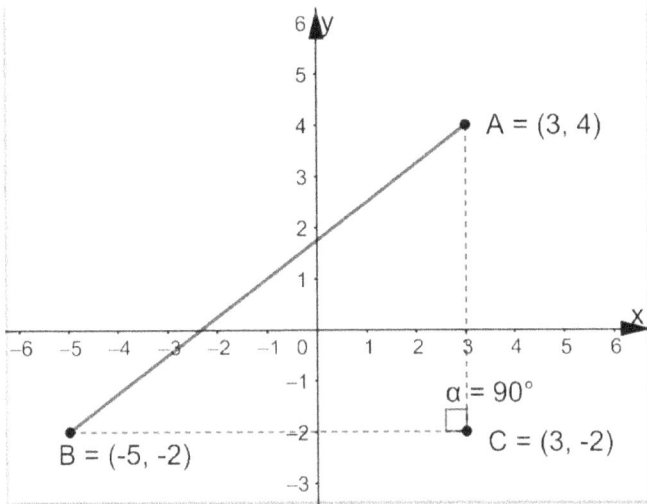

Let's suppose that B is point 2 and A is point 1

$$Slope = m = \frac{Rise}{Run} =$$

$$\frac{Vertical\ distance}{Horizontal\ distance} = \frac{y_2 - y_1}{x_2 - x_1} = \frac{-2-4}{-5-3} =$$

$$\frac{-6}{-8} = \frac{6}{8} = \frac{3}{4}$$

This ratio can be interpreted as showing that the vertical distance increases by 3 units as the horizontal distance increases by 4 units.

Which will be the values of y as x increases from -5 to 3.

Using the relation for the slope that we just calculated, for x_2=-4 we get:

$$\frac{3}{4} = \frac{y_2 - y_1}{x_2 - x_1} = \frac{y_2 - (-2)}{-4 - (-5)} = \frac{y_2 + 2}{1}$$

So,

$\frac{3}{4} = \frac{y_2 + 2}{1}$ using the cross-multiplication property

$3 = 4(y_2 + 2)$

$3 = 4y_2 + 8$

$3 - 8 = 4y_2$

$-5 = 4y_2$

$y_2 = \frac{-5}{4} = -1.25$

EXAMPLE

In this same way, we can calculate the values for y for x increasing by 1 unit starting with   -5.

X	-5	-4	-3	-2	-1	0	1	2	3
Y	-2	-1.25	-0.5	0.25	1	1.75	2.5	3.25	4

As you can see, as the values of x increase, the values of y increase as well with the same ratio $\frac{3}{4}$.

Any increase of an unknown number or variable (for example x) is symbolized Δx, where Δ is the Greek letter Delta.

So, the slope can be written as how fast the values of y change as the values of x simultaneously change.

$slope = \frac{\Delta y}{\Delta x}$ or can be thought of as the <u>average rate of change</u>.

1.D Equation of a straight line

a. Non-vertical and non-horizontal line

You have seen in 1C that the slope of a straight line is calculated with the formula:

$Slope = m = \frac{Rise}{Run} = \frac{Vertical\ distance}{Horizontal\ distance} = \frac{y_2 - y_1}{x_2 - x_1}$

So, if we have a point on a straight-line $M(x_1, y_1)$, then the slope of the line that passes through point M and any other point $K(x, y)$ will be written as:

$m = \frac{y - y_1}{x - x_1}$

So, if we isolate y, we get:

$y - y_1 = m(x - x_1)$ **(Called point-slope form)**

$y = y_1 + mx - mx_1$

So,

$y = mx + y_1 - mx_1$

But, $y_1 - mx_1$ is a constant (b)

So, we have the equation of a straight line

$y = mx + b$ **(Called general form)**

Where:

m is the slope

b is the intersection of the line with y axis. It is called the <u>y-intercept</u>

$y - y_1 = m(x - x_1)$ is called the <u>point slope equation</u>.

EXAMPLE

Find the equation of a line that passes through the point M (2,3) and has a slope m=3

We start with the point slope equation;

$$y - y_1 = m(x - x_1)$$

We plug the coordinates of point M and the slope.

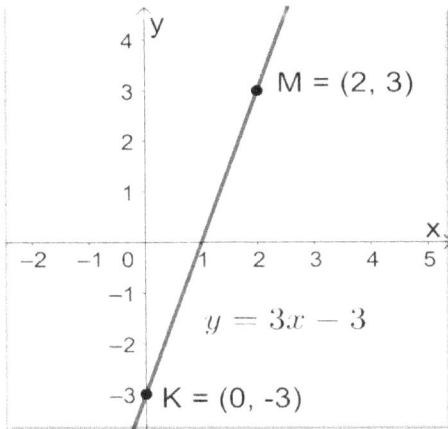

$$y - 3 = 3(x - 2)$$
$$y = 3 + 3x - 6$$
$$\mathbf{y = 3x - 3}$$

Here, the slope is 3 and the intersection with the y axis is the point K (0,-3)

Remember that the intersection of the line with the x axis will have a y coordinate equal to zero.

This is called the <u>x-intercept</u>

1.D Equation of a straight line

b. Vertical and horizontal lines

You have seen in 1C that the slope of a straight line is calculated with the formula:

$$Slope = m = \frac{Rise}{Run} = \frac{Vertical\ distance}{Horizontal\ distance} = \frac{y_2 - y_1}{x_2 - x_1}$$

So, if we have a point on a straight-line M(x_1, y_1), then the slope of the line that passes through point M and any other point K(x, y) will be written as:

$$m = \frac{y - y_1}{x - x_1}$$

Vertical line

The equation for a vertical line is $x = constant$

This means that for any values of y in the formula $m = \frac{y - y_1}{x - x_1}$, the x coordinates will always be the same. But, if x is the same, the denominator of the slope formula will become zero. For the constant x, the <u>slope m is infinite, or undefined</u>.

Horizontal line

The equation for a horizontal line is $y = constant$

In this case, the y coordinates of any point on the horizontal line will be the same. From here we can infer that the difference $y - y_1$ will be zero.

For the constant y, the <u>slope m is equal to zero</u>.

EXAMPLE

Find the slope of a line with the equation y = 3

Let's say we have point one: N (5,3) and point two: M (9,3) situated on the line. Next, let's calculate the slope.

$$m = \frac{y_2 - y_1}{x_2 - x_1} = \frac{3-3}{9-5} = \frac{0}{4} = 0$$

1.E Parallel and perpendicular lines

a. Parallel lines

Remember that the formula for a straight line is:

$y = mx + b$

In this context, two lines are parallels when they have the same slope (variable m). However, the y intercept of the two lines will be different.

EXAMPLE

Graph the following straight lines.

$y = 2x + 1$

$y = 2x + 4$

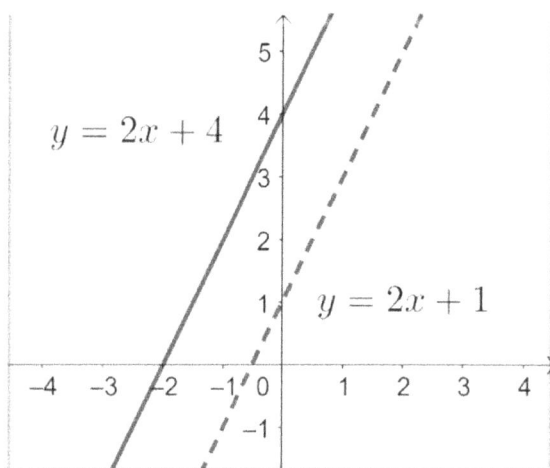

As you can see in the graph, the equations have the same slope, in this case, 2. They also have different y intercepts, in this case, 1 and 4 respectively.

Question:
What happens when the slopes are the same, and the y intercepts are the same as well?

EXAMPLE

We have the lines.

$y_1 = 2x + 1$

$y_2 = 2x + 1$

We can see that these lines y_1 and y_2 are identical. Indeed, if we represent them on the cartesian system, it will be exactly the same line. We will have only one line.

1.E Parallel and perpendicular lines

b. Perpendicular lines

Remember that, for two lines that are not vertical, they are perpendicular (have an angle of 90^0 between them) if the product of their slopes is minus 1.

This means that, if the first line's slope is m_1, and the second line's slope is m_2, then these lines are perpendicular if $m_1 \times m_2 = -1, or\ m_1 = \frac{-1}{m_2}$

EXAMPLE

We have the line given by the equation:

$y_1 = 2x + 1$

Show that the line given by the equation:

$y_2 = -0.5x + 4$

Is perpendicular to the first line.

So,

$m_1 = 2, and\ m_2 = -0.5$

Then,

$m_1 \times m_2 = 2 \times (-0.5) = -1$

From here we see that y_1 and y_2 are perpendicular lines.

These two lines are represented in the graph below.

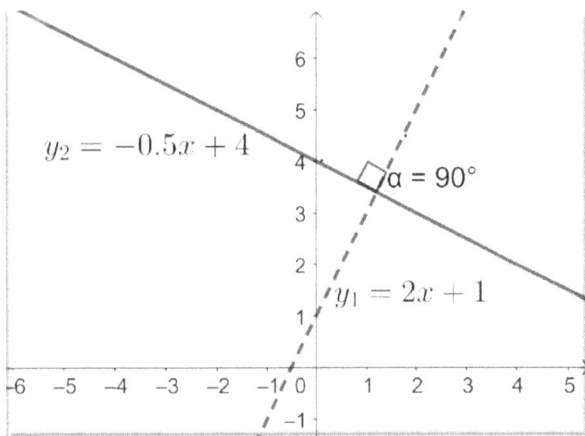

1.F Symmetry

a. About a line

We know that two points are symmetric to a line "L", if they are part of a line "K" that is perpendicular to the line "L", and they are at the same distance from line "L" each point on the other side of line "L" compared with the other point (like in a mirror).

Let's suppose we have a vertical line (y axis), and we want to find the symmetric point on the y axis to the point M (x, y). The horizontal distance from this point to the y axis is x. The same distance x from the y axis on the other side will be the point N $(-x, y)$.

EXAMPLE

Represent the symmetric point of M (2,3) by the y axis. Let's call this symmetric point N.

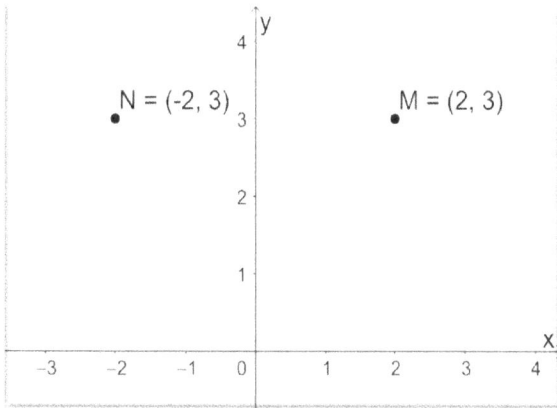

As you can see, the symmetric point N has the same y coordinate, but the x is the same only in value. The sign is negative (-).

That tells us that the distance between these two points and the y axis is the same. They are at the same distance from the y axis, just on opposite sides of it.

Let's suppose we have a horizontal line (x axis), and we want to find the symmetric point on the x axis to the point M (x, y). The vertical distance from this point to the x axis is y. The same distance y from the x axis on the other side will be the point N $(x, -y)$.

EXAMPLE

Represent the symmetric point of P (1,-2) by x axis. Let's call this symmetric point R.

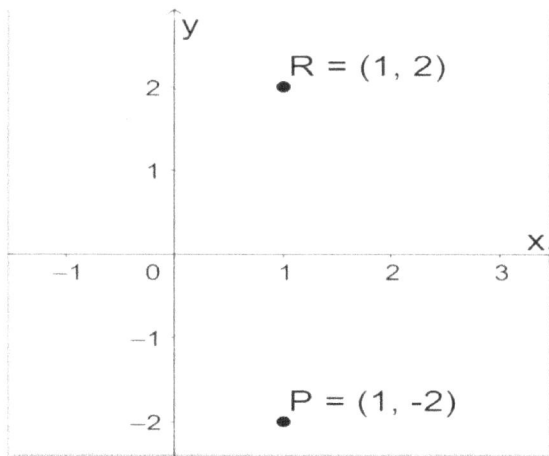

As we can see, the symmetric point R has the same x coordinate but the y is the same only in value. The sign is positive (+).

That tells us that the distance between these two points and the x axis is the same. They are at the same distance from the x axis.

Let's suppose we have a vertical line (x=1), and we want to find the symmetric point on the y axis to the point M (x, y). The horizontal distance from this point to x=1 axis is x+1. The same distance x from the y axis on the other side will be the point N $(x + 1 - x, y)$.

EXAMPLE

Represent the symmetric point of Q (3,-2) by the x=1 line. Let's call this symmetric point S.

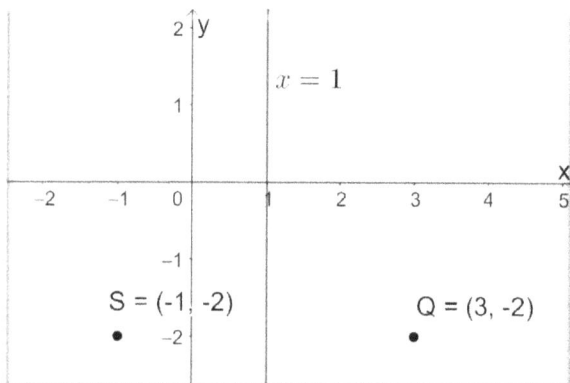

As we can see, the symmetric point S has the same y coordinate but the x coordinate is the same distance of 2 units to the left from the line x=1. Point Q has the x coordinate two units to the right of line x=1, as x=1+2=3

The symmetric point S, has the x coordinate two units to the left of line x=1, as x=1-2=-1.

1.F Symmetry

b. About a point

Let's see what happens when we want to find the symmetric point to a point not on a line.

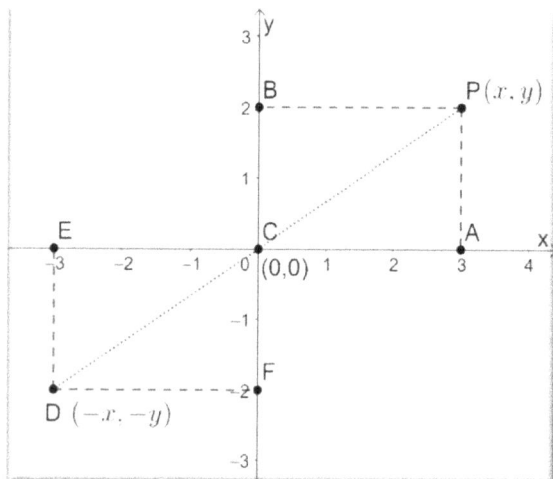

We have point P (x , y) situated in the first quadrant.

The symmetric point by the origin to point P--in this case, point D--is situated at the same distance from the origin.

We can interpret this by saying that points P and D, are symmetric to a point C if point C is the midpoint of the line DP.

Segment CP equals segment DC.

The coordinates of point D are: (-x, -y)

EXAMPLE

Find the symmetric to point R (4,2) by the point S (1,1)

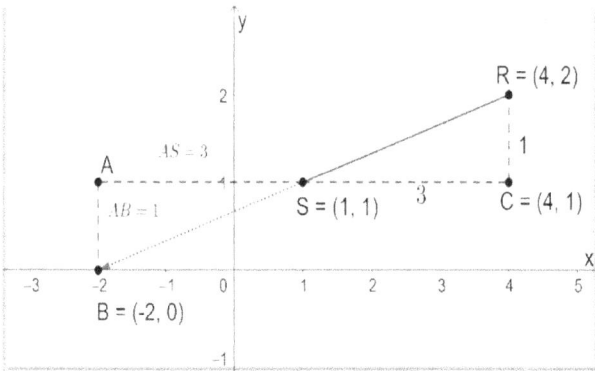

To find the symmetric point of R by S, we extend the line RS in such way that we get an extension that is as long as RS. Or we can find the coordinates of point B from the midpoint formula;

$$x = \frac{x_2 + x_1}{2}$$

$$y = \frac{y_2 + y_1}{2}$$

Here, point 1 will be B, point 2 will be R, and midpoint will be point S

So,

$x_2 = 4, y_2 = 2$

Midpoint coordinates are:

$x = 1, y = 1$

Substituting in the above formulas we have:

The x coordinate:

$1 = \frac{4 + x_1}{2}$

$2 = 4 + x_1$

$-2 = x_1$

The y coordinate:

$1 = \frac{2 + y_1}{2}$

$2 = 2 + y_1$

$0 = y_1$

So,

The coordinates of the symmetric point to R by the point S is B (-2,0), as can be seen in the figure at the beginning of the example.

CHAPTER 2

GEOMETRY

2.A Area

a. Rectangle, Square

We know that the area of a rectangle is the length times the height. A = L * h

EXAMPLE

Find the area of a rectangle that has a length of 10 and height of 5, and a rectangle that has a length of 7 and a height of 6

A = L x h = 10 x 5 = 50
A = L x h = 7 x 6 = 42

EXAMPLE

Find the area of the figure shown below.

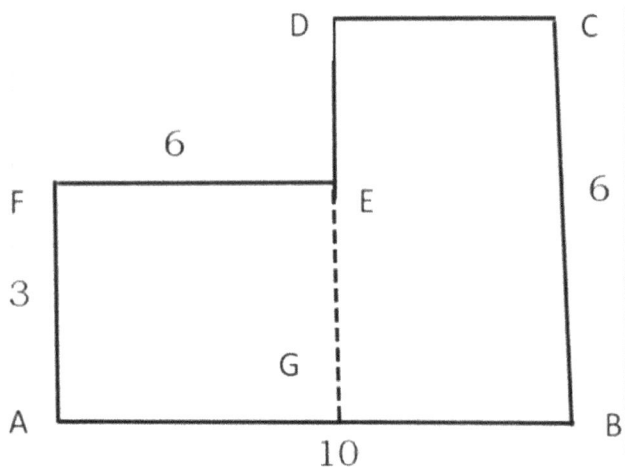

We can split the area into two areas of two rectangles; AFEG and DCBG

Area of AFEG = AG x AF = 6x3=18

GB = 10-6=4

Area of DCBG = DC x CB = 4 x 6 = 24

So,

Area of AFEDCBA = 18 + 24 = 42

EXAMPLE

Calculate the area of the figure shown below.

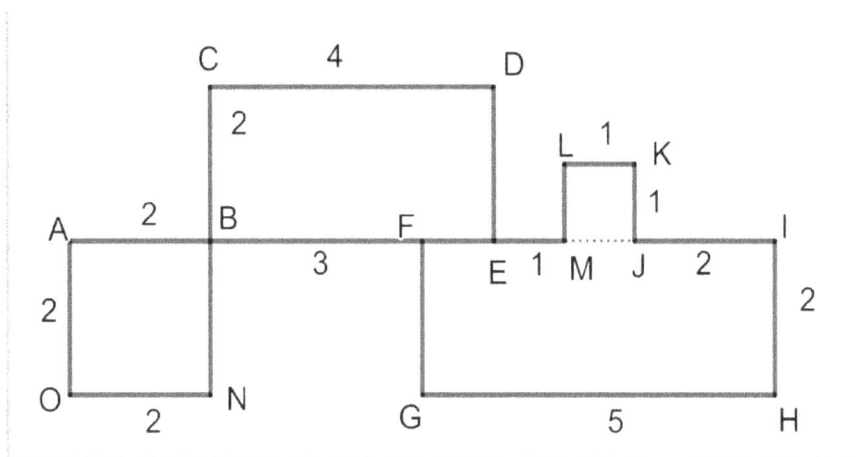

Total Area = Area of OABN +Area of BCDE +Area of FIHG + Area of MLKJ

Area of OABN = 2 x 2 = 4

Area of BCDE = 2 x 4 = 8

Area of FIHG = 2 x 5 = 10

Area of MLKJ = 1 x 1 = 1

Total Area = 4 + 8 + 10 +1 = 23

2.A Area

b. Triangles

As we know, a triangle is a geometrical figure that has three sides.
The area of a triangle is calculated by the formula:

$Area\ of\ triangle = \frac{base \times height}{2}$

To understand this formula, think about the formula of a rectangle;

$Area\ of\ rectangle = base \times height$

EXAMPLE

In the figure below, the $Area\ of\ rectangle = base \times height = DC \times BC$

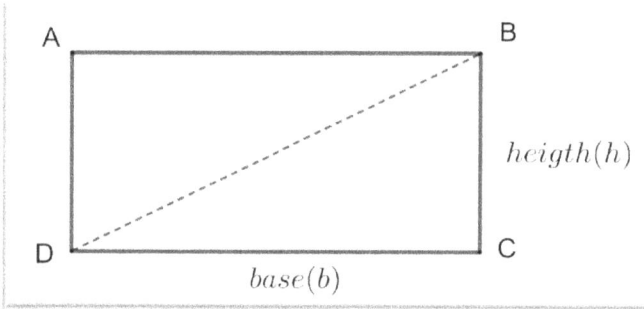

As we can see, the area of the triangle DBC is half of the area of the rectangle ABCD.
So,

$$Area\ of\ triangle = \frac{base \times height}{2}$$

Remember that the height is the perpendicular line to the base, or the line that is extended from the base.

A line that is perpendicular to another line is a line that makes an angle of 90 degrees with the other line it is perpendicular to.

In the example above, BC perpendicular to DC or $BC \perp DC$. $\sphericalangle DCB = 90^0$

What happens when the triangle looks like the one below? What is its area?

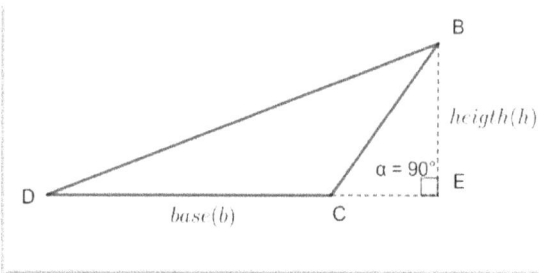

Here, the height of the triangle DCB is the perpendicular BE to the extension from the base DC. In this case DE.

The area of this triangle will be:

$$Area = \frac{base \times height}{2} = \frac{DC \times BE}{2}$$

EXAMPLE

Calculate the area of the triangle shown below.

Here, the base of the triangle is 5.

The height is 4.

The area is calculated with the formula:

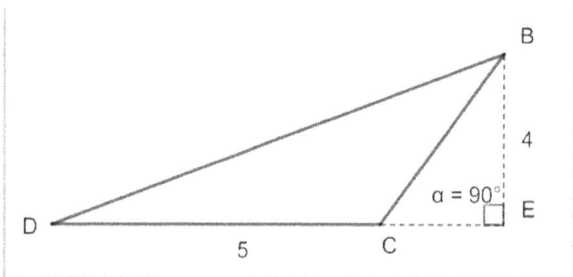

$$Area = \frac{base \times height}{2} = \frac{5 \times 4}{2} = 10$$

EXAMPLE

When the value of the base remains the same, any triangle with the same height will have the same area.

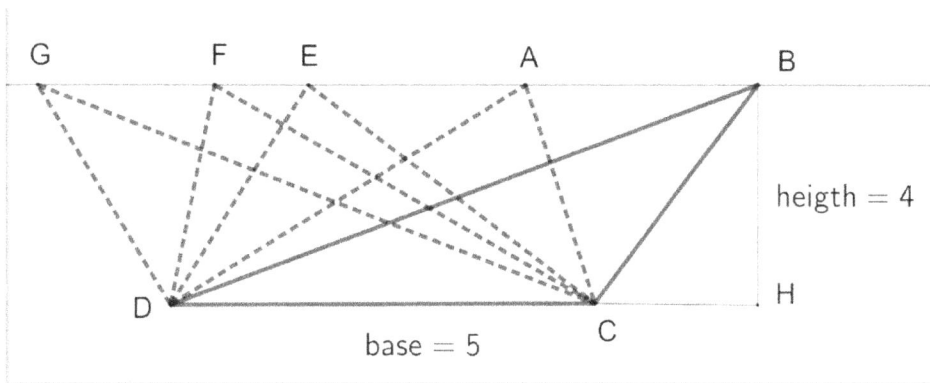

All these triangles have the same area of 10. They have the same base and the same height.

2.A Area

c. Trapezoid, Circle

A trapezoid is a quadrilateral, or a figure with four sides, that has two parallel sides.

EXAMPLE

As we can see, segments AB and DC are parallel.

In the case where the trapezoid does not have any perpendicular side on the base, like the one below, the area of this type of trapezoid is calculated with the formula:

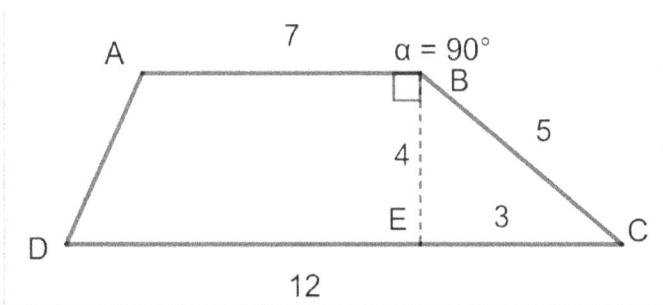

$$Area = \frac{(big\ base + small\ base) \times height}{2}$$

Here:
- The big base is DC
- The small base is AB
- The height is BE

The area of ABCD is $= \frac{(DC + AB) \times BE}{2} = \frac{(12 + 7) \times 4}{2} = 19 \times 2 = 38$

EXAMPLE

As we can see segments AB and DC are parallel.

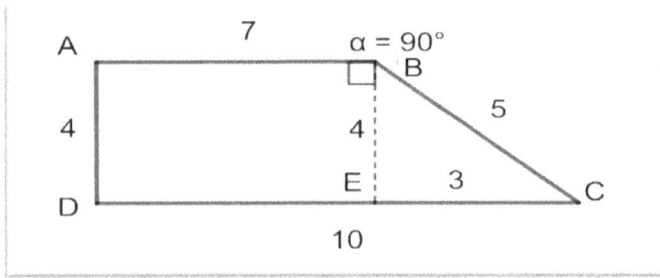

BE is perpendicular on DC. In this particular case AD is perpendicular on DC. The area here could be calculated as Area of the rectangle ABDE plus Area of the right-angle triangle BEC

where angle BEC is right-angle. It means it has 90^0.

So,

Area of the trapezoid ABCD $4 \times 7 + \frac{4+3}{2} = 28 + 6 = 34$

Remember that the Area of a triangle $= \frac{Base\ of\ triangle \times Height\ of\ triangle}{2}$

Area of a circle

Remember that area of a circle with a radius of R, can be calculated using formula:

$Area = \pi R^2$

EXAMPLE

Find the area of the circle with the radius R=5

$Area = \pi R^2 = \pi 5^2 = 25\pi$

EXAMPLE

Calculate the area of the figure shown below. The left half of the circle is missing.

Remember that the radius of a circle is the diameter divided by two. Here, the radius is 2.

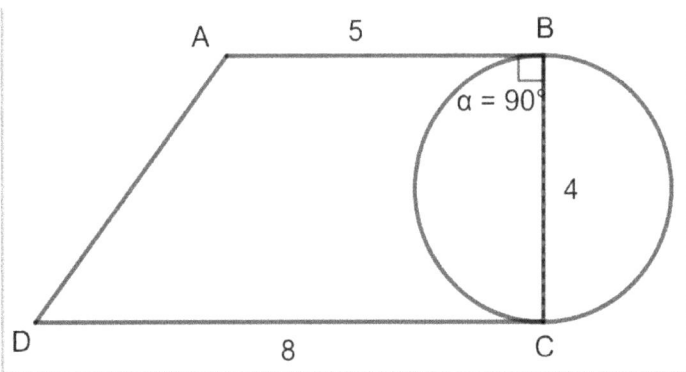

First, we calculate the area of ABCD.

$Area = \frac{(AB+DC)\times BC}{2} = \frac{(5+8)\times4}{2} =$

$(13) \times 2 = 26$

Second, we calculate the area of the half of the circle inside the trapezoid.

$Area = \frac{1}{2}\pi 2^2 = 2\pi$

Third, we calculate the area of the right half of the circle. Area = 2π

Finally, we subtract the area of half a circle from the area of the trapezoid.

$Area = 26 - 2\pi$

Total area = 26 -2π + 2π =26

2.B Volume

a. Prism, cube, cylinder, cone

Volume of a prism and cube

The volume of a prism is calculated using the formula:

$Volume = L \times W \times h = Area\ of\ base \times height$

Where:

L – length

W – width

h – height

EXAMPLE

Calculate the volume of the prism shown below.

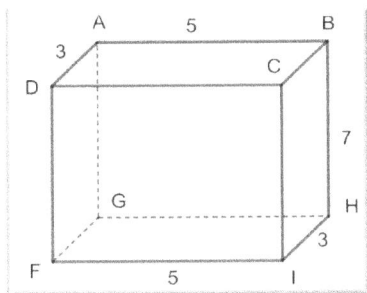

Here,

Length = 5

Weight = 3

height = 7

$Volume = 5 \times 3 \times 7 = 105$

EXAMPLE

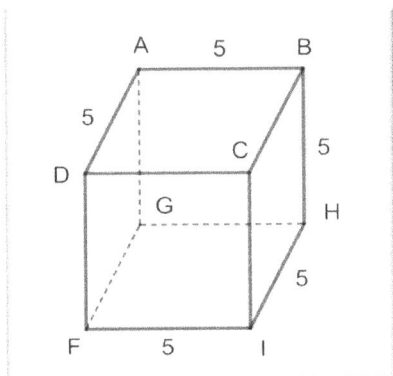

Calculate the volume of the cube shown below.

Here,

Length = Weight = height = 5

$Volume = 5 \times 5 \times 5 = 125$

Volume of a Cylinder

The volume of a cylinder is calculated with the formula:

$Volume = Area\ of\ the\ base \times height = \pi R^2 h$

The base of a cylinder is a circle, so the area is calculated as the area of a circle. (πR^2).

The <u>radius</u> of the circle is the diameter divided by 2.

The <u>diameter</u> of a circle is the line that unites two opposite points on the circle and passes through the center.

EXAMPLE

Calculate the volume of the cylinder shown below.

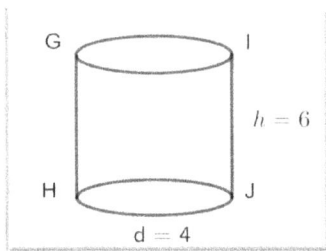

Here,

The radius of the circle is the diameter divided by 2.

The diameter of the circle is 4, so the radius = 2

Radius is 4 divided by 2 = 2

$Volume = \pi R^2 h = \pi(2)^2 \times 6 = \pi \times 4 \times 6 = 24\pi$

Remember that $\pi = 3.14$

So,

If we have to express the volume as a number not in terms of π, we have:

$Volume = \pi R^2 h = \pi(2)^2 \times 6 = \pi \times 4 \times 6 = 24\pi = 24 \times 3.14 = 75.32$

Volume of a Cone

The base of a cone is also a circle.

The volume of a cone is calculated using the formula:

$Volume\ of\ a\ cone = \dfrac{Area\ of\ the\ base \times heigth}{3} = \dfrac{\pi R^2 h}{3}$

EXAMPLE

Calculate the volume of the cone shown below.

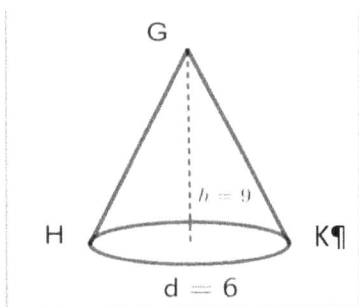

Here,

Radius = 3

height = 9

$Volume = \dfrac{Area\ circle \times heigth}{3} = \dfrac{\pi R^2 h}{3} = \dfrac{\pi 3^2 9}{3} = 27\pi$

EXAMPLE

Calculate the volume of the complex shape shown below

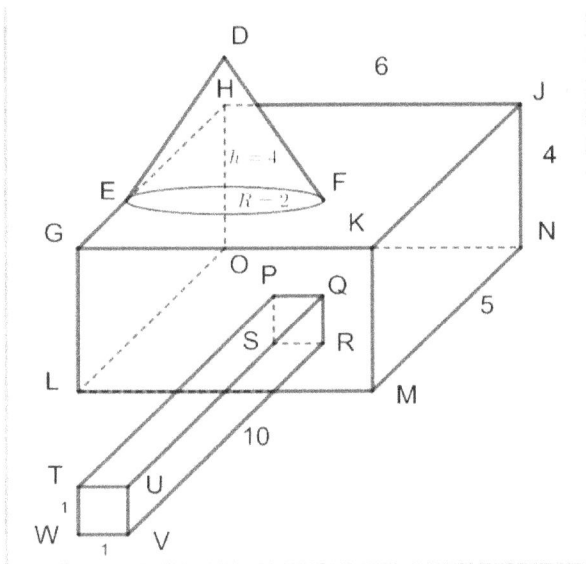

Volume of prism GKMLHJNO =
$6 \times 4 \times 5 = 120$
Volume of prism TUVWPQRS =
$1 \times 1 \times 10 = 10$
Volume of cone EFD = $\pi \times 2^2 \times 4 =$
$16 \times 3.14 = 50.24$

Total volume =
120+10+50.24=180.24

2.C Circle

a. Diameter, Secant, Tangent, Properties of tangents to a circle, Unit circle

The diameter is the line that unites two opposite points on the circle, and passes through the center.

EXAMPLE

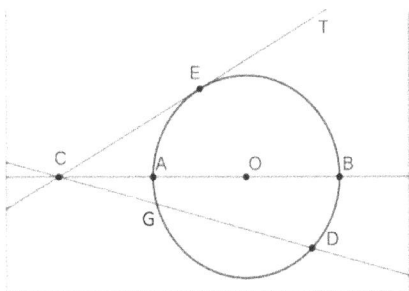

In the figure, the diameter is the line AB. The center of the circle is point O. AO=OB=Radius of the circle. The line that passes through C, G, and D is called the secant. The Secant of a circle is a line that intersects the circle at two points. Here, these points are G

and D.
The segment between G and D is called the chord.

The <u>Tangent</u> to a curve (or a circle in this case) is a line that touches the curve at <u>ONLY ONE</u> point. Here, that point is E. The line passes through C and touches the circle at point E, or line CE.

Properties of tangents to a circle

1. A tangent to a circle is perpendicular to the radius at the point where the tangent touches the circle.

Properties of chords in a circle

1. The line from the centre of a circle that is perpendicular to a chord, bisects the chord or splits the chord in two equal parts.
2. The line that is perpendicular to a chord and bisects the chord, passes through the center of the circle.
3. The line that passes through the centre of a circle and its midpoint is perpendicular to the chord.

EXAMPLE

Determine the size of the line AE that passes through the centre of the circle shown below

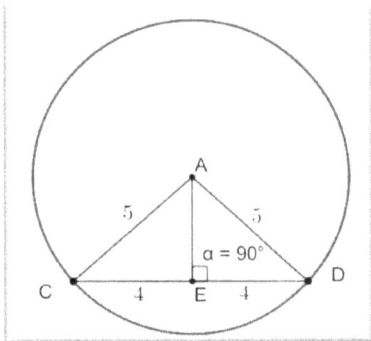

and splits the chord CD in two equal parts.
AE is perpendicular to the chord CD
In the right-angle triangle ACE, $AE^2 = AC^2 - CE^2$
So,

$$AE = \sqrt{AC^2 - CE^2} = \sqrt{5^2 - 4^2} = \sqrt{25 - 16} = \sqrt{9} = 3$$

The circle that has a radius equal to 1 is called <u>unit circle</u>.

2.C Circle

b. Arc length and the radian, properties of angles in a circle

An <u>arc</u> is a part of the circumference of the circle.
In a circle, an <u>angle subtended</u> by an arc is the angle whose arms go through the ends of the arc.

Remember that <u>one radian</u> is the measure of an angle from the center of a circle, <u>subtended</u> by an arc which is equal in length to the radius of the circle as can be seen in the figure below.

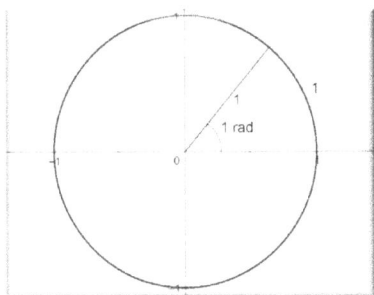

The question is: how many radians are in a circle? The radian is by the above definition the ratio between the length of the subtended arc and the radius of the circle (R).

$$rad = \frac{length\ of\ the\ subtended\ arc}{radius\ of\ the\ circle}$$

If we apply this formula for the entire circle with the total length of $2\pi R$, we have:

$$total\ number\ of\ rad = \frac{2\pi R}{R} = 2\pi$$

So,

In a circle we always have 2π radians

Half a circle has π radians.

A quarter of a circle has $\frac{\pi}{2}$ radians

Remember that π is an irrational number equal with 3.14185........

NOTE

If we divide the circumference of any circle with its diameter, we will always obtain the same number 3.14185 or π.

EXAMPLE

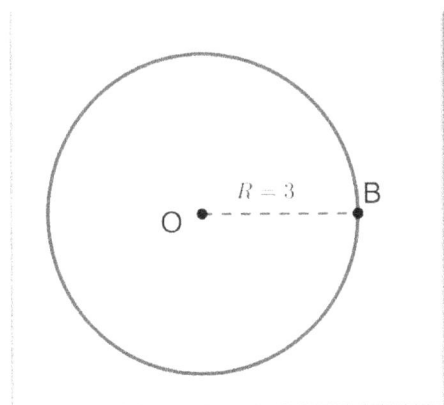

If we consider a circle that has the radius equal with 3, and the circumference equal with: $C = 2\pi R = 2 * \pi * 3$

We have that the ratio between the circumference of the circle and the diameter is:

$$\frac{2\pi R}{Diameter} = \frac{2*\pi*3}{2*R} = \frac{2*\pi*3}{2*3} = \pi$$

As we can see, the ratio doesn't depend on radius, it is always equal with pi.

Angles in a circle

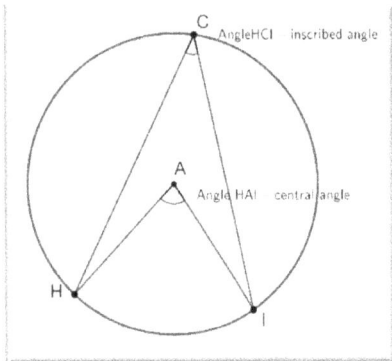

The <u>central angle</u> is an angle whose arms go from the centre of the circle to the circle itself.

The <u>inscribed angle</u> is an angle whose vertex is on the circle and the end of the arms are again on the circle.

The arc that subtends both angles is HI

<u>Property 1</u>

In any circle, the measure of a central angle subtended by an arc is two times the measure of an inscribed angle subtended by the same arc.

EXAMPLE

In the figure above, $\angle HAI = 2 * \angle HCI$

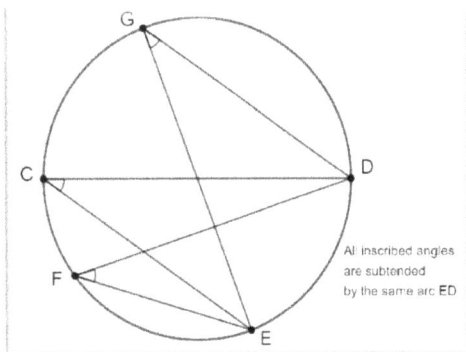

<u>Property 2</u>

In a circle, all the inscribed angles subtended by the same arc are congruent (equal).

EXAMPLE

Here,
$\angle EFD = \angle ECD = \angle EGD$

Note

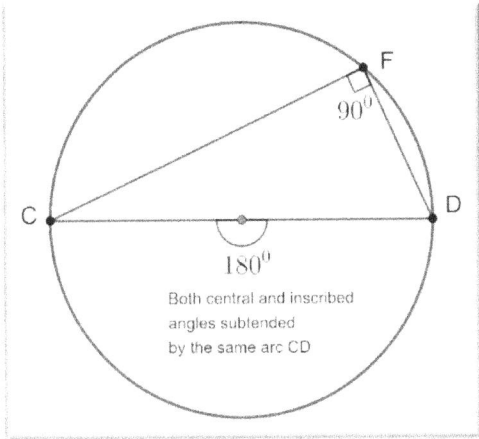

The <u>central angle</u> COD that is the diameter and subtended by half of the circumference (semicircle) is 180^0

The inscribed angle CFD that is subtended by the semicircle is half of the central angle so is 90^0

<u>Property 3</u>

All inscribed angles subtended by a semicircle are right angles.

Both central and inscribed angles subtended by the same arc CD

CHAPTER 3

TRIGONOMETRY

3.A Trigonometric ratios

a. Sine, Cosine, Tangent, Cotangent, solving right-angles triangles, special ratios

We know that the trigonometric ratios are relations between sides of a right-angle triangle. The trigonometric ratios we analyze here are sine, cosine, tangent and cotangent.

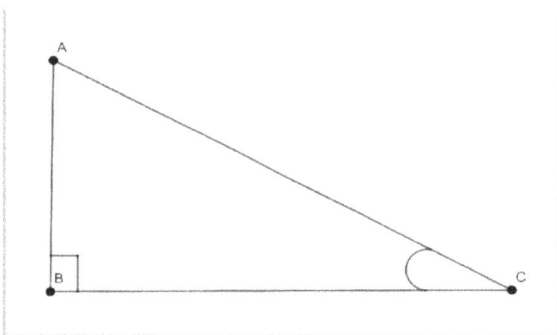

In a right-angle triangle, with $\sphericalangle\phi \neq 90^0$ we have the following *trigonometric ratios*:

$\sin \sphericalangle\phi = \frac{opposite}{hypotenuse}$

$\cos \sphericalangle\phi = \frac{adjacent}{hypotenuse}$

$\tan \sphericalangle\phi = \frac{opposite}{adjacent}$

$\cot \sphericalangle\phi = \frac{adjacent}{opposite}$

EXAMPLE

In the triangle below, calculate the trigonometric ratios sine, cosine, tangent, and cotangent of angle C. In this triangle the angle ABC is right-angle (90^0).

Here,

segment AC is the hypotenuse,

segment AB is the opposite of angle C, and segment BC is the adjacent to angle C.

So, the trigonometric ratios are:

$$\sin \sphericalangle C = \frac{opposite}{hypotenuse} = \frac{AB}{AC}$$

$$\cos \sphericalangle C = \frac{adjacent}{hypotenuse} = \frac{BC}{AC}$$

$$\tan \sphericalangle C = \frac{opposite}{adjacent} = \frac{AB}{BC}$$

$$\cot \sphericalangle C = \frac{adjacent}{opposite} = \frac{BC}{AB}$$

Remember "SOHCAHTOA"?

It comes from:

Sine = **O**pposite ÷ **H**ypotenuse
Cosine = **A**djacent ÷ **H**ypotenuse
Tangent = **O**pposite ÷ **A**djacent

Special trigonometric ratios

Some of the special angles in a right-angle triangle are:

$30^0, 60^0, 45^0$

EXAMPLE

Let's suppose we have an isosceles triangle with the side equal to variable a. We know that the perpendicular BD to the base AC splits the base into half. AD = DC = a/2.

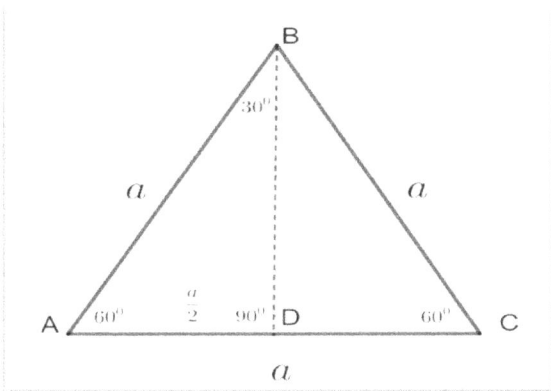

Segment BD is found using the Pythagorean theorem in the triangle ADB.

$$DB^2 = AB^2 - AD^2$$

$$DB^2 = a^2 - \left(\frac{a}{2}\right)^2 = a^2 - \frac{a^2}{4} = \frac{4a^2 - a^2}{4} = \frac{3a^2}{4}$$

$$DB = \sqrt{\frac{3a^2}{4}} = \frac{a\sqrt{3}}{2}$$

The sine, cosine and tangent for angles A and ABD are:

$$sin60^0 = \frac{opposite}{hypotenuse} = \frac{DB}{AB} = \frac{\frac{a\sqrt{3}}{2}}{a} = \frac{\sqrt{3}}{2}$$

$$cos60^0 = \frac{adjacent}{hypotenuse} = \frac{AD}{AB} = \frac{\frac{a}{2}}{a} = \frac{1}{2}$$

$$tan60^0 = \frac{opposite}{adjacent} = \frac{\frac{a\sqrt{3}}{2}}{\frac{a}{2}} = \frac{a\sqrt{3}}{2} * \frac{2}{a} = \sqrt{3}$$

$$sin30^0 = \frac{opposite}{hypotenuse} = \frac{AD}{AB} = \frac{\frac{a}{2}}{a} = \frac{1}{2}$$

$$cos30^0 = \frac{adjacent}{hypotenuse} = \frac{BD}{AB} = \frac{\frac{a\sqrt3}{2}}{a} = \frac{\sqrt3}{2}$$

$$tan30^0 = \frac{opposite}{adjacent} = \frac{\frac{a}{2}}{\frac{a\sqrt3}{2}} = \frac{a}{2} * \frac{2}{a\sqrt3} = \frac{1}{\sqrt3} = \frac{\sqrt3}{\sqrt3\sqrt3} = \frac{\sqrt3}{3}$$

EXAMPLE

Now let's calculate the trigonometric ratios for 45^0.

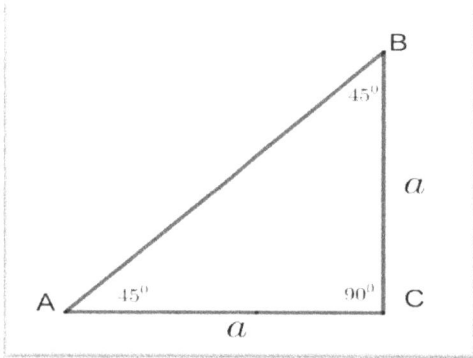

We build a right-angle triangle that is isosceles as well.

The right-angle ABC is isosceles as well. The sides AC and BC are equal with a.

AB can be calculated using the Pythagorean theorem.

$$AB^2 = AC^2 + BC^2 = a^2 + a^2 = 2a^2$$

$$AB = \sqrt{2a^2} = a\sqrt2$$

$$Sin45^0 = \frac{opposite}{hpotenuse} = \frac{a}{a\sqrt2} = \frac{1}{\sqrt2} = \frac{\sqrt2}{\sqrt2\sqrt2} = \frac{\sqrt2}{2}$$

$$cos45^0 = \frac{adjacent}{hpotenuse} = \frac{a}{a\sqrt2} = \frac{1}{\sqrt2} = \frac{\sqrt2}{\sqrt2\sqrt2} = \frac{\sqrt2}{2}$$

$$tan45^0 = \frac{opposite}{adjacent} = \frac{a}{a} = 1$$

Put all together here they are.

	30^0	60^0	45^0
Sin(Φ)	$\dfrac{1}{2}$	$\dfrac{\sqrt3}{2}$	$\dfrac{\sqrt2}{2}$
Cos(Φ)	$\dfrac{\sqrt3}{2}$	$\dfrac{1}{2}$	$\dfrac{\sqrt2}{2}$
Tan(Φ)	$\dfrac{\sqrt3}{3}$	$\sqrt3$	1

CHAPTER 4

ALGEBRA

4.A Introduction to powers

a. Definition, rules

The <u>power</u> is a group of two numbers where one is the base and the other is the exponent. The exponent tells us how many times we multiply the base again and again.

EXAMPLE

The power, in this case, is the group 2^4

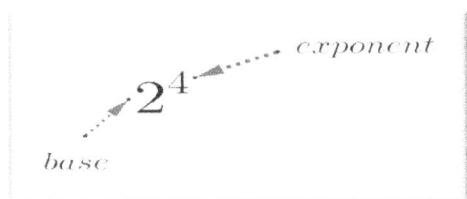

The base is 2, and the exponent is 4. This means that, the base 2 is multiplied with itself 4 times.

$2 \times 2 \times 2 \times 2 = 16$

EXAMPLE

Write the following repeated multiplications as a power.

$3 \times 3 \times 3 \times 3 \times 3 \times 3 = 3^6$

$5 \times 5 \times 5 \times 5 = 5^4$

EXAMPLE

Write the following powers as repeated multiplications.

$7^5 = 7 \times 7 \times 7 \times 7 \times 7$

$8^3 = 8 \times 8 \times 8$

<u>Power rules</u>

We consider a and b as non-zero real numbers, and m and n integers.

1. The product of powers with the same base

$$a^n \times a^m = a^{n+m}$$

EXAMPLE

$$3^5 \times 3^7 = 3^{5+7} = 3^{12}$$

2. The division of powers with the same base

$$a^n \div a^m = a^{n-m}$$

EXAMPLE

$$5^8 \div 5^6 = 5^{8-6} = 5^2$$

3. The product of two numbers at an exponent.

$$(a \times b)^n = a^n \times b^n$$

EXAMPLE

$$(6 \times 3)^4 = 6^4 \times 3^4$$

4. The quotient of two different numbers at an exponent.

$$\left(\frac{a}{b}\right)^n = \frac{a^n}{b^n}$$

EXAMPLE

$$\left(\frac{5}{3}\right)^4 = \frac{5^4}{3^4}$$

5. The negative exponent.

$$a^{-n} = \frac{1}{a^n}$$

EXAMPLE

$$3^{-5} = \frac{1}{3^5}$$

6. Exponent at an exponent

$$(a^m)^n = a^{m \times n}$$

EXAMPLE

$$(2^3)^5 = 2^{3 \times 5} = 2^{15}$$

7. Exponent zero

$$a^0 = 1$$

EXAMPLE

$$(87654789)^0 = 1$$

8. Fractional exponents.

$$a^{\frac{m}{n}} = \sqrt[n]{a^m}$$

EXAMPLE

$$3^{\frac{2}{5}} = \sqrt[5]{3^2}$$

And now, put it all together.

EXAMPLE

$$\frac{3^5 \times 4^4 \times (2 \times 3)^6}{3^3 \times (\frac{4}{3})^2 \times 2^4} = \frac{3^5 \times 4^4 \times 2^6 \times 3^6}{3^3 \times \frac{4^2}{3^2} \times 2^4} = \frac{3^5 \times 4^4 \times 2^6 \times 3^6}{\frac{3^3}{3^2} \times 4^2 \times 2^4} = \frac{3^{5+6} \times 4^4 \times 2^6}{3^{3-2} \times 4^2 \times 2^4} = \frac{3^{11} \times 4^4 \times 2^6}{3^1 \times 4^2 \times 2^4} = \frac{3^{11}}{3^1} \times \frac{4^4}{4^2} \times \frac{2^6}{2^4} =$$

$$3^{11-1} \times 4^{4-2} \times 2^{6-4} = 3^{10} \times 4^2 \times 2^2 = 3^{10} \times (2^2)^2 \times 2^2 = 3^{10} \times 2^4 \times 2^2 = 3^{10} \times 2^{4+2} =$$

$$3^{10} \times 2^6$$

EXAMPLE

$$8^{\frac{1}{3}} \times 5^7 \times \frac{(75-4+256-56+2457)^0}{2^3 \times 5^6} = \sqrt[3]{8} \times 5^7 \times \frac{(2728)^0}{2^3 \times 5^6} = \frac{2 \times 5^7 \times 1}{2^3 \times 5^6} = \frac{2}{2^3} \times \frac{5^7}{5^6} = 2^{1-3} \times 5^{7-6} =$$

$$2^{-2} \times 5 = \frac{5}{2^2} = \frac{5}{4}$$

4.A. Introduction to powers

b. Radicals, rules

Remember that a number multiplied by itself twice is that number squared.

EXAMPLE

$$2 \times 2 = 2^2 = 4$$

The reverse operation from 4 to 2 is the square root.

$$\sqrt{4} = 2$$

Cube root of 8 is 2

$$\sqrt[3]{8} = 2$$

Square roots of a perfect square are rational numbers. Square roots of a non-perfect square are irrational numbers. Cubic roots of a perfect cubic number are rational numbers.

Cubic roots of a non-perfect cubic number are irrational numbers.

Remember that:

$$\sqrt{4} = \sqrt{2} \times \sqrt{2} = \sqrt{2^2} = 2$$

$$\sqrt[3]{8} = \sqrt[3]{2} \times \sqrt[3]{2} \times \sqrt[3]{2} = \sqrt[3]{2^3} = 2$$

So, $\sqrt[n]{x^n} = x$

Rules of radicals

a. $\sqrt[n]{a} \times \sqrt[n]{b} = \sqrt[n]{a \times b}$

EXAMPLE

$\sqrt[3]{5} \times \sqrt[3]{7} = \sqrt[3]{35}$

b. $\sqrt[n]{\dfrac{a}{b}} = \dfrac{\sqrt[n]{a}}{\sqrt[n]{b}} = \sqrt[n]{a} \div \sqrt[n]{b}$

EXAMPLE

$\sqrt{\dfrac{16}{25}} = \dfrac{\sqrt{16}}{\sqrt{25}} = \dfrac{4}{5}$

c. $\sqrt[3]{-27} = -3$

The odd root from negative numbers is a negative number.

The even root from negative numbers is an imaginary number.

EXAMPLE

$\sqrt[3]{-27} = -3$

$\sqrt{-2} = \sqrt{-1 \times 2} = \sqrt{-1} \times \sqrt{2} = i\sqrt{2}$

Here $\sqrt{-1} = i \ or \ imaginary$

4.A Introduction to powers

c. Mixed radicals, conversions of radicals

When the number under the radical is not a perfect root, the result will be a mixed radical. A <u>mixed radical</u> is when we have a term (number and or letters) multiplied with a radical.

EXAMPLE

$\sqrt{180} = \sqrt{4 \times 9 \times 5} = \sqrt{4} \times \sqrt{9} \times \sqrt{5} = 2 \times 3 \times \sqrt{5} = 6\sqrt{5}$

$\sqrt{180}$ is called entire radical

$6\sqrt{5}$ is a mixed radical.

$\sqrt[3]{56x^4} = \sqrt[3]{8 \times 7 \times x^3 \times x} = \sqrt[3]{8} \times \sqrt[3]{7} \times \sqrt[3]{x^3} \times \sqrt[3]{x} = \sqrt[3]{2^3} \times \sqrt[3]{7} \times \sqrt[3]{x^3} \times \sqrt[3]{x} = 2x\sqrt[3]{7x}$

$\sqrt[3]{56x^4}$ is the entire radical

$2x\sqrt[3]{7x}$ is a mixed radical

Conversion of radicals

Sometimes, we need to convert the entire radical into a mixed radical. Sometimes we have to do the reverse, to transform a mixed radical into an entire radical.

a. Entire radical to mixed radical

EXAMPLE

Find the result. Write the result into a mixed radical

$$\sqrt{112} = \sqrt{4^2 \times 7} = \sqrt{4^2} \times \sqrt{7} = 4\sqrt{7}$$

$$\sqrt{20x^5y^3} = \sqrt{4 \times 5 \times x^5 \times y^3} = \sqrt{4 \times x^4 \times x \times y^2 \times y \times 5} = \sqrt{4 \times x^4 \times y^2 \times x \times y \times 5} =$$
$$\sqrt{4} \times \sqrt{x^4} \times \sqrt{y^2} \times \sqrt{5xy} = 2x^2y\sqrt{5xy}$$

b. Mixed radical into a entire radical

EXAMPLE

Find the result. Write the result into an entire radical

$$3\sqrt{5} = \sqrt{3^2 \times 5} = \sqrt{9 \times 5} = \sqrt{45}$$
$$5\sqrt[3]{7} = \sqrt[3]{5^3 \times 7} = \sqrt[3]{125 \times 7} = \sqrt[3]{875}$$
$$7xy^2\sqrt{3} = \sqrt{7^2 \times x^2 \times y^4 \times 3} = \sqrt{49 \times 3 \times x^2 \times y^4} = \sqrt{147x^2y^4}$$

EXAMPLE

Find the result. Write the result into a mixed radical

$$\sqrt{450} + \sqrt{882} - \sqrt{162} = \sqrt{45 \times 10} + \sqrt{2 \times 441} + \sqrt{2 \times 81} = \sqrt{9 \times 5 \times 2 \times 5} +$$
$$\sqrt{2 \times 49 \times 9} - \sqrt{2 \times 81} = \sqrt{9 \times 25 \times 2} + \sqrt{2 \times 49 \times 9} - 9\sqrt{2} = \sqrt{9} \times \sqrt{25} \times \sqrt{2} +$$
$$\sqrt{9} \times \sqrt{49} \times \sqrt{2} - \sqrt{81} \times \sqrt{2} = 3 \times 5 \times \sqrt{2} + 3 \times 7 \times \sqrt{2} - 9 \times \sqrt{2} = 15\sqrt{2} + 21\sqrt{2} - 9\sqrt{2} =$$
$$27\sqrt{2}$$

4.B Fundamental concepts in algebra

a. Associative, commutative and distributive properties

<u>Associative property</u>
In a mathematical expression, the order of operations doesn't matter as long as the operations used in that particular case are only addition or only multiplication.

EXAMPLE

(5+3)+7=5+(3+7)

8+7=15

5+10=15

$(2 \times 5) \times 9 = 2 \times (5 \times 9)$

$10 \times 9 = 90$

$2 \times 45 = 90$

Commutative property

In a mathematical expression, the order of the numbers doesn't matter as long as the operations used in that particular case are only addition or only multiplication.

EXAMPLE

5+7+3=7+5+3=15

$2 \times 9 \times 7 = 7 \times 2 \times 9$

$18 \times 7 = 126$

$14 \times 9 = 126$

Distributive property

Whenever there is a bracket, the multiplication operation will distribute over either addition or subtraction operation that is inside that bracket.

EXAMPLE

$4 \times (5 + 6) = 4 \times 5 + 4 \times 6 = 20 + 24 = 44$

EXAMPLE

$3 \times (7 - 4) = 3 \times 7 - 3 \times 4 = 21 - 12 = 9$

4.B Fundamental concepts in algebra

b. Expanding brackets, cross multiplication property

Expanding brackets means getting rid of them.

$a \times (b + c) = a \times b + a \times c$

EXAMPLE

$2 \times (7 + 4) = 2 \times 7 + 2 \times 4 = 14 + 8 = 22$

EXAMPLE

$6(2 + 5 - 2x) = 6 \times 2 + 6 \times 5 - 6 \times 2x = 12 + 30 - 12x = 42 - 12x$

FOIL (First, Outer, Inner, Last)

First, we multiply the first terms in each binomial. Then, we multiply the Outer which means that we multiply the outermost terms in the product. Then Inner terms and Outer ones.

EXAMPLE

$(a + b)(c + d) = ac + ad + bc + bd$

We multiply the <u>First</u> terms $\boldsymbol{a} \times \boldsymbol{c}$

then we multiply the <u>Outer</u> terms $\boldsymbol{a} \times \boldsymbol{d}$,

We multiply the <u>Inner</u> terms $\boldsymbol{b} \times \boldsymbol{c}$,

and then we multiply the <u>Last</u> terms $\boldsymbol{b} \times \boldsymbol{d}$

<u>Cross multiplication</u> means that when two fractions are equal the product between the numerator of first fraction and the denominator of the second fraction equals the product between the numerator of the second fraction and the denominator of the first fraction.

If we have the fractions:

$\frac{a}{b} = \frac{c}{d}$, and b and d are non-zero numbers,

Then,

$a \times d = c \times b$

EXAMPLE

$\frac{4}{5} = \frac{12}{15}$

So, $4 \times 15 = 5 \times 12 = 60$

Application of cross multiplication.

Find x using cross multiplication property.

$\frac{6(x-3)}{2} = \frac{3(x+4)}{5}$

$5 \times 6(x - 3) = 2 \times 3(x + 4)$; $30x - 90 = 6x + 24$; $30x - 6x = 90 + 24$

$24x = 114$

$x = \frac{114}{24} = \frac{57}{12} = 4\frac{9}{12} = 4\frac{3}{4}$

4.C Introduction to logarithms

a. Definition, rules

Remember that $2^4 = 16$. It means that we multiply 2 by itself four times to get the result 16. Now, let's suppose we want to write the exponent 4 in terms of the result and the base.

This <u>exponent</u> will be written as the logarithm of base 2 from 16.

$4 = \log_2 16$

The logarithm is the exponent at which we have to rise the base to obtain the argument of the logarithm.

EXAMPLE

$\log_3 27 = 3$

Rising 3 at exponent 3 gives us a result of 27.

Rules of logarithms

1. Multiplication into addition

$\log_a(b \times c) = \log_a b + \log_a c$

EXAMPLE

$\log_4(5 \times 7) = \log_4 5 + \log_4 7$

2. Division into subtraction

$\log_a(b \div c) = \log_a b - \log_a c$

EXAMPLE

$\log_2(9 \div 3) = \log_2 9 - \log_2 3$

3. The argument at an exponent

$\log_a b^w = w \times \log_a b$

EXAMPLE

$\log_5 7^9 = 9 \times \log_5 7$

4. Changing the base

$\log_a b = \dfrac{\log_c b}{\log_c a}$

EXAMPLE

$\log_5 7 = \dfrac{\log_2 7}{\log_2 5}$

And now, let's put it all together.

Simplify:

$$\log_3 5 + \log_3(x + 2) - \log_3(x^2 + 4x + 4) = \frac{\log_3 5(x+2)}{\log_3(x^2+4x+4)} = \frac{\log_3 5(x+2)}{\log_3(x+2)^2}$$

Change the base to base 7.

$$\log_5 9 = \frac{\log_7 9}{\log_7 5}$$

4.C introduction to logarithms

b. Exponential and logarithmic equations

An <u>exponential expression</u> is one where the variable x is at the exponent, like 3^x.

We will have to use the laws of logarithms and powers to be able to solve exponential and logarithmic equations.

EXAMPLE

Evaluate:

$$4\log_2 32 - 4\log_2 2 = \frac{4\log_2(4\times8)}{4\times1} = \log_2 4 + \log_2 8 = \log_2 2^2 + \log_2 2^3 = 2 \times 1 + 3 \times 1 =$$

$2 + 3 = 5$

Solve for x:

a. $5^x = 125$

$5^x = 5^3$

$x = 3$

b. $21^{3x-4} = 441^{5x+3}$

$21^{3x-4} = (21^2)^{5x+3}$

$21^{3x-4} = 21^{2(5x+3)}$

$3x + 4 = 2(5x + 3)$

$3x + 4 = 10x + 6$

$-7x = 2$

$x = -\frac{2}{7}$

c. $\log_3 x = \log_3 27 - \log_3 6$

$\log_3 x = \log_3 \frac{27}{6} = \log_3 \frac{9\times3}{2\times3} = \log_3 \frac{9}{2}$

$x = \frac{9}{2} = 4.5$

d. $\log_3(2x - 1) + \log_3 2 = \log_3 x$

$\log_2 2(2x - 1) = \log_3 x$

$2(2x - 1) = x$

$4x - 2 = x$

$3x = 2$

$x = \dfrac{2}{3}$

Remember the natural logarithm written $\ln x$ where the base is e or Euler's number

e = **2.71828.....**

4.D Introduction to polynomials

a. Definition, operations with polynomials

A <u>polynomial</u> is, simply put, a mathematical expression that includes constants and one or several variables. These <u>terms</u> are making a finite number of additions, subtractions and multiplications.

A term is either a constant, or a multiplication between a constant and at least a variable.

A <u>monomial</u> is a polynomial with only one term. Ex: 2, $3x$, $67xy^2$

A <u>binomial</u> is a polynomial with two terms. Ex: $2 + x$, $5x - 3y$, $7xy^2 + 5x^2$

A <u>trinomial</u> is a polynomial with three terms. Ex: $5 + 3x - 3x^2$, $6x - 3y + 5$, $9y^2 + 3x^2 - 11$

Operations with polynomials

1. Addition of polynomials

EXAMPLE

$(2x + 3) + (3x + y - 5) = 2x + 3 + 3x + y - 5 = 5x + y - 2$

2. Subtraction of polynomials

EXAMPLE

$(2x + 3) - (3x + y - 5) = 2x + 3 - 3x - y + 5 = -x - y + 8$

3. Multiplication of polynomials

EXAMPLE

$(-7x^2y + 3x + 2)(4x - 5xy^2) = -28x^3y + 35x^3y^3 + 12x^2 - 15x^2y^2 + 8x - 10xy^2$

4. Division of polynomials

a. Division of a polynomial by a monomial

EXAMPLE

$$\frac{16x^2+4x-1}{2x} = \frac{16x^2}{2x} + \frac{4x}{2x} - \frac{1}{2x} = 8x + 2 - \frac{1}{2x}$$

b. Division of a polynomial by a binomial

EXAMPLE

$$\frac{15x^2+35x-25}{5x+5} = \frac{5(3x^2+7x-5)}{5(x+1)} = \frac{3x^2+7x-5}{x+1} = 3x + 10 + \frac{5}{x+1}$$

$$3x^2 + 7x - 5 \div x + 1 = 3x + 10$$
$$\underline{-(3x^2 + 3x)}$$
$$+10x - 5$$
$$\underline{-(10x + 10)}$$
$$5$$

4.D Introduction to polynomials

b. Rationalizing the denominator, special binomial products

Remember that when we multiply the a radical by itself as many times as the root, the result is the number under the radical.

EXAMPLE

$$\sqrt{3} \times \sqrt{3} = 3$$
$$\sqrt[3]{5} \times \sqrt[3]{5} \times \sqrt[3]{5} = 5$$

Rationalizing the denominator

a. When we have a fraction that has a radical at the denominator, we can transform the radical into a non-radical by multiplying the same radical by itself as many times as the root. The result is the number under the radical.

EXAMPLE

Rationalize the denominator.

$$\frac{5}{\sqrt{2}} = \frac{5\times\sqrt{2}}{\sqrt{2}\times\sqrt{2}} = \frac{5\sqrt{2}}{2}$$

$$\frac{2x}{\sqrt{x+1}} = \frac{2x\times\sqrt{x+1}}{\sqrt{x+1}\times\sqrt{x+1}} = \frac{2x\sqrt{x+1}}{x+1}$$

b. Using the special binomial product; the difference of squares

$$(a + b) \times (a - b) = a^2 - ab + ab - b^2 = a^2 - b^2$$

So,

$(a + b) \times (a - b) = a^2 - b^2$

EXAMPLE

Rationalize the denominator.

$$\frac{\sqrt{3}}{\sqrt{5}-\sqrt{3}} = \frac{\sqrt{3}\times(\sqrt{5}+\sqrt{3})}{(\sqrt{5}-\sqrt{3})\times(\sqrt{5}+\sqrt{3})} = \frac{\sqrt{15}+\sqrt{3}\times\sqrt{3}}{[(\sqrt{5})^2-(\sqrt{3})^2]} = \frac{\sqrt{15}+3}{5-3} = \frac{\sqrt{15}+3}{2}$$

Other two special binomial products

$(a + b)^2 = (a + b) \times (a + b) = a^2 + 2ab + b^2$

$(a - b)^2 = (a - b) \times (a - b) = a^2 - 2ab + b^2$

EXAMPLE

$(x + 2)^2 = (x + 2) \times (x + 2) = x^2 + 2x \times 2 + 2^2 = x^2 + 4x + 4$

$(3x - 5)^2 = (3x - 5) \times (3x - 5) = 9x^2 - 2 \times 3x \times 5 + 5^2 = 9x^2 - 30x + 25$

EXAMPLE

Rationalize the denominator.

$$\frac{1}{\sqrt{x^2+4x+4}} = \frac{1}{\sqrt{(x+2)^2}} = \frac{1}{x+2}$$

4.E Linear equations

a. Straight-line graph

a. From equation to the graph

We have seen in 1E that the slope of a straight line is calculated with the formula:

$$Slope = m = \frac{Rise}{Run} = \frac{Vertical\ distance}{Horizontal\ distance} = \frac{y_2-y_1}{x_2-x_1}$$

So, if we have a point on a straight-line M(x_1, y_1), then the slope of the line that passes through point M and any other point K(x, y) will be written as:

$$m = \frac{y-y_1}{x-x_1}$$

So, if we isolate y, we have:

$$y = mx + b$$

Where:

m – slope

b – intersection of the line with the y axis. This is called the y-intercept

EXAMPLE

Graph the straight-line represented by $y = 2x + 3$

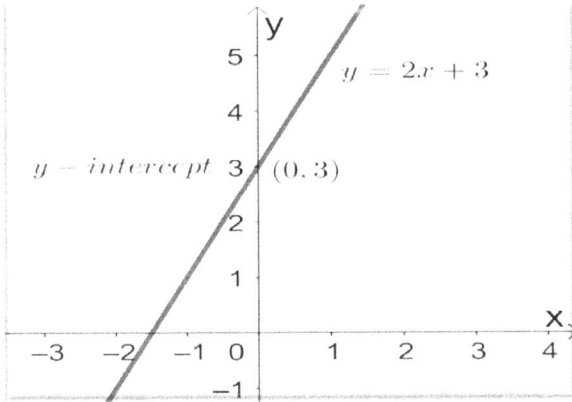

Here, the slope is 2, and the y intercept is the point (0,3)

EXAMPLE

Graph the following lines on the same graph.

a) $y = \frac{1}{2}x$

b) $y = -3x + 2$

c) $y = -x - 1$

d) $y = 3$

e) $x = -2$

Let's analyze them separately.

a) $y = \frac{1}{2}x$

The slope here is 0.5 and the y intercept is (0,0).

b) $y = -3x + 2$

Whenever the slope is negative, the line is tilted from down right to up left. Here, the slope is -3 and the y intercept is (0,2).

c) $y = -x - 1$

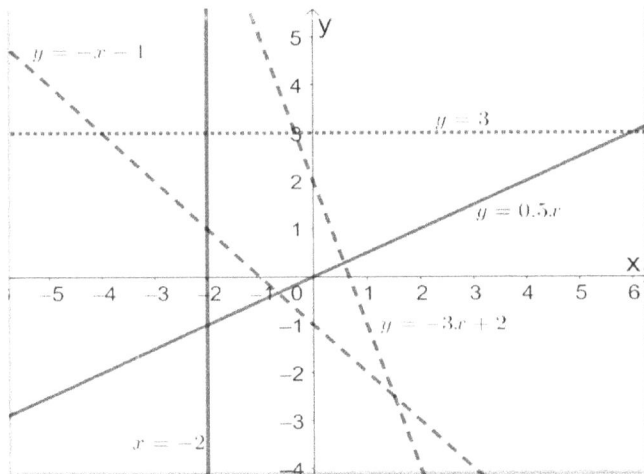

Here, the slope is negative (-1), so the line is tilted from the lower right to the upper left. The y intercept is (0,-1)

d) $y = 3$

Here, there is no x. this line is a horizontal line that goes through point (0,3) and is parallel with the x axis.

e) $x = -2$

This is a vertical line that that goes through point (-2,0) and is parallel with the y axis.

b. From graph to equation

Next is the case in which we are given the line, and we have to determine the mathematical equation of the line.

EXAMPLE

Find the equations of lines a) and b).

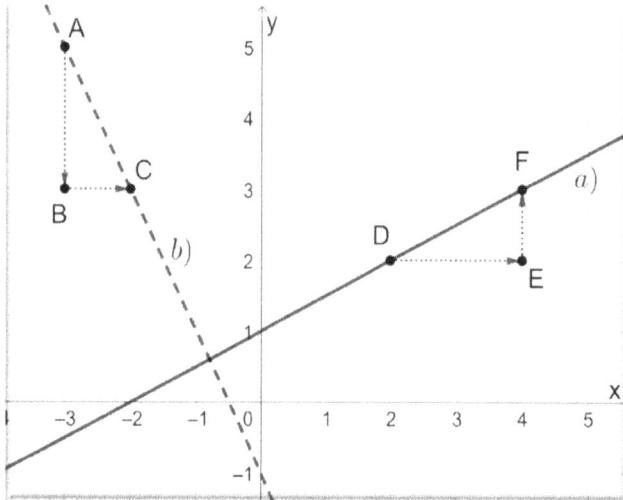

Line a)

We notice that the y intercept is 1, so the constant in the equation is known.

To find the slope, we start at point D, go horizontally (run) for +2 units to the right, then go vertically (rise) up +1 unit. So, the slope is calculated as:

$$slope = \frac{rise}{run} = \frac{1}{2}$$

The equation is $y = \frac{1}{2}x + 1$

Line b)

We notice that the y intercept is -1, so the constant in the equation is known. To find the slope we start at point A, go vertically (rise) for -2 units, then go horizontally (run) plus 1 unit. So, the slope is calculated as:

$$slope = \frac{rise}{run} = \frac{-2}{1} = -2$$

The equation is $y = -2x - 1$

4.F Quadratic equations

a. Factoring used in quadratic equations

Remember that the terms of a product are called <u>factors</u>.

EXAMPLE

We have the product:

$6 \times 8 = 48$

6 and 8 are the <u>factors</u> of the multiplication.

Remember that the prime factors are the numbers that can be divided only by 1 and themselves.

EXAMPLE

Some prime factors are: 2, 3, 5, 7, 11, 13, 17, 19, 23, 29 etc.
Many times, it is useful to be able to factor numbers into their prime factors.

EXAMPLE

$540 = 2^2 \times 3^3 \times 5$
When we are dealing with mathematical expressions that include numbers and letters (variables), one method used to help us is the Greatest Common Factor (GCF).

EXAMPLE

Find the GCF.
$4x^4y^3 + 8x^2y = 4x^2y(x^2y^2 + 2)$
Methods used in the factorization of quadratic expressions.

a. *form where the coefficient of x^2 is 1*.

$x^2 + bx + c$

1. by inspection
We have to find two integers that have their <u>product equal to c</u>, and their <u>sum equal to b.</u>

EXAMPLE

Factor $x^2 + 5x + 6$.
the product $= 2 \times 3 = 6$ *and the sum* $= 2 + 3 = 5$
So, $x^2 + 5x + 6 = (x + 2)(x + 3)$
Remember:
- If the product of the integers is positive, then the integers are either both positive or both negative.
- If the product of the integers is negative, then one integer is positive and the other is negative.

2. the factoring of the form: $x^2 + bxy + cy^2$
Remember that when we have: $(x + 3y)(x + 2y),$ *we get*

$$x^2 + 2xy + 3xy + 6y^2 = x^2 + 5xy + 6y^2$$

If we compare this relation with the one above $x^2 + 5x + 6 = (x + 2)(x + 3)$, we notice that in this case we have the same product and the same sum, so we only have to multiply y to both constants 3 and 2 respectively. $x^2 + 5xy + 6y^2 = (x + 3y)(x + 2y)$

EXAMPLE

Factor $x^2 + 9xy + 20y^2$
$x^2 + 9xy + 20y^2 = (x + 4y)(x + 5y)$

3. the factoring of the difference of squares
Remember that $x^2 - c^2 = (x - c)(x + c)$

EXAMPLE

Factor $x^2 - 5^2$
$x^2 - 5^2 = (x - 5)(x + 5)$

b. *The form where the coefficient of x^2 is different from 1.*

The form is $ax^2 + bx + c$

1. The factoring by grouping
Here, we are splitting the middle term of the trinomial into two terms, in such a way that we can then group first two terms and second two terms together.

EXAMPLE

$6x^2 + 19x + 15 = 16x^2 + 9x + 10x + 15 = 3x(2x + 3) + 5(2x + 3) = (2x + 3)(3x + 5)$

2. The factoring using the decomposition method

Here, we will find two integers so that their product equals a times c, and their sum equals b. In the next step we are splitting the middle term of the trinomial into two terms so that we can then group first two terms and second two terms together.

EXAMPLE

Factor the trinomial: $8x^2 + 10x + 3$

Here, a=8; b=10; c=3 so,

$a \times c = 24$

So, $6 \times 4 = 24 \, ; and \, 6 + 4 = 10$

Then,

$8x^2 + 4x + 6x + 3 = 4x(2x + 1) + 3(2x + 1) = (2x + 1)(4x + 3)$

3. The factoring of the form $ag^2 + bg + c$, where g is monomial

EXAMPLE

Factor $5x^4 + 25x^2 + 30$

Here, the monomial is x^2 which we will substitute with A.

So, $A = x^2$

The original trinomial becomes:

$5A^2 + 25A + 30 = 5(A^2 + 5A + 6) = 5(A + 2)(A + 3)$

Now, we substitute back $A = x^2$ so, we have:

$5(x^2 + 2)(x^2 + 3)$

4. The factoring of the form $a^2g^2 - c^2$, where g is monomial

EXAMPLE

Factor $x^4 - 81$

Here, we substitute $x^2 = A$

$A^2 - 81 = (A - 9)(A + 9)$

Now, we substitute A with x^2, so we have:

$(x^2 - 9)(x^2 + 9) = (x - 3)(x + 3)(x^2 + 9)$

c. Application of the factoring.

Finding the roots of the quadratic equations.

EXAMPLES

Find the roots of $x^2 + 5x + 6 = 0$.

the product $= 2 \times 3 = 6 \, and \, the \, sum = 2 + 3 = 5$

So, $x^2 + 5x + 6 = (x + 2)(x + 3) = 0, x = -2, or \, x = -3$

3. the factoring of the difference of squares

Find the roots of $x^2 - 5^2 = 0$

$x^2 - 5^2 = (x - 5)(x + 5) = 0 \; so, x = 5 \; or \; x = -5$

b. *The form where the coefficient of x^2 is different from 1.*

The form is $ax^2 + bx + c$

1. The factoring by grouping

$6x^2 + 19x + 15 = 16x^2 + 9x + 10x + 15 = 3x(2x + 3) + 5(2x + 3) = (2x + 3)(3x + 5)$

$6x^2 + 19x + 15 = 0$

$(2x + 3)(3x + 5) = 0$

So, $x = -\dfrac{3}{2}, or \; x = -\dfrac{5}{3}$

2. The factoring using the decomposition method

$8x^2 + 10x + 3 = 0$

$(2x + 1)(4x + 3) = 0 \; so, x = -\dfrac{1}{2}, or \; x = -\dfrac{3}{4}$

3. The factoring of the form $ag^2 + bg + c$, where g is monomial

Find the roots of $5x^4 + 25x^2 + 30 = 0$

Here, the monomial is x^2 that we will substitute with A.

So, $A = x^2$

The original trinomial becomes:

$5A^2 + 25A + 30 = 5(A^2 + 5A + 6) = 5(A + 2)(A + 3)$

Now, we substitute back $A = x^2$ so, we have:

$5x^4 + 25x^2 + 30 = 0$

$5(x^2 + 2)(x^2 + 3) = 0$ so, there is no real solutions.

4. The factoring of the form $a^2g^2 - c^2$, where g is monomial

Find the solutions of $x^4 - 81 = 0$

Here, we substitute $x^2 = A$

$A^2 - 81 = (A - 9)(A + 9)$

Now, we substitute A with x^2, so we have:

$(x^2 - 9)(x^2 + 9) = (x - 3)(x + 3)(x^2 + 9)$

$x^4 - 81 = 0$

$(x - 3)(x + 3)(x^2 + 9) = 0$

$x = 3, or\ x = -3$

4.F Quadratic equations

b. Completing the square

In this case, we will try to arrive at a form: $x^2 + 2cx + c^2 = (x + c)^2$

EXAMPLE

Find the roots of the equation $3x^2 + 6x - 9 = 0$

$3(x^2 + 2x - 3) = 0$

We take 3 as common factor.

$3(x^2 + 2x + 1 - 3 - 1) = 0$

Here, $x^2 + 2x + 1 = (x + 1)^2$

We take -4 out of the bracket by multiplying it with 3.

$3(x^2 + 2x + 1) - 4 \times 3 = 0$

We move 6 on the other side of the equation.

$3(x^2 + 2x + 1) = +12$

We divide by 3

$(x + 1)^2 = \frac{12}{3}$

We square both sides.

$\sqrt{(x + 1)^2} = \mp\sqrt{4}$

$x + 1 = \mp\sqrt{4}$

So,

$x = -1 - 2 = -3$

$x = -1 + 2 = 1$

Check

We substitute -3 and 1 respectively in the equation to see if it is satisfied.

If x=-3

$3(-3)^2 + 6(-3) - 9 = 27 - 18 - 9 = 0$

If x=1

$3(1)^2 + 6(1) - 9 = 3 + 6 - 9 = 0$

So,

$x = -3$ and $x = 1$ are the roots of the equation $3x^2 + 6x - 9 = 0$.

TEST 1

1.

1B.a. 1) The horizontal distance between point A and point C is 10

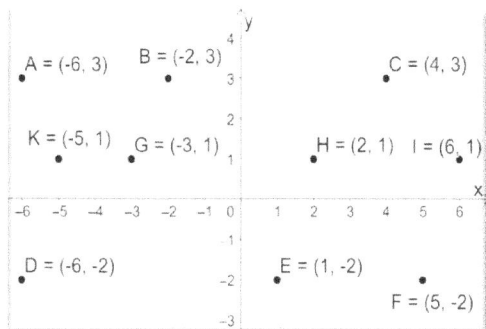

2.

1B.a. 2) The distance between point A and point B is:

3.

1B.b. 1) The vertical distance between point A and point K is:

4.

1B.b. 2) The vertical distance between point A and point D is:

5.

1B.c. 1) The distance between point B and point H is:

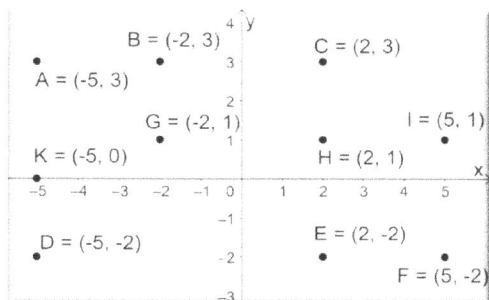

6.

1B.c. 2) The distance between point K and point C is:

7.

1B.d. 1) The mid-point coordinates of segment AB are:

8.

1B.d. 2) The mid-point coordinates of segment BC are:

9.

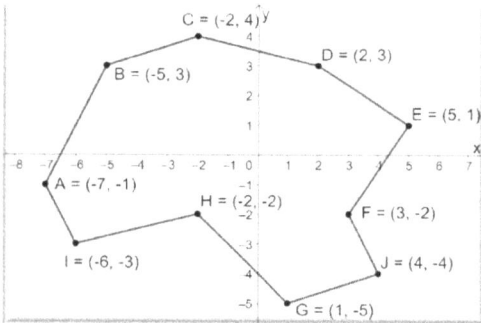

1C. 1) The slope of segment AB is:

10.

1C. 2) The slope of segment BC is:

Mark yourself

1	2	3	4	5	6	7	8	9	10
# of Good Answers (NGA)=				NGA/total number of questions=Ratio				Percent=Ratio*100	
				YOUR PERCENT IS:					%

TEST 2

1.

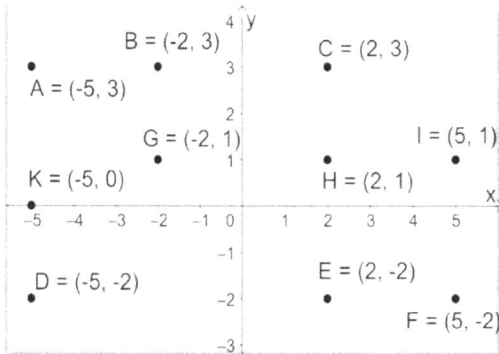

1B.a. 3) The horizontal distance between point D and point E is:

2.

1B.a. 4) The horizontal distance between point E and point F is:

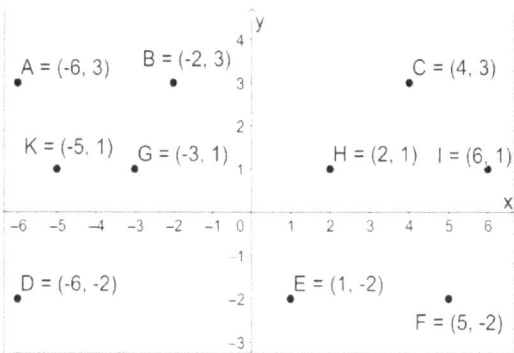

3.

1B.b. 3) The vertical distance between point B and point G is:

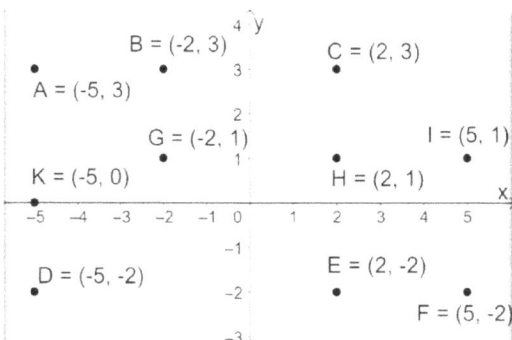

4.

1B.b. 4) The vertical distance between point C and point H is:

5.

1B.c. 3) The distance between point A and point G is:

6.

1B.c. 4) The distance between point A and point H is:

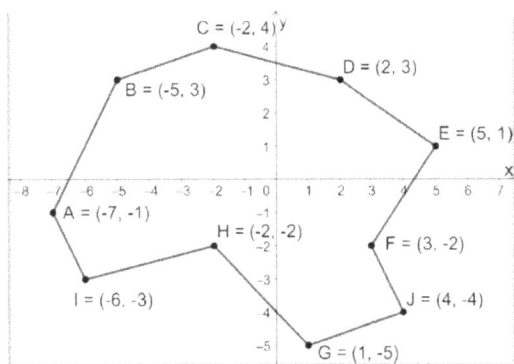

7.

1B.d. 3) The mid-point coordinates of segment CD are:

8.

1B.d. 4) The mid-point coordinates of segment DE are:

9.

1C. 3) The slope of segment CD is:

10.

1C 4) The slope of segment DE is: 4

11.

1D.a. 1) The equation of the line through M (-3,1) and slope -2 is $y = -2x - 5$

12.

1D.a. 2) The y intercept of the line $y = -2x - 5$ is: y=

13.

1D.b. 1) The equation of horizontal line through M (3,4) is:

14.

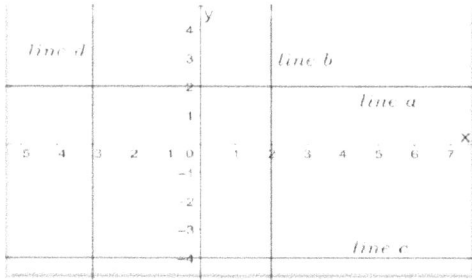

1D.b. 2) The equation of line a is $y = 2$

15.

1E.a. 1) The equation of the parallel line with $y = x - 1$ that intersects y axis at point M (0,5) is:

16.

1E.a. 2) The equation of the parallel line with $y = -3x + 2$ that intersects y axis at point M (0,-3) is:

17.

1E.b. 1) The lines that have the equations $y = 3x + 4$ and $y = -\frac{1}{3}x - 5$ are:

_____

18.

1E.b. 2) The lines that have the equations $y = 3x - 3$ and $y = 3x + 3$ are not

_____.

19.

1F.a. 1) The point symmetric to M (3,4) about y axis has x coordinate equal -3

20.

1F.a. 2) The point symmetric to M (3,4) about x axis has y coordinate equal 2

21.

1F.b. 1) The symmetric point to B (2,3) by the origin is C (-2,-3)

22.

1F.a. 2) The symmetric point to R (0,2) by the origin is B (0,-2)

Mark yourself

1	2	3	4	5	6	7	8	9	10
11	12	13	14	15	16	17	18	19	20
21	22								

# of Good Answers (NGA)=	NGA/total number of questions=Ratio	Percent=Ratio*100
	YOUR PERCENT IS:	%

TEST 3

1.

A = (-6, 3)   B = (-2, 3)   C = (4, 3)

K = (-5, 1)   G = (-3, 1)   H = (2, 1)   I = (6, 1)

D = (-6, -2)   E = (1, -2)   F = (5, -2)

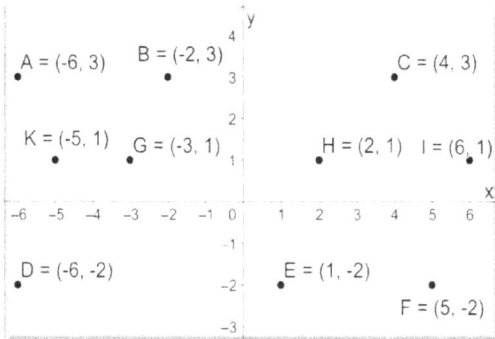

1B.a. 5) The horizontal distance between point K and point G is:

2.

B = (-2, 3)   C = (2, 3)

A = (-5, 3)

G = (-2, 1)   I = (5, 1)

K = (-5, 0)   H = (2, 1)

D = (-5, -2)   E = (2, -2)   F = (5, -2)

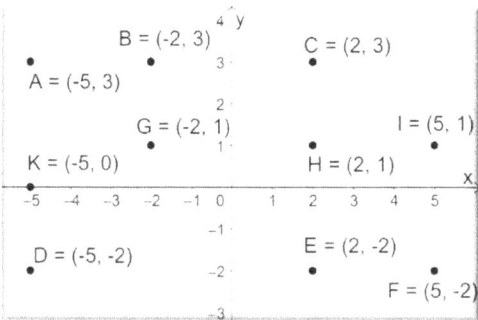

1B.b. 5) The vertical distance between point C and point E is:

3.

1B.c. 5) The distance between point A and point I is:

4.

1B.d. 5) The mid-point coordinates of segment EF are:

5.

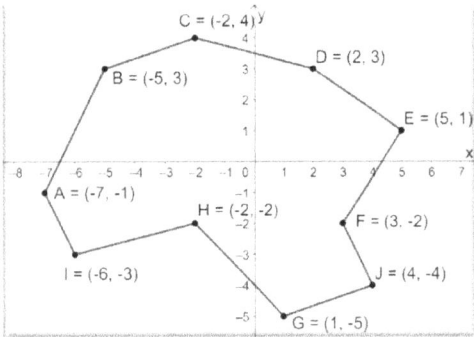

C = (-2, 4)
D = (2, 3)
B = (-5, 3)
E = (5, 1)
A = (-7, -1)
H = (-2, -2)
F = (3, -2)
I = (-6, -3)
J = (4, -4)
G = (1, -5)

1C. 5) The slope of segment EF is:

6.

1D.a. 3) In the slope relation, $m = \dfrac{y-5}{x+4}$, the y intercept in terms of the slope m, is $b =$

7.

1D.a. 4) The equation of the parallel line with $y = 3x + 1$ that passes through the point M (5,6) is:

8.

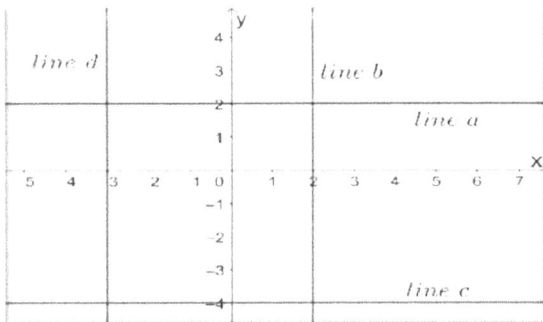

line d
line b
line a
line c

1D.b. 3) The equation of line b is $x = 3$

9.

1D.b. 4) The equation of line c is $y = -4$

10.

1E.a. 3) The line $y = -5x + 3$ is not parallel with $y = -4x + 3$

11.

1E.a. 4) The $y = 3x - 1$ is the same with $y = 3x + 1$

12.

1E.b. 3) The line perpendicular to the line $y = -2x + 7$ has the slope $m = 2$.

13.

1E.b. 4) The equation of the line perpendicular to $y = 5x - 1$ that passes through M (3,4) is:

14.

1F.a. 3) The point symmetric to M (5,4) about the vertical x=1 has the x coordinate equals:

15.

1F.a. 4) The point symmetric to M (4,3) about the vertical x=2 has the x coordinate equals:

16.

1F.b. 3) The symmetric point to R (3,0) by the point S (-1,3) is:

17.

1F.b. 4) The symmetric point to P (-2,-3) by the point S (-1,2) is L (1,2)

18.

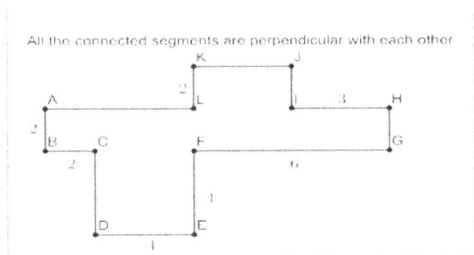

All the connected segments are perpendicular with each other

2A.a. 1) The area of the rectangle ABFL is 12

19.

2A.a. 2) The area of the square CDEF is 160

20.

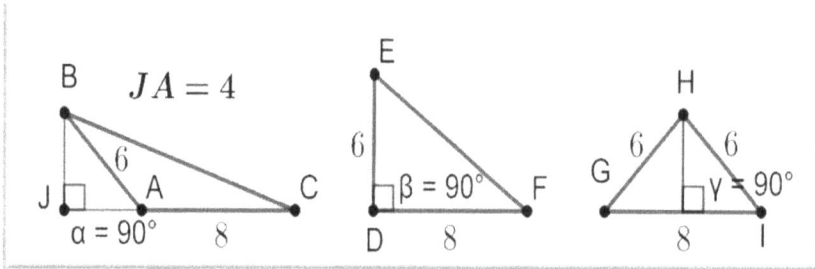

2A.b. 1) The area of the triangle BAC is:

21.

2A.b. 2) The area of the triangle DEF is:

22.

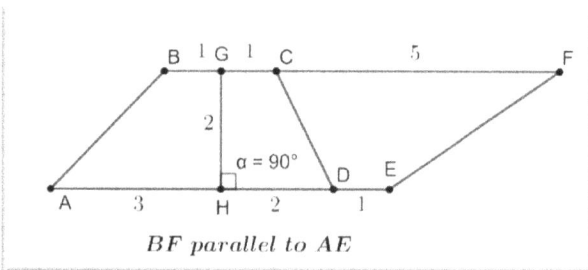

BF parallel to AE

2A.c. 1) The area of the trapezoid ABGH is:

23.

2A.c. 2) The area of the trapezoid ABCD is:

24.

2B.a. 1) The volume of the prism with a rectangular base where Length =5, Width = 4 and Height = 3 units is:

25.

2B.a. 2) The volume of a right triangular prism with the base a triangle with area = 20 and height = 5 is:

26.

2C.a. 1) The secant is the line that intersects the circle in _____ points.

27.

2C.a. 2) The length of the line EA is 11.66

28.

2C.b. 1) In a circle we have _____ radians

29.

2) The <u>central angle</u> is an angle whose arms go from the _____ of the circle to the circle itself.

Mark yourself

1	2	3	4	5	6	7	8	9	10
11	12	13	14	15	16	17	18	19	20
21	22	23	24	25	26	27	28	29	
# of Good Answers (NGA)=				NGA/total number of questions=Ratio				Percent=Ratio*100	
				YOUR PERCENT IS:					%

TEST 4

1.

A = (-6, 3) B = (-2, 3) C = (4, 3)
K = (-5, 1) G = (-3, 1) H = (2, 1) I = (6, 1)
D = (-6, -2) E = (1, -2) F = (5, -2)

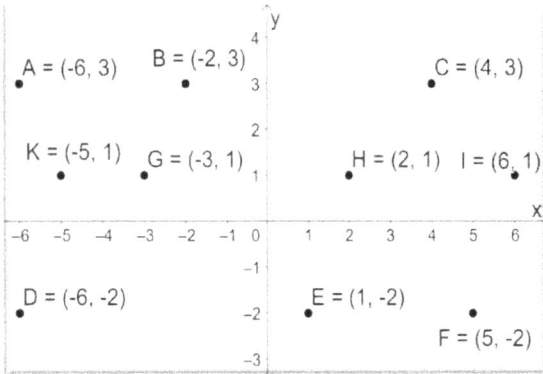

1B.a. 6) The horizontal distance between point K and point H is:

2.

B = (-2, 3) C = (2, 3)
A = (-5, 3)
G = (-2, 1) I = (5, 1)
K = (-5, 0) H = (2, 1)
D = (-5, -2) E = (2, -2) F = (5, -2)

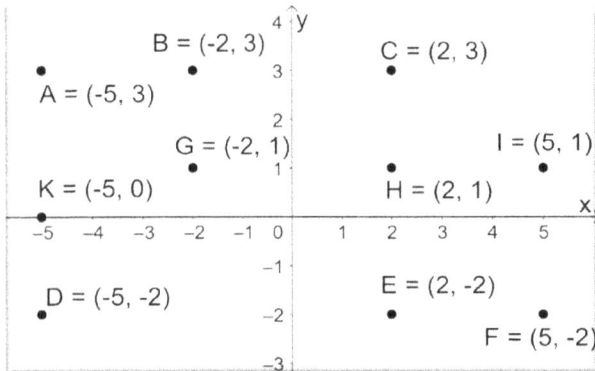

1B.b. 6) The vertical distance between point K and point D is:

3.

1B.c. 6) The distance between point A and point F is:

4.

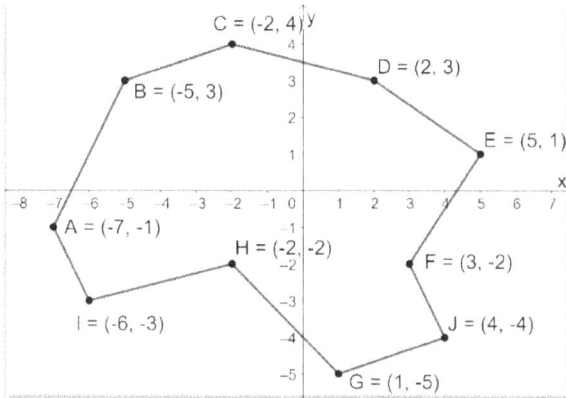

1B.d. 6) The mid-point coordinates of segment FJ are:

5.

1C. 6) The slope of segment FJ is:

6.

1D.a. 5) Y intercept of the parallel line with $y = 3x + 1$ in problem 4 is b=

7.

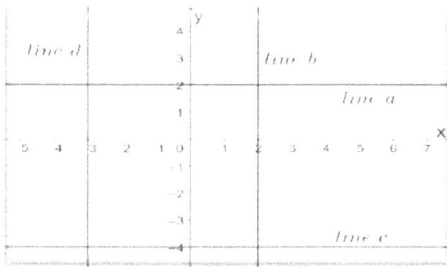

1D.b. 5) The equation of line d is $x = +3$

8.

1E.a. 5) The line $y = 4x + 3$ is parallel with $y = x - 25$

9.

1E.b. 5) The equation of the line perpendicular to $y = -2x + 3$ through M (-2,-3) is:

10.

1F.a. 5) The point symmetric to M (-1,2) about the vertical x=3 has the x coordinate equals:

11.

1F.b. 5) The symmetric point to P (-3,4) by the point S (0,0) is L (   )

12.

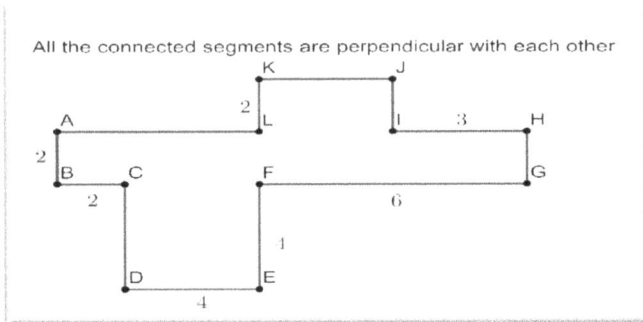

All the connected segments are perpendicular with each other

2A.a. 3) The area of FKJIHGF is:

13.

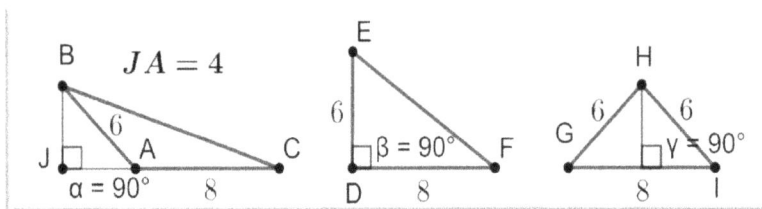

$JA = 4$

$\alpha = 90°$

$\beta = 90°$

$\gamma = 90°$

2A.b. 3) The area of triangle GHI is:

14.

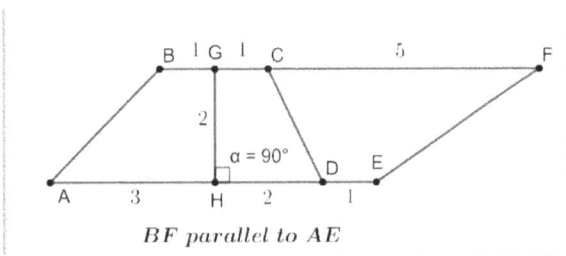

$\alpha = 90°$

BF parallel to AE

2A.c. 3) The area of the trapezoid CDEF is:

15.

2B.a. 3) The volume of a cube with the side of 5 is:

16.

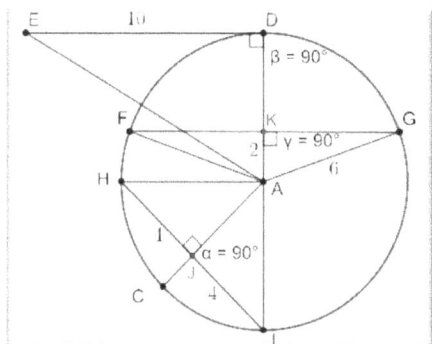

2C.a. 3) **The length of the segment KD is:**

17.

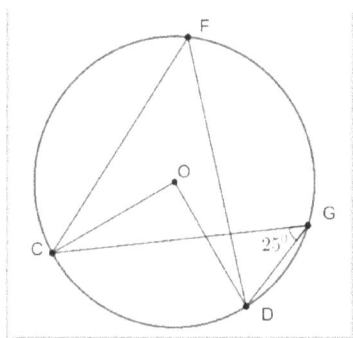

2C.b. 3) **The measure of the angle CFD is:**

18.

3A.a. 1) **The shadow of a 20 m tall tree with the light rays of 23^0 with the vertical is:**

19.

4A.a. 1) The simplified expression of $\frac{5x^6y^9}{15x^4y^3} ; x, y \neq 0$ is:

20.

4A.b. 1) The root of $\sqrt[4]{81x^4}$ is:

21.

4A.c. 1) The mixed radical of $\sqrt{192}$ is:

22.

4B.a. 1) $(6+7)+4$ is equal with $(6+4)+7$

23.

4B.b. 1) $(3 + x)(x - 1)$ is equal with:

24.

4C.a. 1) The expression $\log_3(4 \times 6)$ will become $\log_3 4 + \log_3 6$

25.

4C.b. 1) The solution of the equation $620 = 2 * 3^{x+1}$ is $x =$

26.

4D.a. 1) The simplified polynomial of $(2x^2 - 3x - 5) + (-4x + 3)$ is:

27.

4D.b. 1) The rationalized expression of $\frac{2}{\sqrt{7}}$ is:

Mark yourself

1	2	3	4	5	6	7	8	9	10
11	12	13	14	15	16	17	18	19	20
21	22	23	24	25	26	27			
# of Good Answers (NGA)=				NGA/total number of questions=Ratio				Percent=Ratio*100	
				YOUR PERCENT IS:					%

TEST 5

1.

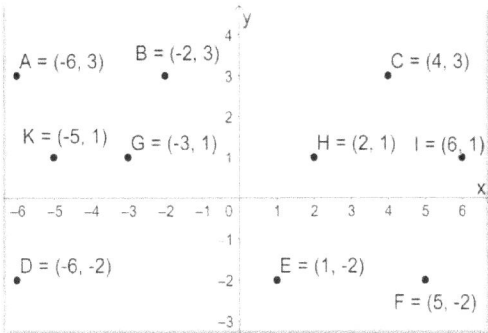

1B.a. 7) The distance between point K and point I is:

2.

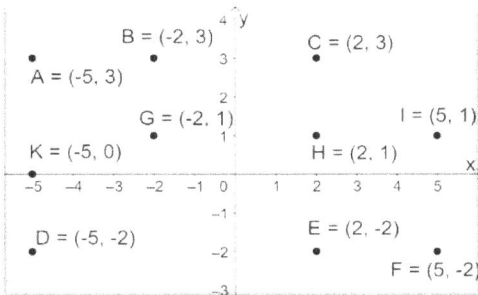

1B.b. 7) The distance between point H and point E is:

3.

1Bc7) The distance between point A and point E is:

4.

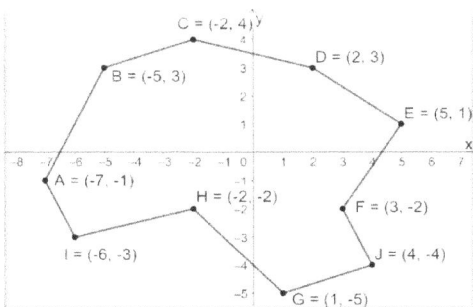

1B.d. 7) The mid-point coordinates of segment JG are:

5.

1C. 7) The slope of segment JG is: $\frac{1}{3}$

6.

1D.a. 6) The intersection to x axis of $y = 4x - 8$ *is* P (   )

7.

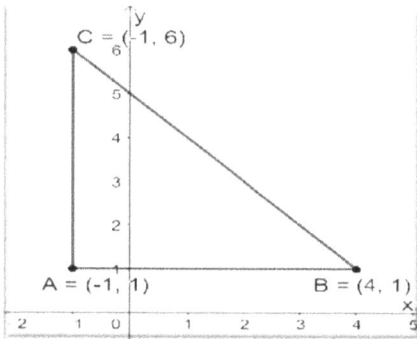

1D.b. 6) The equation of the line that passes through the points C and A in the figure below is $x =$

8.

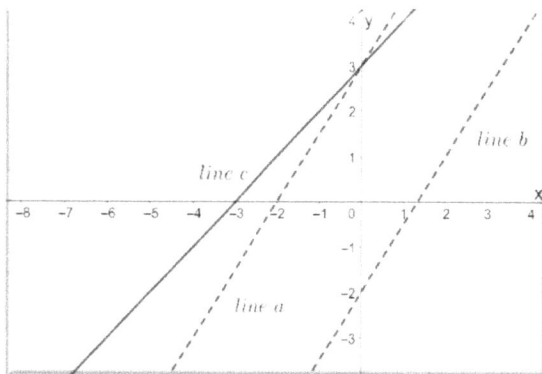

1E.a. 6) The line a is _____ to line b

9.

1E.b. 6) The equation of the line perpendicular to $y = x + 2$ through K (1,2) intersects the y axis in M (   )

10.

1F.a. 6) The point symmetric to M (-1,1) about the horizontal y=2 has y=

11.

1F.b. 6) The symmetric point to P (3,4) by the point S (0,0) is L (   )

12.

2A.a. 4) The area of the whole figure is:

All the connected segments are perpendicular with each other

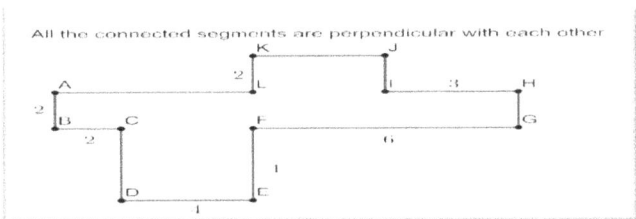

13.

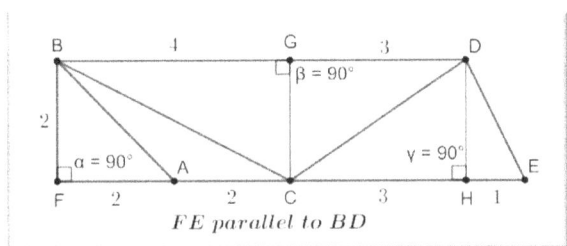

FE parallel to BD

2A.b. 4) The area of triangle BFA is:

14.

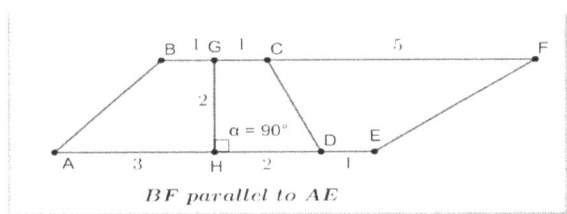

BF parallel to AE

2A.c. 4) The area of the trapezoid GHEF is:

15.

2B.a. 4) The volume of a cylinder with radius = 3 and height = 6 is:

16.

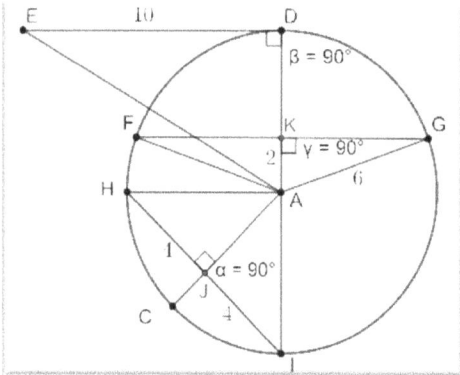

2C.a. 4) The length of the segment KG is:

17.

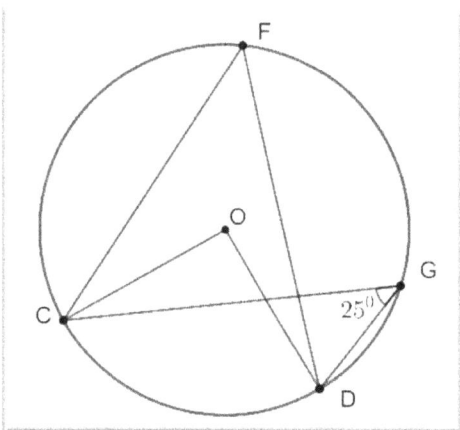

2C.b. 4) The measure of the angle COD is 50^0

18.

3A.a. 2) Eiffel Tower in Paris has the line between top and the end of the shadow 400 m long. The height of the tower is 324 m. The angle of the light rays and soil is:

19.

4A.a. 2) The simplified expression of $(6x^4y)(xy^2)(-3x^4y^3)$ is:

20.

4A.b. 2) The simplified expression of $\sqrt[3]{x^2y^3} * \sqrt[3]{x^4y^6}$ is:

21.

4A.c. 2) The mixed radical of $\sqrt{160}$ is:

22.

4B.a. 2) 2+(3+6)+5 is equal with (2+6)+(2+5)-5

23.

4B.b. 2) $(-5 + 3x)(x + 6)$ is equal with $3x^2 + 13x - 30$

24.

4C.a. 2) The expression $\frac{1}{3}\log_6 a + 5 \log_6 b - 7 \log_6 c$ will become

$\log_6(\frac{a^{\frac{1}{3}} * b^5}{c^7})$; $a, b, c > 0$

25.

4C.b. 2) The solution of the equation $5 = \log_5 x + \log_5(x - 3)$ is $x = 36.54; x > 3$

26.

4D.a. 2) The simplified polynomial of $(-4x^2 + 2x) - (-3x^2 - 5x + 7)$ is:

27.

4E.a. 1) The slope of the line $y = 3x - 1$ is:

28.

4F.a. 1) The factored expression of $x^2 + 2x - 15$ is:

29.

4F.b. 1) The number to be added to $x^2 + 2x$ to make a perfect square is:

Mark yourself

1	2	3	4	5	6	7	8	9	10
11	12	13	14	15	16	17	18	19	20
21	22	23	24	25	26	27	28	29	

# of Good Answers (NGA)=	NGA/total number of questions=Ratio	Percent=Ratio*100
	YOUR PERCENT IS:	%

CHAPTER 5

FUNCTIONS

5.A Parent functions (Review Pre-Calculus 11 and 12)

a. Determine: if a relation is a function, the values of a function, the range

A relation is a <u>function</u> when for each value x that belongs to the domain, there is <u>only one value</u> y that belongs to the range.

EXAMPLE

Going through the function f in the picture below, there is only one value in the Range that corresponds to any value in the Domain. For value 2 in the Domain, it corresponds with only one value in the Range (4). For value 4 in the Domain, it corresponds with only one value in the Range (4). For value 3 in the Domain, it corresponds with only one value in the Range (9)..

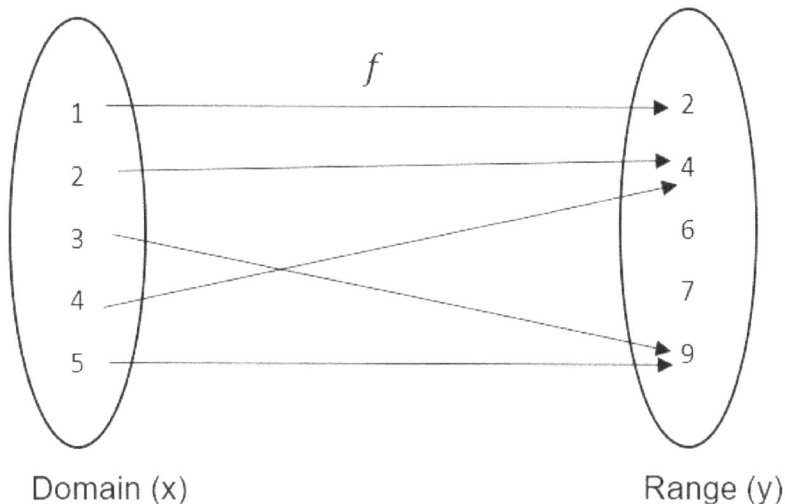

Domain (x)   Range (y)

The value of a function is the value in the Range that is the result of a value in the Domain going through the function f.

EXAMPLE

In the figure above one value of the function f is **9.** This number is the result of the domain value of 3 going through f as well as 5 going through f.
The <u>Range</u> is the set of elements (letters or numbers) that are the result of the elements (letters or numbers) in the Domain going through function f.

EXAMPLE

The set of numbers {2,4,6,7,9} in the figure above is the Range of f.
The Domain is the set of numbers {1,2,3,4,5}.

5.A Parent functions (Review Pre-Calculus 11 and 12)

b. Linear and quadratic functions and their graphs

A function is <u>linear</u> when the difference between consecutive values in the Domain is always the same. In the same time, the difference between consecutive values in the Range is always the same, not necessarily the same with the difference between consecutive Domain values.

EXAMPLE

Domain(X)	Range(Y)	Point
1	5	A
2	7	B
3	9	C
4	11	D

Here the difference between consecutive values in the Domain is 1 $x_B - x_A = 2 - 1 = 1$, or $x_D - x_C = 4 - 3 = 1$.
In the same time the difference between consecutive values in the Range is **2.** $y_B - y_A = 7 - 5 = 2$, or $y_D - y_C = 11 - 9 = 2$.

When we graph a linear function, the graph is a straight line.

The pairs (x,y) represent points in the Cartesian system of axes.

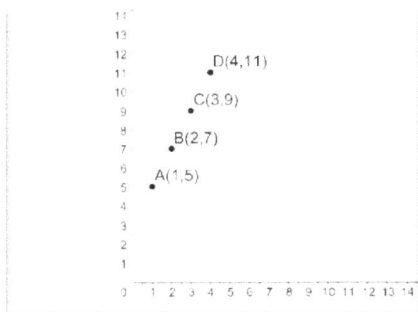

We can see here that if we connect these points it will create a straight line.

The equation that represents this graph is $y = 2x + 3$ and it is called a linear equation. (the grade of the equation is either 0 or 1).

A function is *quadratic* when the relation between x and y is a polynomial of second degree.

$y = x^2 - 5x + 4$ or $xy + 2x^2 - 3y = 5$

EXAMPLE

$y = x^2 - 5x + 4$

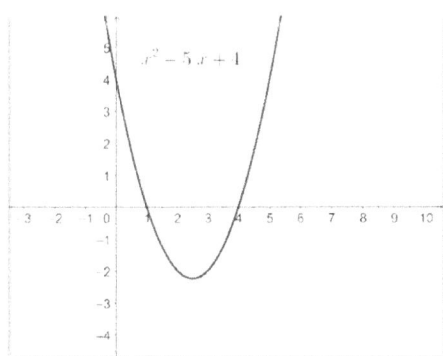

Here the quadratic function $y = x^2 - 5x + 4$ is represented.

VERTICAL LINE TEST

If we draw a vertical line through a graph and this line **intersects the graph** at **only one point**, then the graph represents a **function**.

If this line **intersects the graph** at **more than one point**, then the graph DOES NOT represent a Function.

5.A Parent functions (Review Pre-Calculus 11 and 12)

c. Inverse functions and their graphs

An <u>inverse</u> function represented by $f^{-1}(x)$ is the function that has the Domain equal to the Range of the original function $f(x)$. The Range of the inverse function equals the Domain of the original function.

EXAMPLE

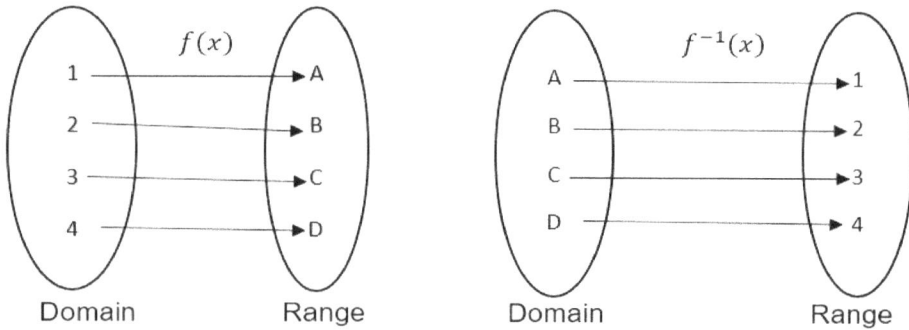

EXAMPLE

We obtain the equation of the inverse function in few steps:

1. Write the equation of the original function.

The original function is $f(x) = y = 4x + 7$

2. Switch the variables x and y in the original formula.

$x = 4y + 7$

3. Solve for y

$x - 7 = 4y$ so, $y = f^{-1}(x) = \frac{x-7}{4}$

The graph of an inverse function is always a reflection of the graph of the original function by y=x.

EXAMPLE

$f(x) = 4x + 1 \text{ and } f^{-1}(x) = \frac{x-1}{4}$

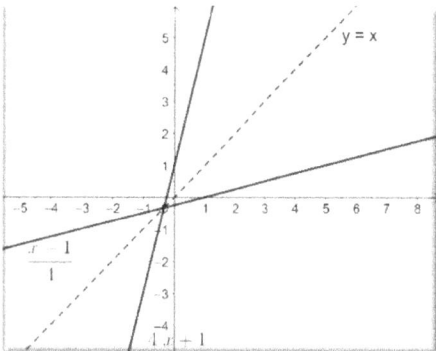

PROPERTY OF INVERSE FUNCTIONS

$$f\left(f^{-1}(x)\right) = f^{-1}(f(x)) = x$$

The function of the inverse function for a value x, equals the inverse function of the same function for value x, and equals value x.

5.B Piecewise functions

A *piecewise function* behaves differently on certain intervals of the Domain. It can be split into subsections.

EXAMPLE

$$f(x) = \begin{cases} x \text{ for } x \leq -2 \\ 1 \text{ for } -2 < x < 3 \\ 2x - 2 \text{ for } x \geq 3 \end{cases}$$

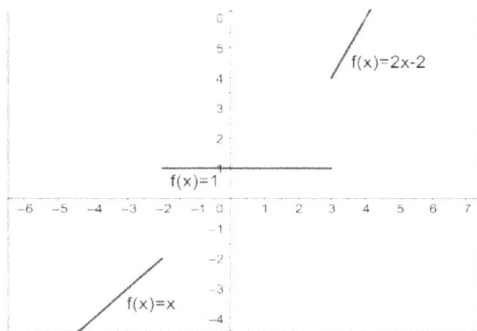

This function has three subsections.

As we can see for all the values,

$x \leq -2 \; f(x) = x$.

For $-2 < x < 3$ the function is a horizontal line

that crosses y axis at y=1.

For all the values of x greater or equal to 3, the function is a straight line with slope equal with 2 and

y-intercept equal with -2.

EXAMPLE

$$f(x) = \begin{cases} 2 - x \text{ for } x < 1 \\ 3 \text{ for } x \geq 1 \end{cases}$$

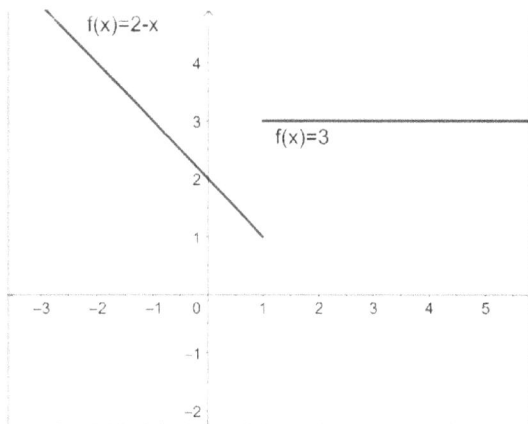

This function has two subsections.

As we can see for all the values,

$x \leq 1 \; f(x) = 2 - x$.

The straight line with slope of -1 crosses the y axis

at y=2.

For all the values of x greater or equal to 1, the

function is a straight line with slope equal to 2.

5.C Trigonometric functions

The 6 trigonometric functions are:

$$\sin(x) \, ; \cos(x) \, ; \tan(x) = \frac{\sin(x)}{\cos(x)} \, ; \cot(x) = \frac{\cos(x)}{\sin(x)} = \frac{1}{\tan(x)} \, ; \sec(x) = \frac{1}{\cos(x)} \, ; \csc(x) = \frac{1}{\sin(x)}$$

The trigonometric functions are *periodic* functions. This means they repeat themselves periodically.

EXAMPLE

The graph of sin(x) is represented below.

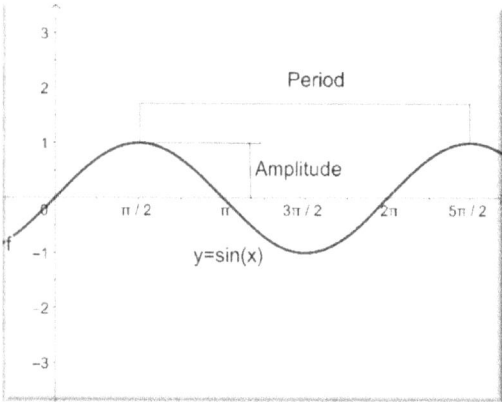

One <u>radian</u> is the measure of an angle subtended at the center of a circle by an arc which is equal in length to the radius of the circle as can be seen in figure below.

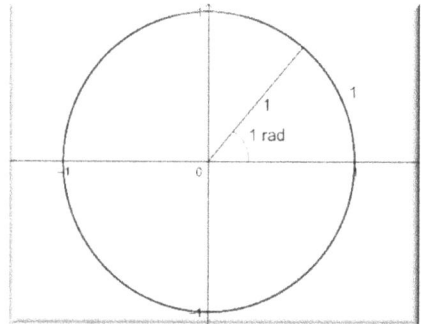

$$2\pi \; radians = 360^0 so, \pi \; radians = 180^0$$

As can be seen below is the period for

$f(x) = \sin(x) \; and \; g(x) = \cos(x) \; is \; 2\pi, instead \; h(x) = \tan(x)$ has a period of π.

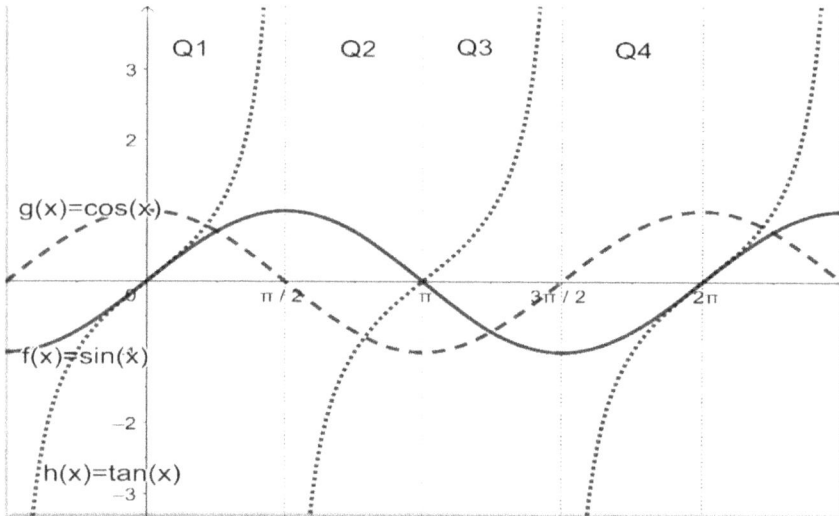

Q1 – first quadrant
Q2 – second quadrant
Q3 – third quadrant
Q4 – fourth quadrant

The sign of functions sin(x); cos(x) and tan(x) in each quadrant is shown below.

	Q1	Q2	Q3	Q4
Sin(x)	+	+	-	-
Cos(x)	+	-	-	+
Tan(x)	+	-	+	-

EXAMPLE

Tan(x) is negative in quadrants Q2 and Q4.

In a right-angle triangle, with $\sphericalangle\phi \neq 90^0$ we have the following *trigonometric ratios*:

$$\sin \sphericalangle\phi = \frac{opposite}{hypotenuse}$$

$$\cos \sphericalangle\phi = \frac{adjacent}{hypotenuse}$$

$$\tan \sphericalangle\phi = \frac{opposite}{adjacent}$$

$$\cot \sphericalangle\phi = \frac{adjacent}{opposite}$$

EXAMPLE

In the right-angle triangle ΔABC with $\sphericalangle B = 90^0$.

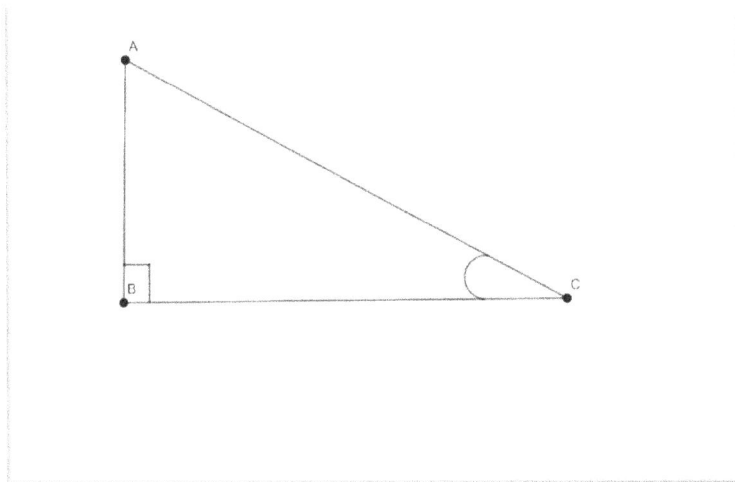

$$\sin \sphericalangle C = \frac{opposite}{hypotenuse} = \frac{AB}{AC}$$

$$\cos \sphericalangle C = \frac{adjacent}{hypotenuse} = \frac{BC}{AC}$$

$$\tan \sphericalangle C = \frac{opposite}{adjacent} = \frac{AB}{BC}$$

$$\cot \sphericalangle\phi = \frac{adjacent}{opposite} = \frac{BC}{AB}$$

5.D Graphs of trigonometric functions

a Graphing sine and cosine functions

Sine and cosine trigonometric functions have the domain: all real values of x, and the range: interval [-1, 1].

EXAMPLE

The graph of $f(x) = \sin(x)$ is shown below.

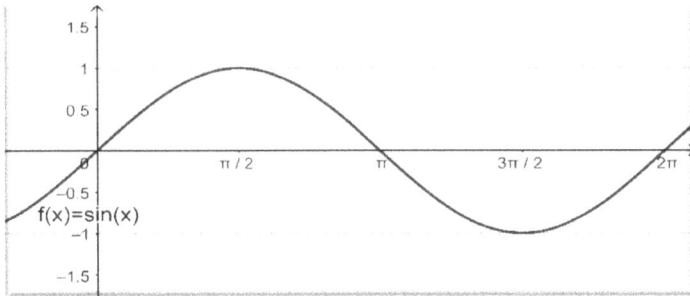

The value of $\sin\left(\frac{3\pi}{2}\right) = -1 \mp 2n\pi, n$ *is an integer.*

The value of $\sin(2\pi) = 0 \mp 2n\pi, n$ *is an integer.*

EXAMPLE

The graph of $f(x) = \cos(x)$ is shown below.

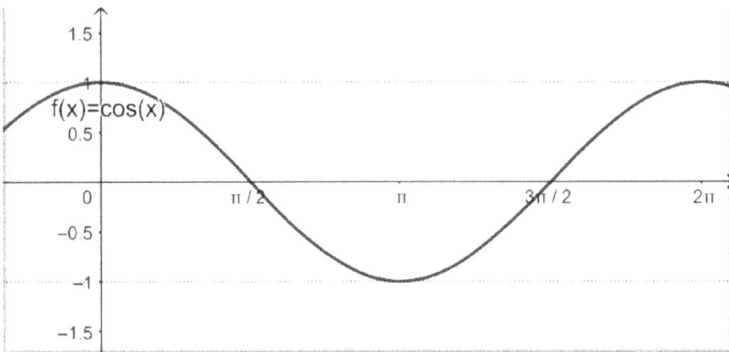

The value of $\cos\left(\frac{3\pi}{2}\right) = 0 \mp 2n\pi, n$ *is an integer.*

The value of $\sin(2\pi) = 1 \mp 2n\pi, n$ *is an integer.*

5.D Graphs of trigonometric functions

b Graphing tangent and cotangent functions

Tangent and cotangent trigonometric functions have the domain all the real values for x but the ones where either sine or cosine functions are zero.

For values of $x = \frac{\pi}{2} \mp k\pi, k - integer$ the tangent is undefined. For these values of x there are vertical asymptotes. The range is: all real numbers.

Asymptotes are lines where graphs of the functions go towards zero and the distance between the graph and asymptote approaches zero but never becomes zero.

The tangent graph is shown below.

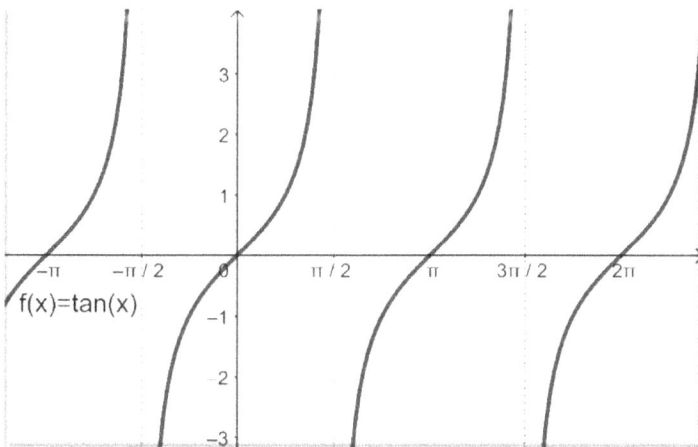

f(x)=tan(x)

EXAMPLE

The tangent of zero is zero.

We know that $\tan(x) = \frac{\sin(x)}{\cos(x)}$ so, the tangent function is zero when the sine function is zero. The tangent function is <u>undefined</u> when the cosine function is zero.

EXAMPLE

Sine function is zero for $x = 0, and \; tangent \; is \; zero \; for \; x = \mp k\pi \; where \; k - integer.$

We know that $\cot(x) = \frac{\cos(x)}{\sin(x)}$. The cotangent function is undefined for $x = \mp k\pi, k - integer.$

The sign functions Sin(Φ), Cos(Φ), and Tan(Φ) in each quadrant is shown below.

	Q1	Q2	Q3	Q4
Sin(x)	+	+	-	-
Cos(x)	+	-	-	+
Tan(x)	+	-	+	-

The values for the special angles in a right-angle triangle are given below.

	30^0	60^0	45^0
Sin(Φ)	$\dfrac{1}{2}$	$\dfrac{\sqrt{3}}{2}$	$\dfrac{\sqrt{2}}{2}$
Cos(Φ)	$\dfrac{\sqrt{3}}{2}$	$\dfrac{1}{2}$	$\dfrac{\sqrt{2}}{2}$
Tan(Φ)	$\dfrac{\sqrt{3}}{3}$	$\sqrt{3}$	1

EXAMPLE

If $\sin (30^0) = \frac{1}{2}$ and, $\cos(30^0) = \frac{\sqrt{3}}{2}$ then:

$$\tan (30^0) = \frac{\sin (30^0)}{\cos (30^0)} = \frac{\frac{1}{2}}{\frac{\sqrt{3}}{2}} = \frac{1}{2} \div \frac{\sqrt{3}}{2} = \frac{1}{2} \times \frac{2}{\sqrt{3}} = \frac{1}{\sqrt{3}} = \frac{\sqrt{3}}{3}$$

The cotangent graph is shown below.

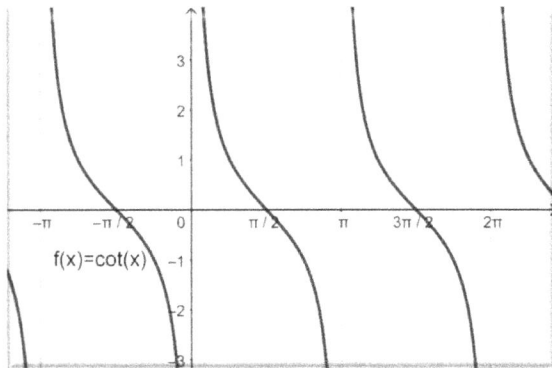

For values of $x = \pi \mp k\pi, k - integer$ the cotangent is undefined. For these values of x, there are vertical asymptotes.

EXAMPLE

If $\sin (30^0) = \frac{1}{2}$ and, $\cos (30^0) = \frac{\sqrt{3}}{2}$ then:

$$\cot (30^0) = \frac{\cos (30^0)}{\sin (30^0)} = \frac{\frac{\sqrt{3}}{2}}{\frac{1}{2}} = \frac{\sqrt{3}}{2} \div \frac{1}{2} = \frac{\sqrt{3}}{2} \times \frac{2}{1} = \sqrt{3}$$

If sin $(60^0) = \frac{\sqrt{3}}{2}$ and, cos $(60^0) = \frac{1}{2}$ then:

cot $(60^0) = \frac{\cos (60^0)}{\sin (60^0)} = \frac{\frac{1}{2}}{\frac{\sqrt{3}}{2}} = \frac{1}{2} \div \frac{\sqrt{3}}{2} = \frac{1}{2} \times \frac{2}{\sqrt{3}} = \frac{1}{\sqrt{3}} = \frac{\sqrt{3}}{3}$

5.E Inverse trigonometric functions

Remember that an _inverse_ function $f^{-1}(x)$ is the function that has the Domain equal to the Range of the original function $f(x)$. The Range of the inverse function equals the Domain of the original function.

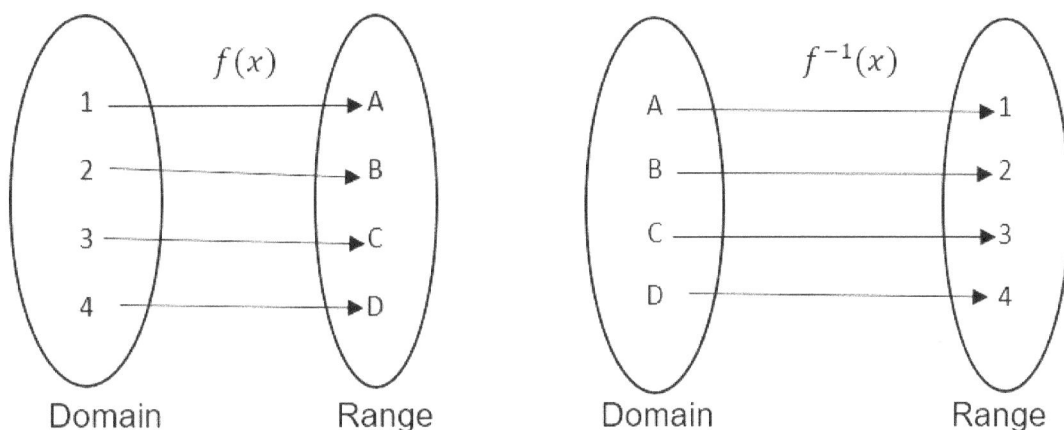

EXAMPLE

Because we are talking about trigonometric functions like sine, cosine and tangent, in the notation for inverse trigonometric function,
we substitute $f^{-1}(x)$ with $\Phi = sin^{-1}(x)$, where:
 x is value of the original trigonometric function.
 Φ is the angle we are looking for.

EXAMPLE

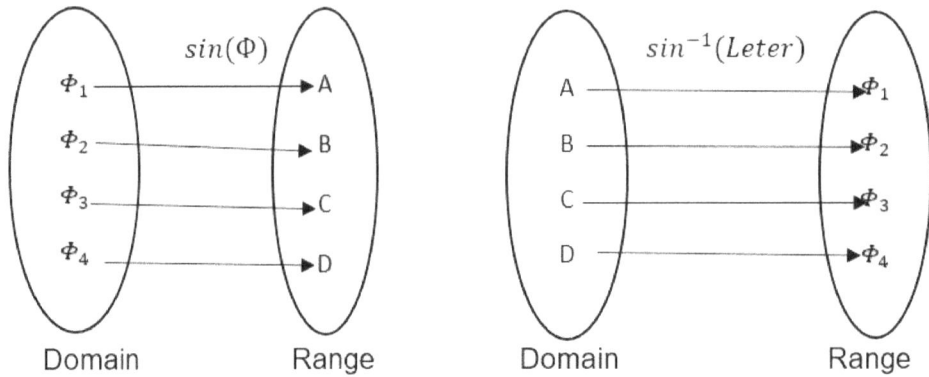

Domain	Range	Domain	Range

EXAMPLE

If $\sin(\Phi) = 0.345$, then $\Phi = sin^{-1}(0.345) = 20.18^0$

If $\tan(\Phi) = 1.89$, then $\Phi = tan^{-1}(1.89) = 62.11^0$

5.F Graphs of inverse trigonometric functions

The inverse function of sine is $f^{-1}(x) = sin^{-1}(x) = \arcsin(x)$.

The graph of $f^{-1}(x) = sin^{-1}(x) = \arcsin(x)$ is shown below.

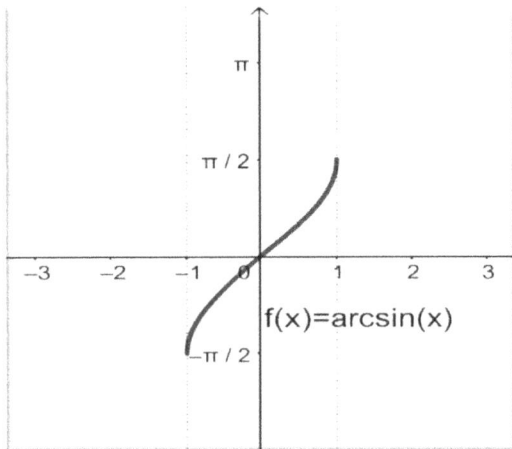

The domain is the interval [-1, 1].

The range is the interval $\left[-\frac{\pi}{2}, \frac{\pi}{2}\right]$.

EXAMPLE

$f(0) = \arcsin(0) = 0$

$f(-1) = \arcsin(-1) = -\frac{\pi}{2}$

The graph of $f^{-1}(x) = cos^{-1}(x) = \arccos(x)$ is shown below.

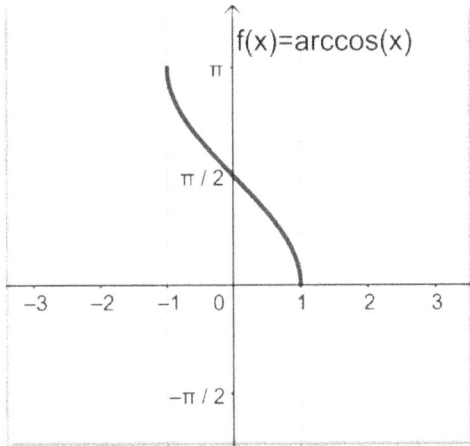

The domain is the interval [-1, 1].
The range is the interval $[0, \pi]$.

EXAMPLE

$f(-1) = \arccos(-1) = \pi$

The graph of $f^{-1}(x) = tan^{-1}(x) = \arctan(x)$ is shown below.

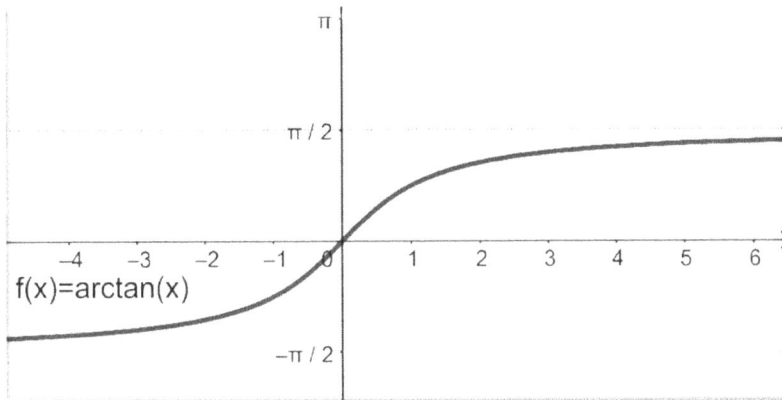

The domain is the real numbers.
The range is the interval $\left[-\frac{\pi}{2}, \frac{\pi}{2}\right]$.

EXAMPLE

$f(1) = \arctan(1) = \frac{\pi}{4}$

$f(-4) = \arctan(-4) = -0.47\,\pi$

The graph of $f^{-1}(x) = cot^{-1}(x) = \text{arccot}(x)$ is shown below.

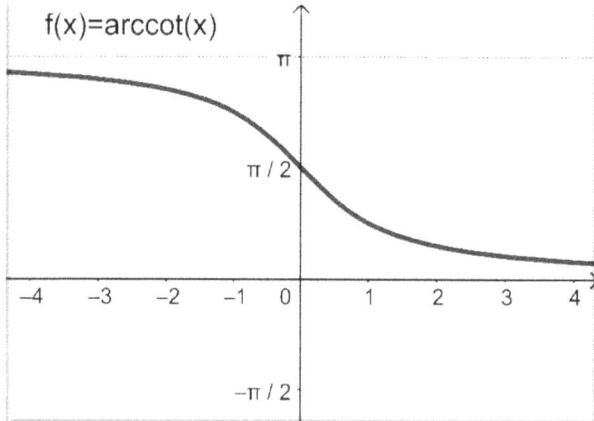

f(x)=arccot(x)

The domain is the real numbers.
The range is the interval $[0, \pi]$.

EXAMPLE

$f(-1) = \text{arccot}(-1) = 0.75\pi$

$f(-4) = \text{arccot}(-4) = 0.92\,\pi$

We could calculate the arccot(x) as $\frac{\pi}{2} - [\arctan(x)]$.

EXAMPLE

$f(-1) = \text{arccot}(-1) = \frac{\pi}{2} - [\arctan(x)] = \frac{\pi}{2} - [\arctan(-1)] = \frac{\pi}{2} - \left[-\frac{\pi}{4}\right] = \frac{\pi}{2} + \frac{\pi}{4} = \frac{3\pi}{4} = 0.75\pi$

$f(-4) = \text{arccot}(-4) = \frac{\pi}{2} - [\arctan(x)] = \frac{\pi}{2} - [\arctan(-4)] = \frac{\pi}{2} - [-0.42\pi] = 0.92\,\pi$

TEST 6

Includes Functions

1.

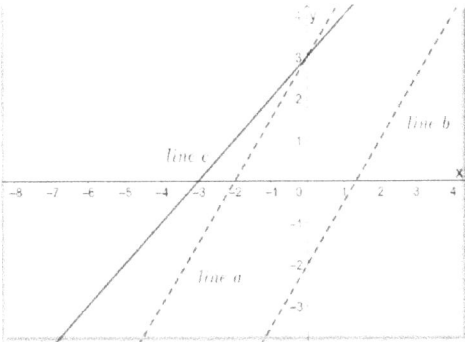

1E.a. 7) The equation of the line a is:

2.

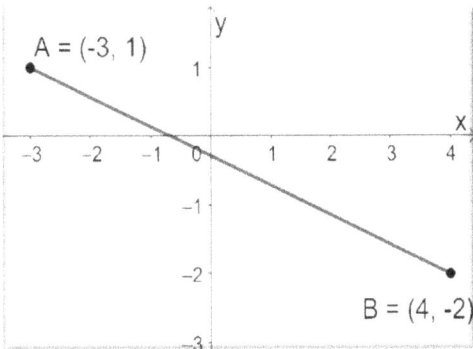

1E.b. 7) The equation of the line perpendicular to AB in point B, is:

3.

1F.a. 7) The point symmetric to M (-3,-2) about the vertical y=3 has the y coordinate =

4.

1F.b. 7) The point L (-1, 1.5) is not symmetric to P (2,-3) by the point S (   )

5.

All the connected segments are perpendicular with each other

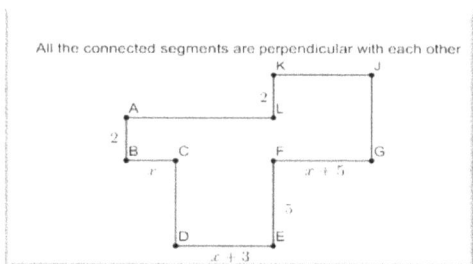

2A.a. 5) The area of the square CDEF is

$5x + 15$

6.

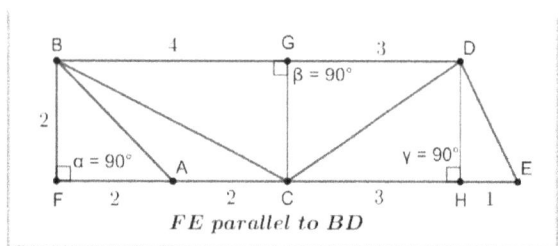

β = 90°

α = 90°

γ = 90°

FE parallel to BD

2A.b. 5) The area of triangle BFC is:

7.

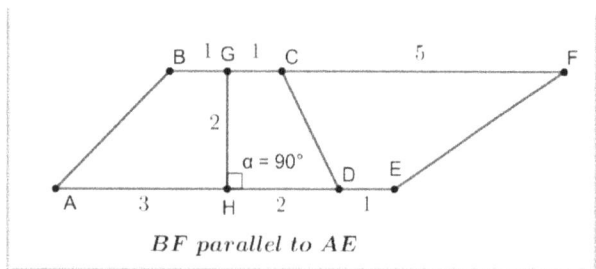

α = 90°

BF parallel to AE

2A.c. 5) The area of the trapezoid GHCD is:

8.

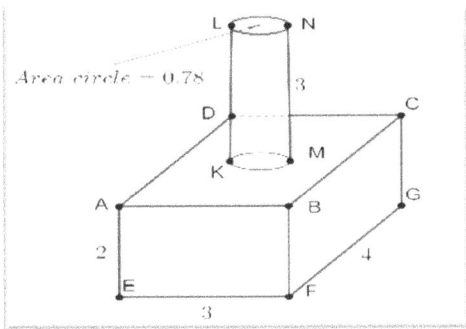

2B.a. 5) The volume of the cylinder is:

9.

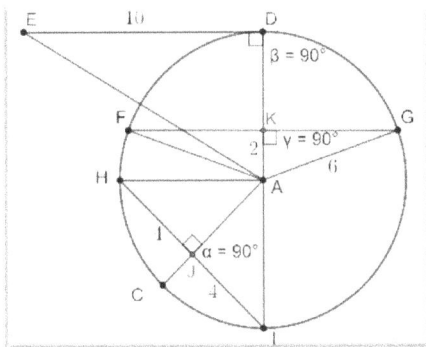

2C.a. 5) The length of the segment CJ is:

10.

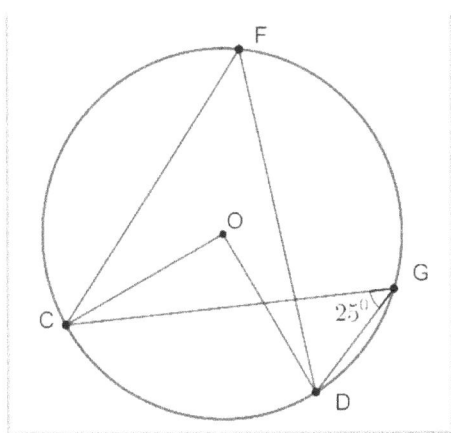

2C.b. 5) The angle CFD is called inscribed angle.

11.

3A.a. 3) The 55.86 m tall Tower of Pisa, Italy, is lean with almost four degrees. The distance from the vertical is:

12.

4A.a. 3) The simplified expression of $(-3a^2b^3)^3$ is:_____; $a \neq 0$

13.

4A.b. 3) The simplified expression of $\sqrt{\dfrac{9x^4y}{y^{-2}}} * \sqrt{\dfrac{4xy^2}{x^{-2}}}$ is:

14.

4A.c. 3) The mixed radical of $\sqrt{175}$ is:

15.

4B.a. 3) $(5 \times 7) \times 3 = 5 \times (7 \times 3)$

16.

4B.b. 3) $(2x^2 - 3x)(4x + 5)$ is equal with $8x^3 + 2x^2 - 15x$

17.

4C.a. 3) The expression $\log_3 5^7$ will become:

18.

4C.b. 3) The solution of the equation $5^{x-4} = 25$ is $x =$

19.

4D.a. 3) The perimeter of a rectangle with Length $=2x - 3$ and Width $= 3x + 7$ is:

20.

4D.b. 2) The rationalized expression of $\dfrac{2\sqrt{3}}{\sqrt{11}}$ is:

21.

4E.a. 2) The y intercept of the line $2x + 3y = 5$ is:

22.

4F.a. 2) The factored expression of $x^2 - xy - 6y^2$ is:

23.

4F.b. 2) The number to be added to $x^2 + 5x$ to make a perfect square is:

24.

5A.a. 1) Determine if the relation below is a function.

{(-3, 6), (-2, 10), (3, 3), (3, -12), (7,12)}

25.

5Ab 1) Determine if the following table of pairs of x and y represents a linear function.

X	Y
5	5
6	7
6	8
9	10

26.

5Ac 1) If $f(x) = y = 2x + 3$ the inverse function is: $f^{-1}(x) = \frac{x-3}{2}$

27.

5C. 1) The length of the part of a trigonometric function that repeats, measured along the x axis is called_____.

28.

5D.b. 1) For each of values of α, the values of $\sin(\alpha)$ are:

α	0	$\frac{\pi}{6}$	$\frac{\pi}{2}$	$\frac{\pi}{3}$	π
$\sin(\alpha)$	0	$\frac{1}{2}$	1	$\frac{\sqrt{3}}{2}$	0

29.

5E. 1) Inverse trig functions do the _____ of the "regular" trig functions.

Mark yourself

1	2	3	4	5	6	7	8	9	10
11	12	13	14	15	16	17	18	19	20
21	22	23	24	25	26	27	28	29	
# of Good Answers (NGA)=				NGA/total number of questions=Ratio				Percent=Ratio*100	
					YOUR PERCENT IS:			%	

CHAPTER 6

LIMITS

Fundamental Concepts

UNDERSTANDING LIMITS:

To understand limits better let's analyze this example. We are watching a soccer game. There is a penalty kick. The player is approaching the ball. The TV transmission is interrupted for 2 seconds. When the transmission is back, the soccer player already scored. The ball is rolling towards the center of the post. What happened while the transmission was off? What did the soccer player do to score? What path did the ball take? Using limits, we can approximate the path the ball followed. Let's split the time interval into very small parts on both images, before and after the player scores. We then compare the path of the ball before and after the transmission interruption. In this way, the path the ball took can be approximated easier.

Now,

If we have a function f(x) that gets very close to a value L for a value of x that gets very close to a value "a", we call L the limit of function f for x getting very close to a certain value "a".

EXAMPLE

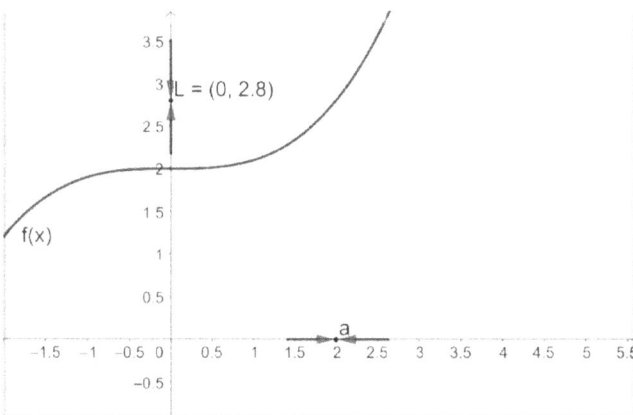

The function f(x) is getting very close to the value L for x approaching very close to a value "a".

Point L (0,2.**8.** is the limit of function f(x) when x is going very close to value a=2.

EXAMPLE

Suppose $f(x) = 0.1x^3 + 2$

We want to see what happens when x is very close to value **2**. We create a table of values in such a way that we take values for x closer and closer to value 2, on each side of 2, smaller and bigger than 2.

The table below shows the values for x around 2 and for f(x) around the limit L=2.8.

As we can see, as we approach x=2 from both sides f(x) is approaching L=2.8

X	$f(x) = 0.1x^3 + 2$
1.5	2.3375
1.9	2.6859
1.99	2.7881
1.999	2.7988
2.001	2.8012
2.01	2.8121
2.1	2.9261

A geometric illustration

Let's suppose we have a circle. The chord AB is the longest chord compared with the others in this example. As we are approaching point A going around the circle from point B towards point A, through points B, C, D, E, the length of the chord is becoming smaller and smaller. When we are at a point which is extremely close to point A, the length of the chord is extremely small. The moment we are in point A the chord becomes a tangent to the circle, line AT.

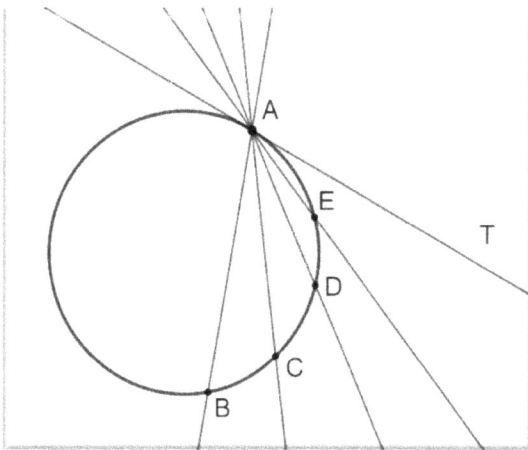

The word tangent comes from the Latin word **tangens** which means "touching". Any tangents to any curve, touch that particular curve in one and *only* one point.

6.A Limits

b. Algebraically

The symbol used when we calculate the limit L of a function f(x) for a value of x approaching a value "a" is:

$$\lim_{x \to a} f(x) = L$$

Simplistically, when the function has values (Range) for each value of real values of x (Domain), that function is called <u>continuous</u>.

When x equals a certain value where the function exists and is continuous, we just substitute that value of x in the expression of the limit.

EXAMPLE

$$\lim_{x \to 2} \frac{x^2+2x+4}{x+3} = \lim_{x \to 2} \left[\frac{(x+2)^2}{x+3} \right] = \frac{\lim_{x \to 2}(x+2)^2}{\lim_{x \to 2}(x+3)} = \frac{(2+2)^2}{(2+3)} = \frac{16}{5} = 3\frac{1}{5}$$

There are **three cases** for when we have to find the limits of a rational expression algebraically for x approaches plus or minus infinity.

1. When the degree of the polynomial at the numerator **is less** than the degree of the polynomial at the denominator. In this case, the limit will always be zero for x going towards $\mp\infty$. The graph will have a horizontal asymptote x=0.

EXAMPLE

Find $\lim_{x \to \infty} \frac{x+1}{x^2-2}$

$$\lim_{x \to \infty} \frac{x+1}{x^2-2} = \lim_{x \to \infty} \frac{x+1}{x^2-2} = \lim_{x \to \infty} \left[\frac{x^2(\frac{1}{x}+\frac{1}{x^2})}{x^2(1-\frac{2}{x^2})} \right] = \frac{\lim_{x \to \infty}(\frac{1}{x}) + \lim_{x \to \infty}(\frac{1}{x^2})}{\lim_{x \to \infty} 1 - \lim_{x \to \infty}(\frac{2}{x^2})} = \frac{0+0}{1-0} = \frac{0}{1} = 0$$

2. When the degree of the polynomial at the numerator **is the same** as the degree of the polynomial at the denominator. In this case, the limit will always equal the ratio between the coefficients of the terms with the highest exponent for x going towards $\mp\infty$. The graph will have a horizontal asymptote at x= the ratio between the coefficients of the terms with the highest exponent.

EXAMPLE

Find $\lim_{x \to \infty} \frac{3x^2+1}{x^2-2}$

$$\lim_{x\to\infty}\frac{3x^2+1}{x^2-2}=\lim_{x\to\infty}\frac{3x^2+1}{x^2-2}=\lim_{x\to\infty}\left[\frac{x^2(3+\frac{1}{x^2})}{x^2(1-\frac{2}{x^2})}\right]=\frac{\lim_{x\to\infty}(3)+\lim_{x\to\infty}(\frac{1}{x^2})}{\lim_{x\to\infty}1-\lim_{x\to\infty}(\frac{2}{x^2})}=\frac{3+0}{1-0}=\frac{3}{1}=3$$

3. When the degree of the polynomial at the numerator **is greater** than the degree of the polynomial at the denominator. In this case, the graph will tend to go either upward or downward, depending on the signs of the terms with the highest exponents from the numerator and denominator.

EXAMPLE

Find $\lim_{x\to\infty}\frac{3x^3+1}{x^2-2}$

$$\lim_{x\to\infty}\frac{3x^3+1}{x^2-2}=\lim_{x\to\infty}\frac{3x^3+1}{x^2-2}=\lim_{x\to\infty}\left[\frac{x^2(3x+\frac{1}{x^2})}{x^2(1-\frac{2}{x^2})}\right]=\frac{\lim_{x\to\infty}(3x)+\lim_{x\to\infty}(\frac{1}{x^2})}{\lim_{x\to\infty}1-\lim_{x\to\infty}(\frac{2}{x^2})}=\frac{\infty+0}{1-0}=\infty$$

As we can see, both terms $3x^3$ and x^2 are positive. In this case, the result will be positive so, the graph will go toward plus infinity.

EXAMPLE

$$\lim_{x\to\infty}\left(\sqrt{3x^2-2}-\sqrt{7x}\right)=\lim_{x\to\infty}\left(\sqrt{3x^2-2}-\sqrt{7x}\right)\left(\frac{\sqrt{3x^2-2}+\sqrt{7x}}{\sqrt{3x^2-2}+\sqrt{7x}}\right)=\lim_{x\to\infty}\left(\frac{3x^2-2-7x}{\sqrt{3x^2-2}+\sqrt{7x}}\right)=$$

$$\lim_{x\to\infty}\frac{3x^2-7x-2}{\sqrt{3x^2+1}+\sqrt{7x}}=\lim_{x\to\infty}\frac{x(3x-7-\frac{2}{x})}{x\left(\sqrt{3+\frac{1}{x^2}}+\sqrt{\frac{7}{x}}\right)}=\lim_{x\to\infty}\frac{3x-7-\frac{2}{x}}{\sqrt{3+\frac{1}{x^2}}+\sqrt{\frac{7}{x}}}=\frac{\lim_{x\to\infty}(3x-7-\frac{2}{x})}{\lim_{x\to\infty}\left(\sqrt{3+\frac{1}{x^2}}+\sqrt{\frac{7}{x}}\right)}=\frac{\infty-7-0}{\sqrt{3+0}+0}=\infty$$

EXAMPLE

Find $\lim_{h\to0}\frac{f(x+h)-f(x)}{h}$ when $f(x)=x^2-2x$

$$\lim_{h\to0}\frac{f(x+h)-f(x)}{h}=\lim_{h\to0}\frac{[(x+h)^2-2(x+h)]-(x^2-2x)}{h}=\lim_{h\to0}\frac{x^2+2hx+h^2-2x-2h-x^2+2x}{h}\lim_{h\to0}\frac{2hx+h^2-2h}{h}=$$

$$\lim_{h\to0}\frac{h(2x-h-2)}{h}=\lim_{h\to0}(2x-h-2)=2x-2$$

6.B One side versus two sides

As we saw before, when we talk about limits towards a certain value "a", we have to take into consideration that the values x approaching a certain value "a" could be bigger or smaller compared with that value "a". As we can see in the graph to the left.

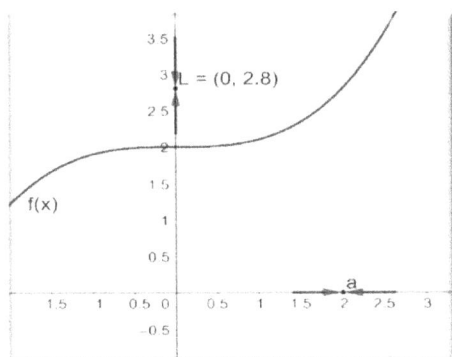

There are cases when we could have the function defined only for values bigger than the value where we want to know the limit.

EXAMPLE

The radical function.

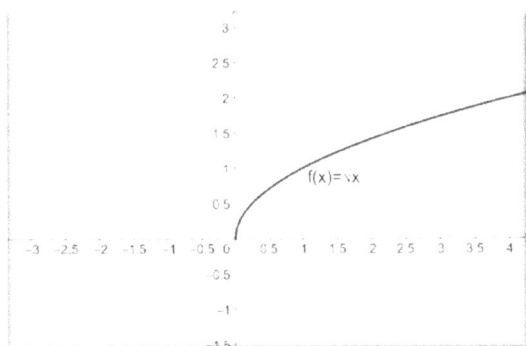

Here the values of x are bigger that zero. The limit exists only for x bigger than zero. We call this limit **one side limit.**

The notation for this limit is: $\lim_{x \to 0^+} f(x)$

There are cases when we could have the function defined for values larger and smaller than the value where we want to know the limit, but the limits are different.

EXAMPLE

In this case the left limit of f(x) for values of x<1 is 1

$$\lim_{x \to 1^-} f(x) = 1$$

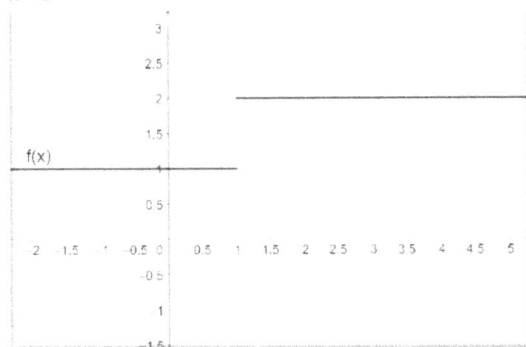

The right limit of f(x) for values of x>1 is 2

$$\lim_{x \to 1^+} f(x) = 2$$

There is no limit for x=1

As we can see,

$$\lim_{x \to 1^-} f(x) \neq \lim_{x \to 1^+} f(x)$$

6.C End behavior

This section of the chapter is all about what happens with the graph of the function when x is approaching plus or minus infinity. The question will be: what is the limit of the function when x approaches $\pm\infty$?

As x goes to $\pm\infty$, the limit of the function could be infinite, finite or does not exit.

In case the function is polynomial, there are a few scenarios.

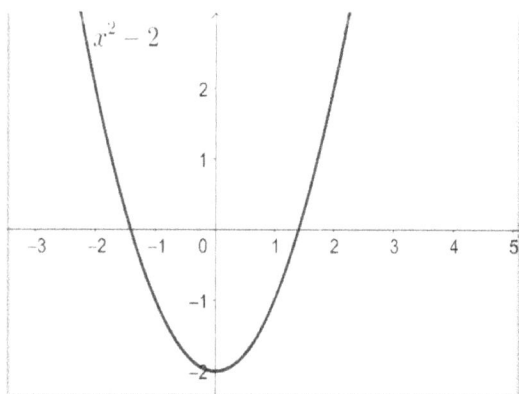

a) The degree of the function is even and the coefficient of the term with the highest degree is positive.

The graph goes upward in quadrant I and II.

b) The degree of the function is even and the coefficient of the term with the highest degree is negative.

The graph goes downward in quadrant III and IV.

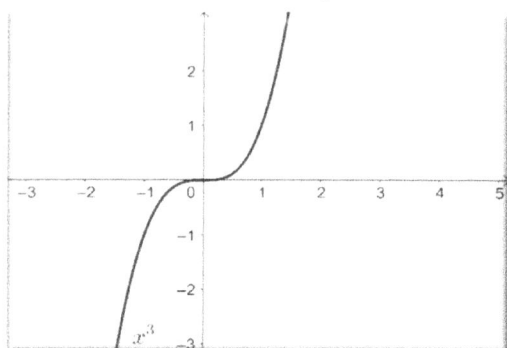

c) The degree of the function is odd and the coefficient of the term with the highest degree is positive, the graph goes upward in quadrant I and downward in quadrant III.

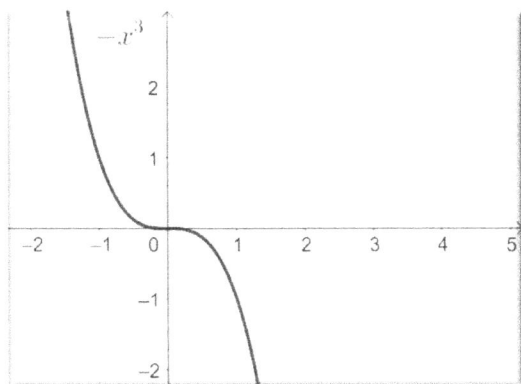

d) The degree of the function is odd and the coefficient of the term with the highest degree is negative, the graph goes upward in quadrant II and downward in quadrant IV.

6.D Intermediate limits theorem

A <u>continuous</u> function is a function that has no sudden changes for all the values of the domain.

Intuitive *EXAMPLE*

Whenever we draw the graph of a continuous function, we don't have to take the pencil off the paper. As we draw the graph, the pencil remains on the paper for the entire graph.

The <u>Intermediate limits theorem</u> states in a nutshell, that for any closed interval of real numbers [a,b], a continuous function f(x) with the interval [a,b] as domain, and a point M between f(a) and f(b), $f(a) \neq f(b)$, there will be at least a point c between a and b for which f(c)=M.

EXAMPLE

As it can be seen in the graph below, the interval [a,b] is [-2,2]. The function is a continuous function of the relation:

$$f(x) = 0.1x^3 + 1.5.$$

There is a point M between f(-2)=0.7 and f(2)=2.3 for which the horizontal line will intersects the graph in a point of a value [c,f(c)] or [c,M].

In this case M is 1.4 and c=-1

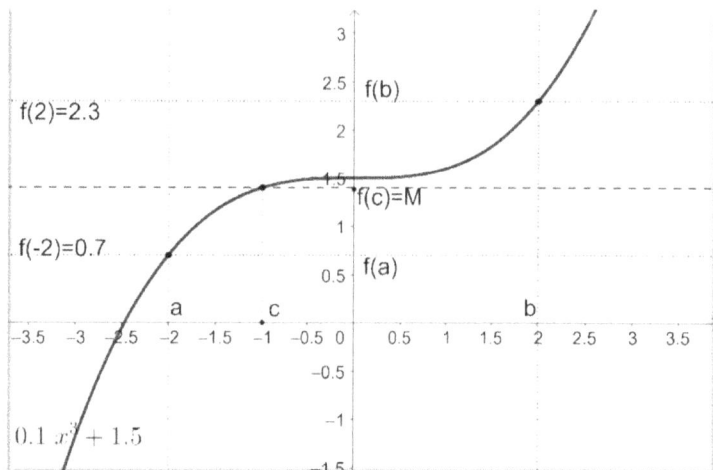

f(2)=2.3

f(b)

f(c)=M

f(-2)=0.7

f(a)

$0.1\,x^3 + 1.5$

6.E Left and right limits

The left limit exists for values of x smaller than a certain value "a".

The notation for this limit is: $\lim\limits_{x \to a^-} f(x)$

The right limit exists for values of x bigger than a certain value "a".

The notation for this limit is: $\lim\limits_{x \to a^+} f(x)$

EXAMPLE

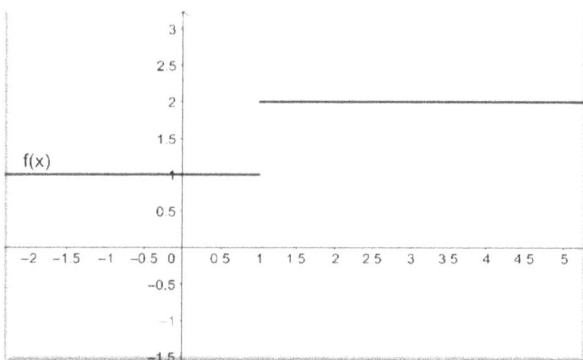

f(x)

In this case the left limit of f(x) for values of x<1 is 1

$$\lim\limits_{x \to 1^-} f(x) = 1$$

The right limit of f(x) for values of x>1 is 2

$$\lim\limits_{x \to 1^+} f(x) = 2$$

As we can see, the limits are finite but are not the same. The limit for x=1 does not exist. In this case the function f(x) is not continuous in x=1.

EXAMPLE

The function in this case is:

$$f(x) = \begin{cases} 0.5x^3 + 1 \ if \ x < 0.5 \\ -x + 0.8 \ if \ x > 0.5 \end{cases}$$

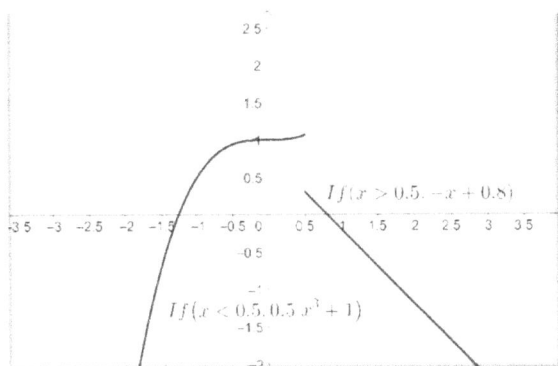

To find the left limit of the function around 0.5, we use the branch for which x values are smaller than 0.5.

$$\lim_{x \to 0.5^-} f(x) = 0.5(0.5)^3 + 1 = 1.0625$$

To find the right limit of the function around 0.5, we use the branch for which x values are bigger than 0.5.

$$\lim_{x \to 0.5^+} f(x) = -0.5 + 0.8 = 0.3$$

6.F Limits to infinity

We have three cases when we have to find the limits of a rational expression algebraically for x approaches plus or minus infinity.

1. When the degree of the polynomial at the numerator <u>is less</u> than the degree of the polynomial at the denominator. In this case, the limit will always be zero for x going towards $\mp\infty$. The graph will have a horizontal asymptote x=0.

EXAMPLE

Find $\lim\limits_{x \to \infty} \dfrac{x+1}{x^2-2}$

$$\lim_{x \to \infty} \frac{x+1}{x^2-2} = \lim_{x \to \infty} \frac{x+1}{x^2-2} = \lim_{x \to \infty} \left[\frac{x^2(\frac{1}{x}+\frac{1}{x^2})}{x^2(1-\frac{2}{x^2})} \right] = \frac{\lim_{x \to \infty}(\frac{1}{x}) + \lim_{x \to \infty}(\frac{1}{x^2})}{\lim_{x \to \infty} 1 - \lim_{x \to \infty}(\frac{2}{x^2})} = \frac{0+0}{1-0} = \frac{0}{1} = 0$$

The graph of $f(x) = \dfrac{x+1}{x^2-2}$ is represented below.

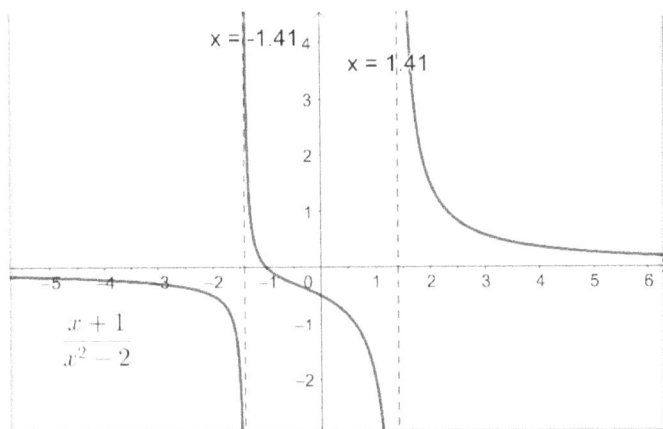

103

2. When the degree of the polynomial at the numerator <u>is the same</u> as the degree of the polynomial at the denominator. In this case, the limit will always equal the ratio between the coefficients of the terms with the highest exponent for x going towards $\mp\infty$. The graph will have a horizontal asymptote at x= the ratio between the coefficients of the terms with the highest exponent.

EXAMPLE

Find $\lim\limits_{x\to\infty} \dfrac{3x^2+1}{x^2-2}$

$$\lim_{x\to\infty} \frac{3x^2+1}{x^2-2} = \lim_{x\to\infty} \frac{3x^2+1}{x^2-2} = \lim_{x\to\infty}\left[\frac{x^2(3+\frac{1}{x^2})}{x^2(1-\frac{2}{x^2})}\right] = \frac{\lim\limits_{x\to\infty}(3)+\lim\limits_{x\to\infty}(\frac{1}{x^2})}{\lim\limits_{x\to\infty}1-\lim\limits_{x\to\infty}(\frac{2}{x^2})} = \frac{3+0}{1-0} = \frac{3}{1} = 3$$

The graph of $f(x) = \dfrac{3x^2+1}{x^2-2}$ is represented below.

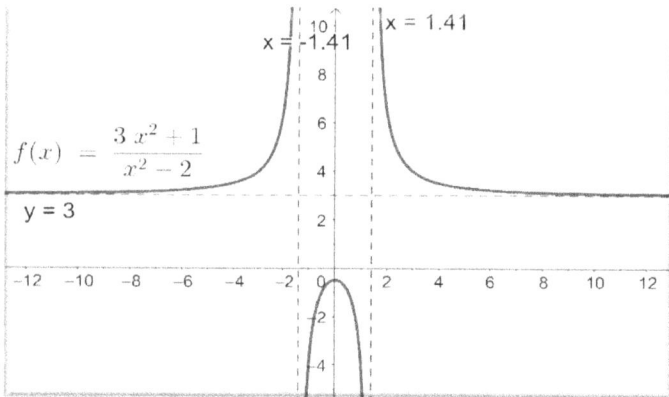

3. When the degree of the polynomial at the numerator <u>is bigger</u> than the degree of the polynomial at the denominator. In this case, the graph will tend to go either upward or downward, depending of the signs of the terms with the highest exponents from the numerator and denominator.

EXAMPLE

Find $\lim\limits_{x\to\infty} \dfrac{3x^3+1}{x^2-2}$

$$\lim_{x\to\infty} \frac{3x^3+1}{x^2-2} = \lim_{x\to\infty} \frac{3x^3+1}{x^2-2} = \lim_{x\to\infty}\left[\frac{x^2(3x+\frac{1}{x^2})}{x^2(1-\frac{2}{x^2})}\right] = \frac{\lim\limits_{x\to\infty}(3x)+\lim\limits_{x\to\infty}(\frac{1}{x^2})}{\lim\limits_{x\to\infty}1-\lim\limits_{x\to\infty}(\frac{2}{x^2})} = \frac{\infty+0}{1-0} = \infty$$

As we can see, both terms $3x^3$ and x^2 are positive. In this case, the result will be positive so, the graph will go toward plus infinity.

The graph of $f(x) = \frac{3x^3+1}{x^2-2}$ is represented below.

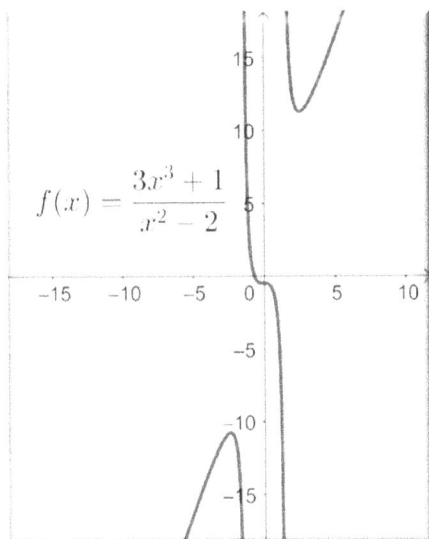

6.G Continuity

Until now we had an intuitive approach to the continuity of functions. If we don't have to take the pencil off the page while we draw the function graph this function is continuous. A rigorous definition of continuity says that:

a function f is continuous at a number c if,

$$\lim_{x \to c} f(x) = f(c)$$

The definition requires 3 things:

a) $f(c)$ is defined. It means that the value "c" belongs to the domain of f.

b) $\lim_{x \to c} f(x)$ exists

c) $\lim_{x \to c} f(x) = f(c)$

EXAMPLE

As it can be seen in the graph below, for a value c belonging to the domain of function f,

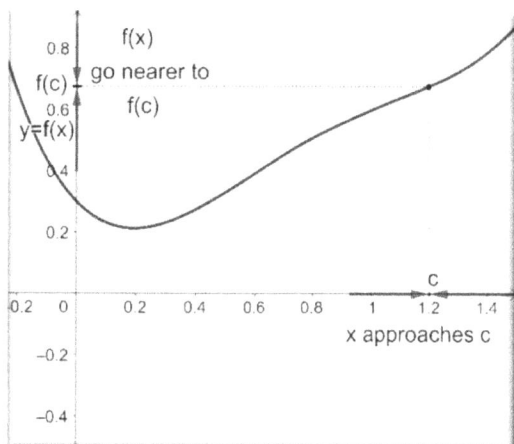

f(c) is defined. As x approaches "c" from left and right, the function f(x) approaches f(c).

$\lim\limits_{x \to c} f(x)$ exists and equals f(c).

A function is <u>continuous coming from the left</u> of a value c if $\lim\limits_{x \to c^-} f(x) = f(c)$

A function is <u>continuous coming from the right</u> of a value c if $\lim\limits_{x \to c^+} f(x) = f(c)$

A function is **not** <u>continuous coming from the right or left</u> of a value c if,

$$\lim\limits_{x \to c^-} f(x) \neq \lim\limits_{x \to c^+} f(x)$$

EXAMPLE

For x approaching from the left, the left limit in the graph below is:

$$\lim\limits_{x \to 1^-} f(x) = f(1) = 2$$

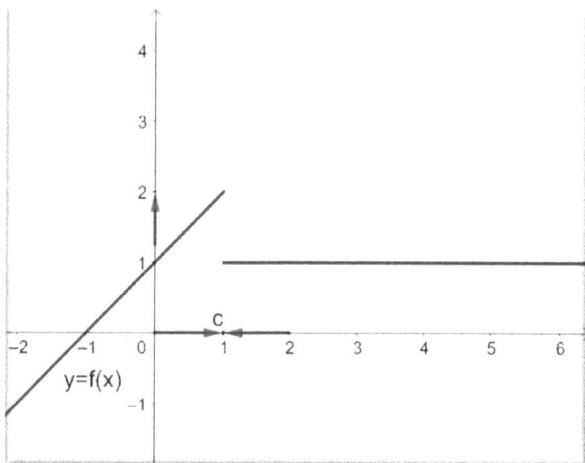

For x approaching from the right, the right limit in the graph below is

$$\lim\limits_{x \to 1^+} f(x) = f(1) = 1$$

The function exists for x=1,
The left and right limits exist but they are not the same.
In this case the function is not continuous for x=1.

Theorem

If functions f and g are continuous at c and b is a constant, the following functions are continuous as well.

f+g ; f-g ; bf ; fg ; f/g if g(c)≠ 0

Theorem

1. Any polynomial function is continuous everywhere (on real numbers).

2. Any rational function is continuous wherever it is defined; it is continuous on its domain.

3. These types of functions are continuous on their domain as well.

Root functions and trigonometric functions are continuous functions.

Theorem

If f is continuous at c and $\lim_{x \to c} g(x) = c$ then $\lim_{x \to c} f[g(x)] = f(c)$

or;

$$\lim_{x \to c} f[g(x)] = f[\lim_{x \to c} g(x)]$$

TEST 7

Includes Limits

1.

3A.a. 4) Big Ben is 96 m tall. The shade is 70 m long. The tangent of the light rays with the ground is:

2.

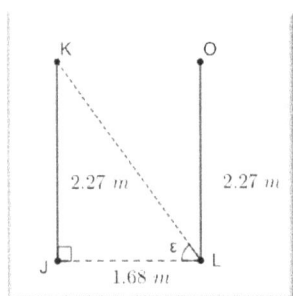

3A.a. 5) The height of the Caryatides is 2.27 m. they are 1.68 m apart. The tangent of the angle that the shade of one touch the next is:

3.

4A.a. 4) The simplified expression of $(\frac{-7x^4y^3}{xy})^2(\frac{x^2yz}{7x^2z})^3; x, y \neq 0$ is:

4.

4A.a. 5) The simplified expression of $(-5x^4y^{-3}z^0)^{-3}$ $x \neq 0$ is:

5.

4A.b. 4) The perimeter of a rectangle with length $L = 3\sqrt{20} + 5\sqrt{5}$ and width $W = 5\sqrt{20}$ is:

6.

4A.c. 4) The mixed radical of $\sqrt{486}$ is:

7.

4B.a. 4) $5\times (9 + 3)$ is equal with $5 + 5\times 9 + 5 \times 3$

8.

4B.b. 4) $(-3x^2 + 4x - 5)(2x^3 + 7)$ is equal with:

9.

4C.a. 4) The expression $\log_3 63 + \log_3 5 - \log_3 35$ equals:

10.

4C.b. 4) The solution of the equation $49^{x-2} = 343^{3x+5}$ is $x =$

11.

4D.a. 4) The perimeter of a rectangle with Length $=2x - 3$ and Width $= 3x + 7$ is $140x + 8$. The value of the above perimeter for $x = 5\ m$ is

12.

4D.b. 3) The rationalized expression of $\frac{\sqrt{7}+3\sqrt{5}}{\sqrt{3}}$ is:

13.

4E.a. 3) The slope of the line $-3x + 5y = -1$ is:

14.

4F.a. 3) The factored expression by grouping of $35x^2 + 8x - 3$ is:

15.

4F.b. 3) The perfect square of $x^2 - 3x$ is:

16.

5A. a 2) Determine if the relation below is a function
{(-3, 6), (-2, 10), (0, 3), (3, -12), (7,22)}

17.

5Ab 2) Determine if the following table of pairs of x and y represent a linear function.

X	Y
1	3
2	5
3	7
4	9

18.

5A.c. 2) If $f(x) = y = 3x - 5$ the inverse function is: $f^{-1}(x) =$

19.

5C. 2) One radian is the measure of an angle subtended at the center of a circle by an arc which is equal in length to the _____.

20.

5D.b. 2) Does the tangent graph looks like the one below for $-\frac{\pi}{2} < \alpha < \frac{\pi}{2}$?

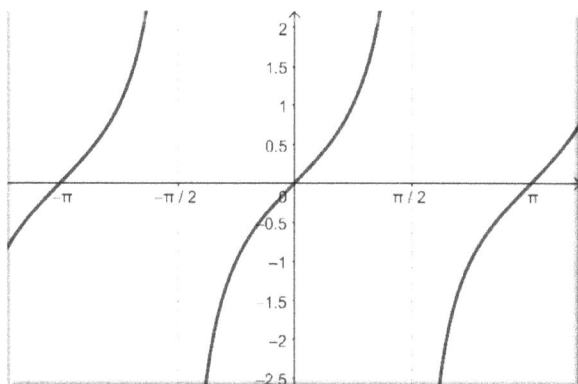

21.

5E. 2) The angle α in right angle triangle PSQ is:

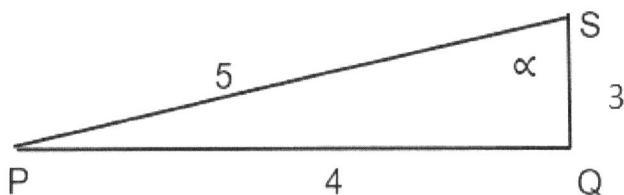

22.

6A.b. 1) $\lim\limits_{x \to 0} \left(\dfrac{6x^2 - 3}{x + 3}\right) =$

23.

6B 1) In the graph below $g(x) = -(0.5x - 3)^2 + 3 \; for \; x < 5$

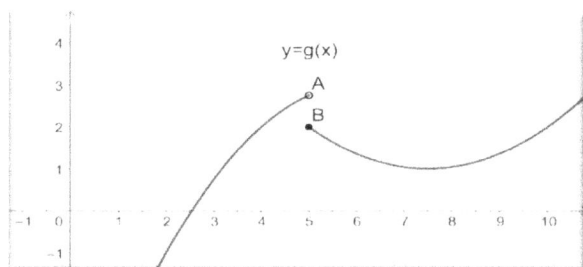

y=g(x)

$g(x) = (0.4x - 3)^2 + 1 \; for \; x \geq 5$

1) $\lim\limits_{x \to 5^-} g(x) = 2.75$

24.

6C 1) $f(x) = 2x^4 - 3x^2 + 15x - 24$

: when x goes to +∞ it goes up in

_____ quadrant

25.

6D. 1) If f is a continuous function on the closed interval [-1,5] where $f(-1) = 2; f(5) = 7$ then; $f(c) = 6$ for at least one value of c that is between -1 and 5

26.

6E 1) Using the graph below,

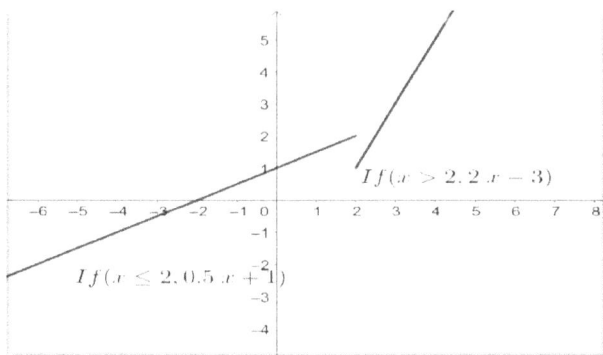

$If(x > 2, 2.x - 3)$

$If(x \leq 2, 0.5.x + 1)$

Where: $f(x) = \begin{cases} 0.5x + 1 \, , x \leq 2 \\ 2x - 3, x > 2 \end{cases}$

$\lim\limits_{x \to 0^-} f(x) =$

27.

6F. 1) $\lim\limits_{x \to \infty} \dfrac{x^2 + 2x - 45}{3x^3 - 5x^2 + 7x - 3} =$

28.

6G. 1) The following function is represented below: $f(x) =$

$$\begin{cases} 2x - 1, x \geq 2 \\ 3x - 3, 1.5 < x < 2 \\ 0.5x, x \leq 1.5 \end{cases}$$

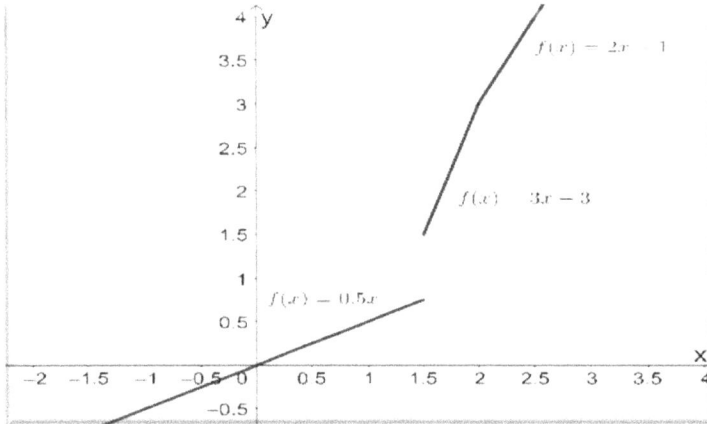

Is the function f(x) is continuous in x=4 ?

Mark yourself

1	2	3	4	5	6	7	8	9	10
11	12	13	14	15	16	17	18	19	20
21	22	23	24	25	26	27	28		

| # of Good Answers (NGA)= | | | | NGA/total number of questions=Ratio | | | | Percent=Ratio*100 | |
| YOUR PERCENT IS: | | | | | | | | % | |

CHAPTER 7

DIFFERENTIATION

Interesting historical facts about Calculus

It is generally accepted that two men founded Calculus; Isaak Newton and Gottfried Leibniz.

Although Newton and Leibniz both had fundamental contributions in the creation of Calculus, their approach was quite different. (1).

Newton took into account that variables are changing with time, **(9)** whereas

Leibniz considered the variables x and y as extending over a number of infinitely close values. He was the first to introduce the notations dx and dy as differences between two successive values of these sequences.

Newton instead used the quantities x' and y', as notations for derivatives, which in his case were finite velocities, used to compute the tangent.

It is interesting to note that neither Leibniz nor Newton worked the concepts in terms of functions, but always in terms of graphs.

For Newton, the calculus was geometrical, while Leibniz took it towards analysis.

Leibniz was very aware of the importance of having a good notation while explaining the new concepts. He was very careful regarding the symbols he used.

Newton, instead, tended to use whatever notation he liked on that particular day.

Leibniz's notation became more used for generalizing calculus to multiple variables.

Leibniz's notation emphasized the operator aspect of the derivative as well as the integral. Much of the notation that we are using in Calculus today is what Leibniz used.

7.B and C Differentiation Definition of derivatives and Notation

We know from Pre-Calculus that the slope of a straight line that passes through points $A(x_1, y_1)$ and $B(x_2, y_2)$ can be calculated as:

$slope = \frac{y_2 - y_1}{x_2 - x_1}$

EXAMPLE

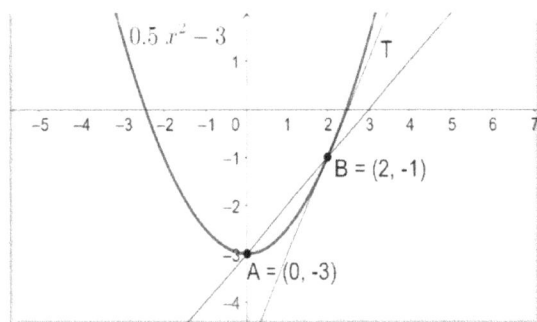

As we can see here, the slope of the line that passes through A(0,-3) and B(2,-1) is:

$slope = \frac{y_2 - y_1}{x_2 - x_1} = \frac{-1 - (-3)}{2 - 0} = \frac{-1 + 3}{2} = \frac{2}{2} = 1$

The slope of the line that only passes through point B and is the tangent to the graph at this point, is defined by the limit of the slope relation when x is approaching **2**.

This limit of the slope when x approaches zero is called derivative.

The formula of the derivative is: $f'(x) = \lim_{h \to 0} \frac{y_2 - y_1}{x_2 - x_1} =$

$\lim_{h \to 0} \frac{f(x+h) - f(x)}{h}$, *where h is a very small number.*

First, we have to calculate the limit of the expression for the difference between x_1 and $x_2 = x_1 + h$, where h is an extremely small value approaching zero.

So, $x_2 - x_1 = x_1 + h - x_1 = h$

The derivative of $y = f(x) = 0.5x^2 - 3$ is $f' = \lim\limits_{h \to 0} \dfrac{f(x+h)-f(x)}{h} = \lim\limits_{h \to 0} \dfrac{0.5(x+h)^2 - 3 - [0.5x^2 - 3]}{h} =$

$\lim\limits_{h \to 0} \dfrac{0.5(x^2+2xh+h^2)-3-0.5x^2+3}{h} = \lim\limits_{h \to 0} \dfrac{0.5x^2+xh+0.5h^2-3-0.5x^2+3}{h} = \lim\limits_{h \to 0} \dfrac{xh+0.5h^2}{h} = \lim\limits_{h \to 0} \dfrac{h(x+0.5h)}{h} =$

$\lim\limits_{h \to 0}(x + 0.5h) = \lim\limits_{h \to 0}(x) + \lim\limits_{h \to 0}(0.5h) = x$

For x=2 $\lim\limits_{h \to 0} \dfrac{f(x+h)-f(x)}{h} = 2$

EXAMPLE

If we consider the very small increase from x as Δx, where $\Delta x \to 0$, then the limit of:

$\dfrac{f(x+h)-f(x)}{h}$ *of* $f(x) = 0.5x^2 - 3$ can be written:

$f' = \lim\limits_{\Delta x \to 0} \dfrac{f(x+\Delta x)-f(x)}{\Delta x} = \lim\limits_{\Delta x \to 0} \dfrac{0.5(x+\Delta x)^2 - 3 - [0.5x^2 - 3]}{\Delta x} =$

$\lim\limits_{\Delta x \to 0} \dfrac{0.5(x^2+2x\Delta x+\Delta x^2)-3-0.5x^2+3}{\Delta x} = \lim\limits_{\Delta x \to 0} \dfrac{0.5x^2+x\Delta x+0.5\Delta x^2-3-0.5x^2+3}{\Delta x} = \lim\limits_{\Delta x \to 0} \dfrac{x\Delta x+0.5\Delta x^2}{\Delta x} =$

$\lim\limits_{\Delta x \to 0} \dfrac{\Delta x(x+0.5\Delta x)}{\Delta x} = \lim\limits_{\Delta x \to 0}(x + 0.5\Delta x) = \lim\limits_{\Delta x \to 0}(x) + \lim\limits_{\Delta x \to 0}(0.5\Delta x) = x$

7.D Differentiation

a Rate of change – Average versus Instantaneous

The **rate of change** tells us how fast y is changing as x is changing.

EXAMPLE

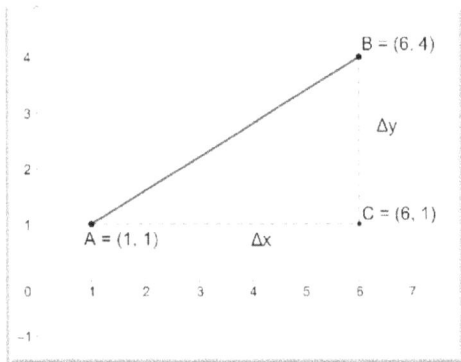

We want to find the rate of change between points A (1,1) and B (6,4). As we go from A to B x varies from 1 to 6, while y varies from 1 to **4.**

The rate of change is then given by the ratio $\frac{\Delta y}{\Delta x}$

In this case $\Delta y = 4 - 1 = 3$

$\Delta x = 6 - 1 = 5$

So, $\frac{\Delta y}{\Delta x} = \frac{3}{5}$

When we have a function f(x), the rate of change between two points $D[x_1, f(x_1)]$ and $E[x_2, f(x_2)]$ that belong to the graph of the function, then the average rate of change between D and E is given by: $\frac{\Delta y}{\Delta x} = \frac{f(x_2) - f(x_1)}{x_2 - x_1}$

EXAMPLE

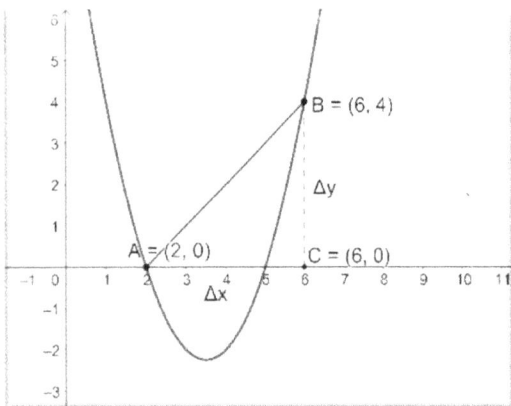

Here, $x_1 = 2 \ and \ x_2 = 6$, so, $f(x_1) = 0 \ and \ f(x_2) = 4$

The average rate of change is:

$\frac{\Delta y}{\Delta x} = \frac{f(x_2) - f(x_1)}{x_2 - x_1} = \frac{4-0}{6-2} = \frac{4}{4} = 1$

The <u>Instantaneous rate of change</u> happens when Δx is extremely small, approaches zero.

The instantaneous rate of change is given by the <u>limit</u> when Δx is extremely small, approaches zero.

EXAMPLE

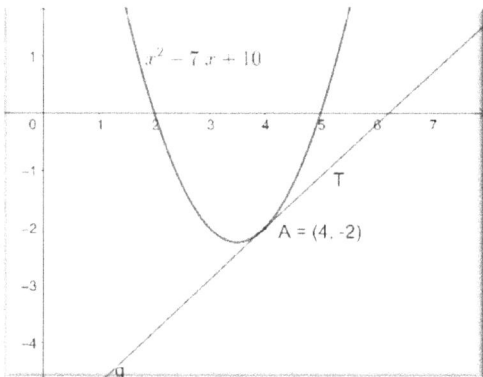

Point B, extremely close to A, has the coordinates $[x + \Delta x, f(x + \Delta x)]$ The formula we use is:

$$f'(x) = \lim_{\Delta x \to 0} \frac{f(x+\Delta x)-f(x)}{\Delta x} =$$

$$\lim_{\Delta x \to 0} \frac{(x+\Delta x)^2 - 7(x+\Delta x)+10-(x^2-7x+10)}{\Delta x} =$$

$$\lim_{\Delta x \to 0} \frac{x^2+2x\Delta x+(\Delta x)^2-7x-7\Delta x+10-x^2+7x-10}{\Delta x} =$$

$$\lim_{\Delta x \to 0} \frac{2x\Delta x+(\Delta x)^2-7\Delta x}{\Delta x} = \lim_{\Delta x \to 0} \frac{\Delta x(2x-\Delta x-7)}{\Delta x} =$$

$$\lim_{\Delta x \to 0} (2x - \Delta x - 7) = 2x - 7$$

So, $f'(x) = 2x - 7$

For x=4 the instantaneous rate of change, called <u>derivative</u> is $f'(4) = 2(4) - 7 = 8 - 7 = 1$

7.D Differentiation

b Rate of change – Slope of secant and tangent lines

A <u>secant</u> is a straight line that passes through a curve and cuts it into two or more parts.

EXAMPLE

The secant DA intersects the graph in points D and A. To be able to calculate the slope,

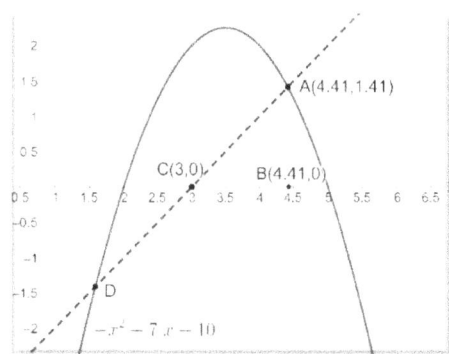

we create our own right-angle triangle ΔABC where angle $\angle CBA$ is 90^0. The slope of the secant DA is calculated using the points C (point 1), and A (point 2), by using the slope relation:

$$slope = \frac{\Delta y}{\Delta x} = \frac{y_2 - y_1}{x_2 - x_1} = \frac{1.41 - 0}{4.41 - 3} = \frac{1.41}{1.41} = 1$$

A **tangent** to a graph is when the line touches the graph in only one point.
In this case, the distance on x axis is between the x coordinates is x+Δx-x=Δx, and on y axis is y+$\Delta y - y = f(x + \Delta x) - f(x) = \Delta y$. We consider Δx extremely small, approaching zero.

EXAMPLE

To be able to calculate the slope of a tangent, we consider the limit of the ratio $\frac{\Delta y}{\Delta x}$ for Δx approaching zero. We know that the function that represents the graph is given by: $f(x) = -x^2 + 7x - 10$

So:

$$slope = \lim_{\Delta x \to 0} \frac{\Delta y}{\Delta x} =$$

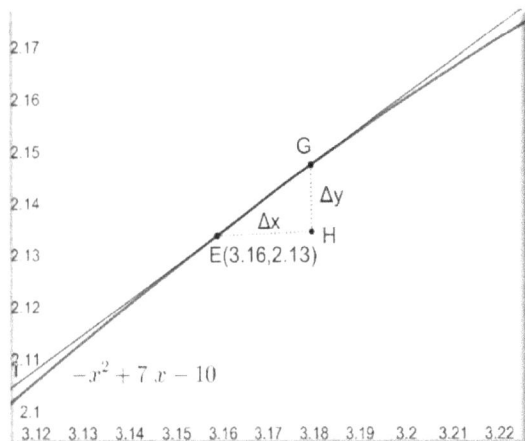

$$\lim_{\Delta x \to 0} \frac{f(x+\Delta x)-f(x)}{\Delta x} = \lim_{\Delta x \to 0} \frac{-(x+\Delta x)^2+7(x+\Delta x)-10-(-x^2+7x-10)}{\Delta x} =$$

$$\lim_{\Delta x \to 0} \frac{-(x^2+2x\Delta x+\Delta x^2)+7(x+\Delta x)-10-(-x^2+7x-10)}{\Delta x} =$$

$$\lim_{\Delta x \to 0} \frac{-x^2-2x\Delta x-\Delta x^2+7x+7\Delta x-10+x^2-7x+10}{\Delta x} = \lim_{\Delta x \to 0} \frac{-2x\Delta x-\Delta x^2+7\Delta x}{\Delta x} = \lim_{\Delta x \to 0} \frac{\Delta x(-2x-\Delta x+7)}{\Delta x} =$$

$$\lim_{\Delta x \to 0} (-2x + 7) = -2x + 7$$

The slope of the tangent at the point E(3.16,2.13) is given by:

$$slope = -2(3.16) + 7 = -6.32 + 7 = 0.68$$

7.E Transcendental Functions – Logarithmic, Exponential, Trigonometric

Formulas used for differentiation of transcendental functions.

a) $f(x) = ln|x|$ *then* $f'(x) = \frac{1}{x}$

EXAMPLE

If $f(x) = 2ln|x| + 3$ *then* $f'(x) = \frac{2}{x}$

b) $f(x) = e^x$ *then* $f'(x) = e^x$ $f(x) = a^x$ *then* $f'(x) = a^x * \ln(a)$

EXAMPLE

If $f(x) = 3^x - 4x + 5$ *then* $f'(x) = 3^x * \ln(3) - 4$

c) $f(x) = sin(x)$ *then* $f'(x) = \cos(x)$ $f(x) = \cos(x)$ *then* $f'(x) = -sin(x)$

EXAMPLE

If $f(x) = 3x - \sin(x)$ then $f'(x) = 3 - \cos(x)$

d) $f(x) = \tan(x)$ then $f'(x) = \sec^2(x)$ $\quad$ $f(x) = \cot(x)$ then $f'(x) = -\csc^2(x)$

EXAMPLE

If $f(x) = \tan(x) + 5$ then $f'(x) = \sec^2(x)$

e) $f(x) = \sec(x)$ then $f'(x) = \sec(x) * \tan(x)$ $\quad$ $f(x) = \csc(x)$ then $f'(x) = -\csc(x) * \cot(x)$

EXAMPLE

If $f(x) = 2\sec(x) + \tan(x)$ then $f'(x) = 2\sec(x) * \tan(x) + \sec^2(x)$

f) $f(x) = \sin^{-1}(x)$ then $f'(x) = \frac{1}{\sqrt{1-x^2}}$ $\quad$ $f(x) = \cos^{-1}(x)$ then $f'(x) = \frac{-1}{\sqrt{1-x^2}}$

EXAMPLE

If $f(x) = 3x - \sin^{-1}(x)$ then $f'(x) = 3 - \frac{1}{\sqrt{1-x^2}} = \frac{3\sqrt{1-x^2}-1}{\sqrt{1-x^2}}$

g) $f(x) = \tan^{-1}(x) = \frac{1}{x^2+1}$ $\quad$ $f(x) = \cot^{-1}(x) = \frac{-1}{x^2+1}$

EXAMPLE

If $f(x) = 7x + \tan^{-1}(x)$ then $f'(x) = 7 + \frac{1}{x^2+1} = \frac{7(x^2+1)+1}{x^2+1} = \frac{7x^2+8}{x^2+1}$

7.F Differentiation

a. Differential rules – Power

When we have a polynomial function like $f(x) = x^n$ where n is a positive integer, the derivative of this function is:

$f'(x) = nx^{n-1}$

EXAMPLE

If $f(x) = x^3$ then, $f'(x) = 3x^{3-1} = 3x^2$

If $f(x) = 2x^4 - 3x^3 + 1$ then, $f'(x) = 2(4)x^{4-1} - 3(3)x^{3-1} + 0 = 8x^3 - 9x^2$

When we have a function where the exponent is a rational number like $f(x) = x^{\frac{m}{n}}$, m,n $\in Z \neq 0$, the derivative of this function is:

$f'(x) = \frac{m}{n}x^{\frac{m}{n}-1} = \frac{m}{n}x^{\frac{m-n}{n}}$

EXAMPLE

If $f(x) = 3x^{\frac{1}{2}}$ then, $f'(x) = \frac{3}{2}x^{\frac{1}{2}-1} = \frac{3}{2}x^{-\frac{1}{2}} = \frac{3}{2}*\frac{1}{x^{\frac{1}{2}}} = \frac{3}{2\sqrt{x}} = \frac{3}{2\sqrt{x}}\frac{\sqrt{x}}{\sqrt{x}} = \frac{3\sqrt{x}}{2x}$

When the unknown is at the denominator.

EXAMPLE

If $f(x) = \frac{3}{x}$; $x \neq 0$ then $f'(x) = 3x^{-1} = (-1)(3)x^{-1-1} = -3x^{-2} = -\frac{3}{x^2}$

In case we have a radical function.

EXAMPLE

If $f(x) = 7\sqrt{x} = 7x^{\frac{1}{2}}$; $x \geq 0$ so, $f'(x) = \frac{7}{2}x^{\frac{1}{2}-1} = \frac{7}{2}x^{-\frac{1}{2}} = \frac{7}{2\sqrt{x}} = \frac{7\sqrt{x}}{2x}$

If $f(x) = \frac{5}{\sqrt{x}} = 5x^{-\frac{1}{2}}; x > 0$

so, $f'(x) = -\frac{5}{2}*x^{-\frac{1}{2}-1} = -\frac{5}{2}x^{\frac{-1-2}{2}} = -\frac{5}{2}x^{\frac{-3}{2}} = -\frac{5}{2x^{\frac{3}{2}}} = -\frac{5}{2\sqrt{x^3}} = -\frac{5\sqrt{x^3}}{2x^3} = -\frac{5x\sqrt{x}}{2x^3} = -\frac{5\sqrt{x}}{2x^2}$

7.F Differentiation

b. Differential rules – Product

If we have two functions f and g, the derivative of the product between f and g is:

$(f * g)' = f'g + fg'$

EXAMPLE

If $f(x) = 2x - 3$ and $g(x) = 5x^3 + 3x^2$

$[f(x) * g(x)]' = [(2x - 3) * (5x^3 + 3x^2)]' = (2x - 3)'(5x^3 + 3x^2) + (2x - 3)(5x^3 + 3x^2)' = 2(5x^3 + 3x^2) + (2x - 3)(15x^2 + 6x) = 10x^3 + 6x^2 + 30x^3 + 12x^2 - 45x^2 - 18x = 40x^3 - 27x^2 - 18x$

EXAMPLE

If $f(x) = \sqrt{x} - 3$ and $g(x) = 3x^2 + 7x$

$[f(x) * g(x)]' = [(\sqrt{x} - 3) * (3x^2 + 7x)]' = (\sqrt{x} - 3)'(3x^2 + 7x) + (\sqrt{x} - 3)(3x^2 + 7x)' = (x^{\frac{1}{2}})'(3x^2 + 7x) + (\sqrt{x} - 3)(6x + 7) = \left(\frac{1}{2}x^{-\frac{1}{2}}\right)(3x^2 + 7x) + 6x\sqrt{x} + 7\sqrt{x} - 18x - 21 = \frac{3}{2}x^{\frac{3}{2}} + \frac{7}{2}\sqrt{x} + 6x\sqrt{x} + 7\sqrt{x} - 18x - 21 = \frac{3}{2}x\sqrt{x} + \frac{7}{2}\sqrt{x} + 6x\sqrt{x} + 7\sqrt{x} - 18x - 21 = \frac{15}{2}x\sqrt{x} + \frac{21}{2}\sqrt{x} - 18x - 21$

EXAMPLE

If $f(x) = ln|x|$ and $g(x) = x^3$

$[f(x) * g(x)]' = [ln|x| * x^3]' = (ln|x|)'(x^3) + ln|x|(x^3)' = \frac{x^3}{x} + ln|x| * 3x^2 = x^2 + 3x^2 ln|x| = x^2(1 + 3ln|x|)$

7.F Differentiation

c. Differential rules – Quotient

If we want to find the derivative of a quotient of f(x) and g(x), we apply this formula:

$[\frac{f(x)}{g(x)}]' = \frac{[f(x)]'g(x) - f(x)[g(x)]'}{[g(x)]^2}$

EXAMPLE

$f(x) = x^2 + 1$, and $g(x) = 2x$

$[\frac{f(x)}{g(x)}]' = \frac{[f(x)]'g(x) - f(x)[g(x)]'}{[g(x)]^2} = \frac{(x^2+1)'(2x) - (x^2+1)(2x)'}{4x^2} = \frac{(2x)(2x) - 2(x^2+1)}{4x^2} = \frac{4x^2 - 2x^2 - 2}{4x^2} =$

$\frac{2(x^2-1)}{4x^2} = \frac{x^2-1}{2x^2}$

EXAMPLE

If $f(x) = \frac{3x^2 - 4x + 5}{6x - 7}$; then $f'(x) = [\frac{s(x)}{m(x)}]' = \frac{[s(x)]'m(x) - s(x)[m(x)]'}{[m(x)]^2} =$

$\frac{[3x^2 - 4x + 5]'(6x-7) - (3x^2 - 4x + 5)[(6x-7)]'}{[(6x-7)]^2} = \frac{(6x-4)(6x-7) - (3x^2 - 4x + 5)6}{(6x-7)^2} =$

$\frac{36x^2 - 42x - 24x + 28 - (18x^2 - 24x + 30)}{(6x-7)^2} = \frac{36x^2 - 42x - 24x + 28 - 18x^2 + 24x - 30}{(6x-7)^2} = \frac{18x^2 - 42x - 2}{(6x-7)^2}$

EXAMPLE

If $f(x) = 2x^3 - 7$, and $g(x) = 2\sqrt{x}$

We can write $\sqrt{x}$ as $x^{\frac{1}{2}}$ so, $(x^{\frac{1}{2}})' = \frac{1}{2}x^{\frac{1}{2}-1} = \frac{1}{2}x^{-\frac{1}{2}} = \frac{1}{2x^{\frac{1}{2}}} = \frac{1}{2\sqrt{x}}$

$[\frac{f(x)}{g(x)}]' = \frac{[f(x)]'g(x) - f(x)[g(x)]'}{[g(x)]^2} = \frac{(2x^3-7)'(2\sqrt{x}) - (2x^3-7)(2\sqrt{x})'}{(2\sqrt{x})^2} = \frac{(6x^2)(2\sqrt{x}) - (2x^3-7)\frac{2}{2\sqrt{x}}}{4x} =$

$\frac{(6x^2)(2\sqrt{x})(\sqrt{x}) - (2x^3-7)}{4x*\sqrt{x}} = \frac{12x^3 - 2x^3 + 7}{4x\sqrt{x}} = \frac{10x^3 + 7}{4x\sqrt{x}}$

7.F Differentiation

d. Differential rules – Chain

We have a function f(u), but u is as well a function of a variable x ; u(x). We will have the situation of a function of a function f(u(x)). To find the derivative of f(u(x)), we apply the "<u>chain</u>" rule:

$$f'(u(x)) = f'(u)[u'(x)]$$

The derivative of a function f of a function u equals the derivative of f(u) times the derivative of u(x).

EXAMPLE

If $f(x) = (6x + 7)^3, u(x) = 6x + 7, and f(u) = u^3$

Here we decide that u(x) is 6x+7 so, f(u) becomes $f(u) = u^3$

Applying the chain rule, we have:

$$f'(x) = f'(u)[u'(x)] = 3(6x + 7)^2(6x + 7)' = 3(6x + 7)^2 * 6 = 18(6x + 7)^2$$

EXAMPLE

If $f(x) = \sqrt{2x^2 + x}, u(x) = 2x^2 + x \ then, f(u) = \sqrt{u}$

We can write $\sqrt{u}$ as $u^{\frac{1}{2}}$ so, $(u^{\frac{1}{2}})' = \frac{1}{2}u^{\frac{1}{2}-1} = \frac{1}{2}u^{-\frac{1}{2}} = \frac{1}{2u^{\frac{1}{2}}} = \frac{1}{2\sqrt{u}}$

Applying the chain rule, we have:

$$f'(x) = f'(u)[u'(x)] = (\sqrt{u})'[u(x)]' = \frac{1}{2\sqrt{u}}(2x^2 + x)' = \frac{4x+1}{2\sqrt{2x^2+x}}$$

EXAMPLE

If $f(x) = \ln(2x - 3), u(x) = 2x - 3 \ then, f(u) = \ln(u)$

We know that $[\ln(u)]' = \frac{1}{u}$

Applying the chain rule, we have:

$$f'(x) = f'(u)[u'(x)] = \frac{1}{u}(2x - 3)' = \frac{2}{2x-3}$$

7.G Higher order differentiation

Until now, we calculated the first derivative of functions. Here, we will calculate second, and third derivative of a function. These higher derivatives are called <u>higher order derivatives</u>.

We have a function f(x) for which we need to calculate the third derivative. For this we calculate the first derivative $f'(x)$. Then we calculate the derivative of the first derivative $f''(x)$. Then we calculate the derivative of the second derivative $f'''(x)$.

EXAMPLE

Find the third derivative of $f(x) = 2x^3 - x^2 + 3x - 73$

If $f(x) = 2x^3 - x^2 + 3x - 73$

First, we calculate the first derivative of f(x).

$f'(x) = (2x^3)' - (x^2)' + (3x)' - (73)' = 6x^2 - 2x + 3$

Second, we calculate the first derivative of $f'(x)$.

$f''(x) = (6x^2)' - (2x)' + (3)' = 12x - 2$

Third, we calculate the first derivative of $f''(x)$.

$f'''(x) = (12x)' - 2' = 12$

EXAMPLE

Find the third derivative of $f(x) = 3x^3 - e^x$

If $f(x) = 3x^3 - e^x$

First, we calculate the first derivative of f(x)

$f'(x) = (3x^3)' - (e^x)' = 9x^2 - e^x$

Second, we calculate the first derivative of $f'(x)$.

$f''(x) = (9x^2)' - (e^x)' = 18x - e^x$

Third, we calculate the first derivative of $f''(x)$.

$f'''(x) = (18x)' - e^x = 18 - e^x$

7.H Implicit differentiation

Sometimes we have expressions like $x^2 - y + 3x + y^2 = 5$. In this case we have to differentiate term by term to find the differential $f'(x) = y'$.

EXAMPLE

We have the expression given above. We want to find $f'(x) = y'$.

$x^2 - y + 3x + y^2 = 5$

We differentiate each term of the relation with regard with x. Y is a function of a function.

$(x^2)' - y' + (3x)' + (y^2)' = 5'$

$2x - y' + 3 + 2yy' = 0$

$y'(-1 + 2y) + 2x + 3 = 0$

$y'(-1 + 2y) = -2x - 3$

$$y' = \frac{-2x-3}{-1+2y}$$

EXAMPLE

If $3x^3 - 3y^3 = 7$; $find\ y''$

We differentiate each term.

$(3x^3)' - (3y^3)' = 7'$

$9x^2 - 9y^2 y' = 0$ From here we have: $y' = \frac{x^2}{y^2}$

We differentiate another time each term.

$(9x^2)' - (9y^2 y')' = 0$

We apply the product rule for $(9y^2 y')'$.

$18x - [18y(y')^2 + 9y^2 y''] = 0$

We substitute y' in the equation above.

$18x - 18y(\frac{x^2}{y^2})^2 - 9y^2 y'' = 0$

$18x - 18\frac{x^4}{y^3} = 9y^2 y''$

$y'' = \frac{2x}{y^2} - \frac{2x^4}{y^5}$

7.I Differentiation - Applications

a. Relating graph of f(x) to f′(x) and f″(x)

We know that the value of the derivative gives the value of the slope of the tangent to the graph.

When the second derivative is positive, this means that the slopes of the graph keep increasing.

When the second derivative is negative, this means that the slopes of the graph keep decreasing.

There are a few rules that we will apply here regarding the relation between the sign of the first and second derivative and the graph of f(x).

	Positive	Negative
First derivative	Graph of f(x) goes UP	Graph of f(x) goes DOWN
Second derivative	Graph is Concave UP	Graph is Concave DOWN

EXAMPLE

Graph the function $f(x) = 2x^3 - 2x^2 - 5x + 4$

First, we calculate the first derivative.

$f'(x) = 6x^2 - 4x - 5$

$f'(x) = 0$ for:

$x_{1,2} = \frac{4 \mp \sqrt{4^2 - 4(6)(-5)}}{2*6} = \frac{4 \mp \sqrt{16+120}}{12} = \frac{4 \mp \sqrt{136}}{12} = \frac{4 \mp 11.66}{12}$ so, $x_1 = -0.63$ and $x_2 = 1.3$

Second, we calculate the second derivative.

$f''(x) = 12x - 4$ so, $f''(x) = 0$ for $x = 0.33$

The table below shows the sign of the derivative and the graph behavior.

	-0.63		0.33		1.3	
Sign of derivative	+		-		-	+
F(x) graph						
Sign of second deriv.	-		-		+	+
Concavity of the graph						

The graph is shown below.

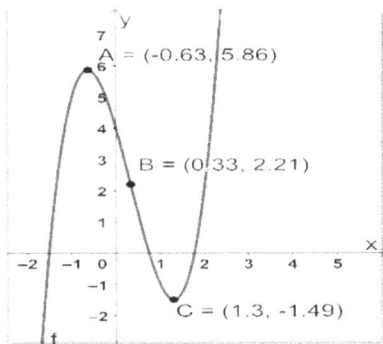

A = (-0.63, 5.86)
B = (0.33, 2.21)
C = (1.3, -1.49)

7.I Differentiation - Applications

b. Differentiability, mean value theorem

A function is called <u>differentiable</u> at a point when it has a derivative at that particular point.

Remember, the <u>derivative of a function</u> at a point is the instantaneous rate of change at that point.

Suppose a continuous function has the graph between f(b) and f(c) where c>b.

<u>The mean value theorem</u> says that there is at least one point on the graph D [d,f(d)] of a continuous and differentiable function, where the slope at that point of the graph equals

the slope of the line that passes through the ends points [b,f(b)] and [c,f(c)]. The formula is:

$$f'(d) = \frac{f(c)-f(b)}{c-b}$$

EXAMPLE

We have the graph of the function: $f(x) = (x-1)^5 + 4x^2 - 6x + 3$ showed below for 0<x<1.

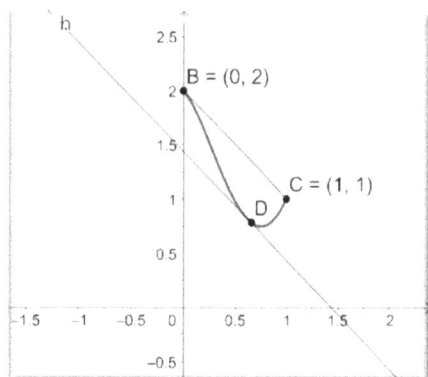

Here, we have the segment that passes through B and C. let's call B point 1 and C point **2**.
The slope of the line is calculated:

$$slope_{BC} = \frac{y_2-y_1}{x_2-x_1} = \frac{1-2}{1-0} = -1$$

By the **mean value theorem**, there is a point D on the graph of f(x) where the tangent to the graph has the same slope as the slope of the segment that passes through B and C.

We have to determine the expression of the tangent to the graph. We know that the expression of the slope of the tangent is given by the first derivative of the function. The derivative of the function is:

$f'(x) = [(x-1)^5]' + (4x^2)' - (6x)' + 3' = 5(x-1)^4 + 8x - 6$

We have to find the value of x for which $f'(x) = -1$

In this case this value is x=0.**65.**

$f(0.65) = (0.65-1)^5 + 4(0.65)^2 - 6(0.65) + 3 = -0.005 + 1.69 - 3.9 + 3 = 0.78$

So, the coordinates of point D are: (0.61,0.78.

7.I Differentiation - Applications

c. Newton's method

In a nutshell, Newton's method approximates the solution of the equation f(x)=0. Let us analyze the figure below.

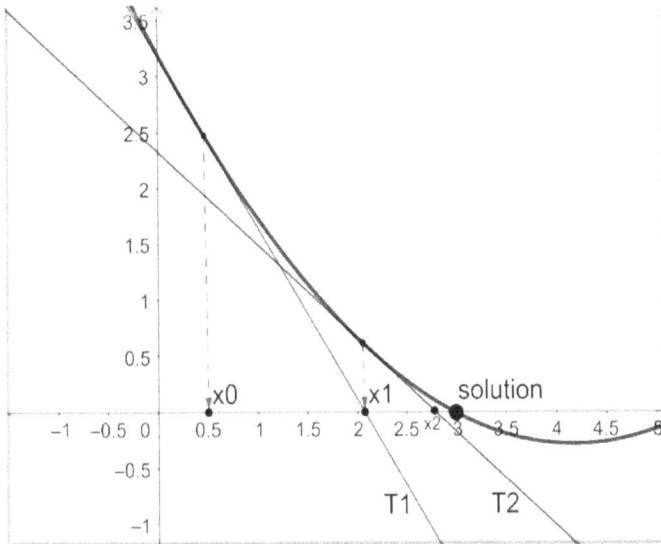

We start by guessing a solution x_0. Here we calculate the tangent at the function.

$$f'(x_0) = \frac{f(x)-f(x_0)}{x-x_0}$$

This tangent (T_1) intersects the x axis in x_1.

x_1 is of coordinates (x_1,0). This point is part of the tangent.

To find x_1, we substitute its coordinates in the tangent relation.

$$f'(x_0) = \frac{f(x_1)-f(x_0)}{x_1-x_0}, where \ f(x_1) = 0$$

$$f'(x_0) * (x_1 - x_0) = -f(x_0)$$

$$x_1 - x_0 = -\frac{f(x_0)}{f'(x_0)}$$

$$x_1 = x_0 - \frac{f(x_0)}{f'(x_0)}$$

So, $x_2 = x_1 - \frac{f(x_1)}{f'(x_1)}$ and so on. We repeat this process until the value of x_n converges to a value. This value is the solution we are looking for.

EXAMPLE

Find the solution of $f(x) = x^3 + 1$

The values used in the next calculations are shown in the table below

x_i	$f(x)$	$f'(x)$	x_{i+1}
-1.5	-2.37	6.75	-1.1418
-1.1418	-0.51	3.95	-1.018
-1.018	-0.055	3.11	-1.00033

Suppose the solution is -1.5

$f'(x) = 3x^2$ so, $f'(-1.5) = 3(-1.5)^2 = 6.75$

$f(-1.5) = -2.37$

$x_1 = x_0 - \dfrac{f(x_0)}{f'(x_0)} = -1.5 - \dfrac{-2.37}{6.75} = -1.5 + 0.35 = -1.1418$

$x_2 = x_1 - \dfrac{f(x_1)}{f'(x_1)} = -1.14 - \dfrac{-0.51}{3.95} = -1.14 + 0.1298 = -1.018$

$x_3 = x_2 - \dfrac{f(x_2)}{f'(x_2)} = -1.018 - \dfrac{-0.055}{3.11} = -1.018 + 0.01797 = -1.00033$

$x_4 = x_3 - \dfrac{f(x_3)}{f'(x_3)} = -1. - \dfrac{-0.00098}{3.00196} = -1 + 0.00033 = -1.$

The solution is x=-1

7.I. Differentiation - Applications

d. Problems in contextual situations, including related rates and optimization problems

We use math all the time whether we realize it or not. Math helps us build things, we use it in science, at the grocery store, when saving money, for time, etc.

EXAMPLE

At what time does a space shuttle change the trajectory from concave down to concave upward?

The equation of the space shuttle height with time for the first 5 minutes is given by:

$h(t) = 2008 - 0.047t^3 + 18.3t^2 - 345t$

Source of the relation: https://spacemath.gsfc.nasa.gov/weekly/5Page40.p

A modified equation with different coefficients is showed in the graph below to be able to visualize the trajectory.

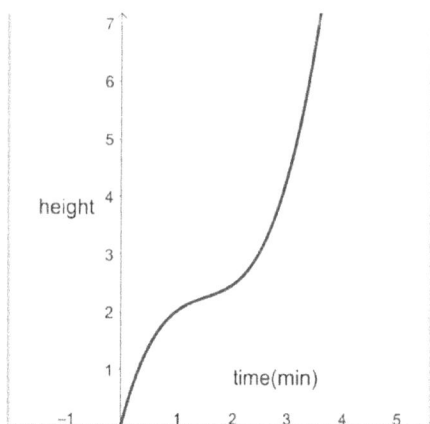

In the calculations we use the real equation used in reality.

To find it we first calculate the first derivative of the equation.

$h'(t) = -3(0.047)t^2 + 2(18.3)t - 345$

Then we calculate the second derivative.

$h''(t) = -3(2)(0.047)t + 2(18.3)$

We equalize the second derivative with zero.

$0 = -3(2)(0.047)t + 2(18.3)$

$0.282t = 36.6$

$t = \frac{36.6}{0.282} = 129.78 \; sec = 2.18 \; min$

At 2.18 minutes, the shuttle is changing the trajectory from concave down to concave upward.

TEST 8

Includes Differentiation

1.

4D.b. 4) The rationalized expression of $\dfrac{\sqrt{5}}{\sqrt{3}+\sqrt{2}}$ is:

2.

4E.a 4) The y intercept of the line $2(x-5)+3(y+2)=2$ is y=

3.

4F.a. 4) The factored by decomposition of $12x^2+11x-5$ is:

4.

4F.b. 4) The term outside the square of expression x^2-7x+6 is:

5.

5Aa 3) Determine if the relation below is a function.
{(-4, 8., (-4, 1), (-2, 3), (0, -12), (1,2), (2, 3)}

6.

5Ab 3) Determine if the following table of pairs of x and y represent a linear function.

X	Y
2	0
2	5
3	10
4	15

7.

5A.c. 3) If $f(x)=y=x^2-1$ the inverse function is:
$f^{-1}(x)=$

8.

5B. 1) Determine if the expressions below represent piecewise functions.

$f(x) = 2x^2 + 5x - 3, x \in R$

9.

5C. 3) $\pi \, radians =$ _____ *degrees*

10.

5D.b. 3) The cotangent function is zero for _____.

11.

5E. 3) The angle $\propto$ in right angle triangle PSQ is:

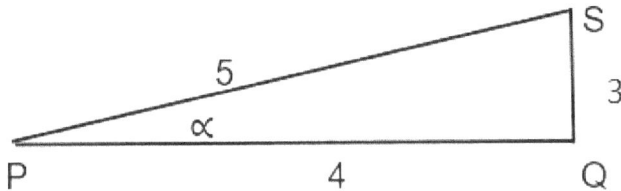

12.

6A.b. 2) $\lim\limits_{x \to 3}(\dfrac{9x^2-1}{3x-1}) =$

13.

6B 2) In the graph below $g(x) = -(0.5x - 3)^2 + 3 \, for \, x < 5$

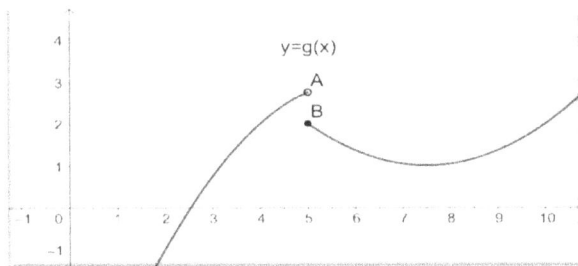

$$\lim\limits_{x \to 5^+} g(x) =$$

14.

6C 2) Determine what is the end behavior of these functions using limits.

$f(x) = -4x^4 + 2x^2 - 5x$: when x goes to -∞ it goes down in the third quadrant

15.

6D 2) If f is a continuous function on the closed interval [-1,5] where

$f(-1) = 2; f(5) = 7$

$f(c) = 2$ for at least one value of c that is between -1 and 5

16.

6E 2) Using the graph below, where: $f(x) = \begin{cases} 0.5x + 1, x \le 2 \\ 2x - 3, x > 2 \end{cases}$

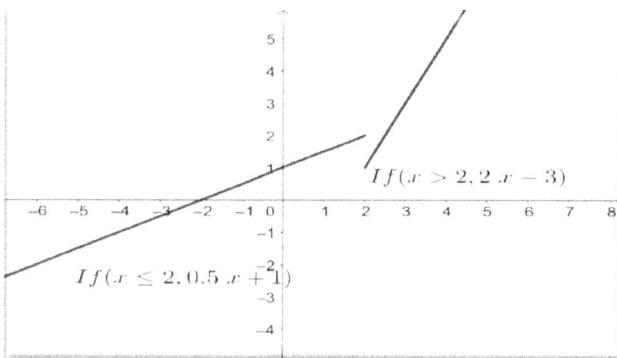

$If(x > 2, 2x - 3)$

$If(x \le 2, 0.5x + 1)$

$$\lim_{x \to 0^+} f(x) =$$

17.

6F 2) $\lim_{x \to -\infty} \dfrac{x^5 + 5x - 5}{2x^3 - 56x^2 + 8x - 3} =$

18.

6G 2) The following function is represented below: $f(x) =$

$\begin{cases} 2x - 1, x \ge 2 \\ 3x - 3, 1.5 < x < 2 \\ 0.5x, x \le 1.5 \end{cases}$

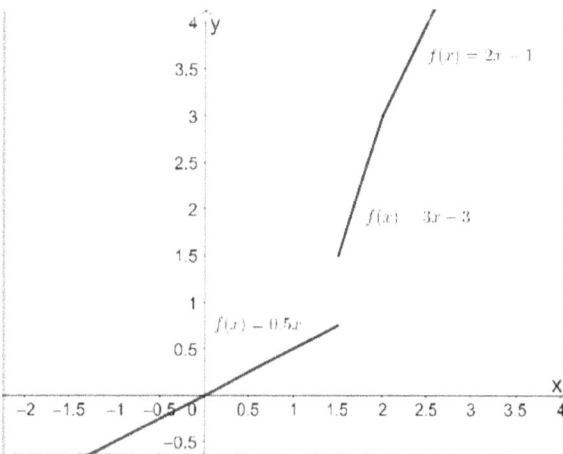

$f(x) = 2x - 1$

$f(x) = 3x - 3$

$f(x) = 0.5x$

The function f(x) is

_____ in x=2

19.

7B 1) The formula of the derivative

is: $f'(x) = \lim_{h \to 0} \dfrac{f(x+h) - f(x)}{h}$

Using this formula, calculate the following derivative.

$f(x) = 2x + 45$ so $f'(x) =$

20.

7C 1) Is Gottfried Leibniz's notation: $\frac{dy}{dx}$ correct?

21.

Distance (km)

$If(x > 0, 2x + 1)$

Time (hours)

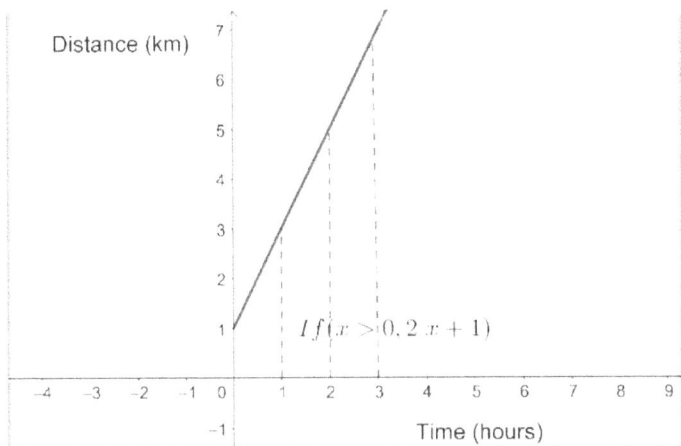

7D.a. 1) The average rate of change (speed) between 1 and 3 hours is: .

22.

7Db 1) The function represented below is:

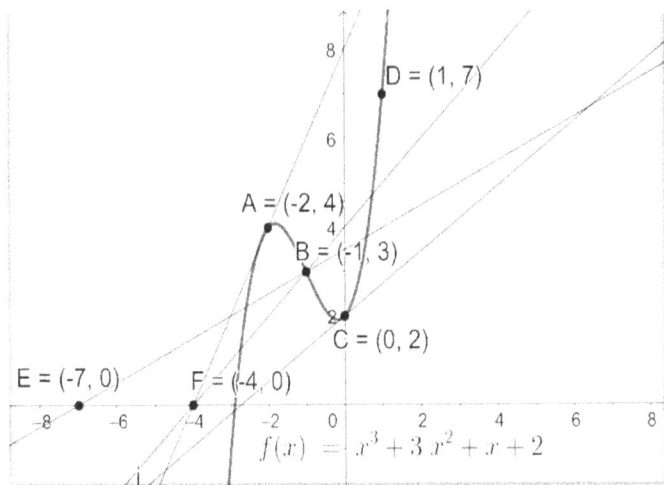

D = (1, 7)

A = (-2, 4)

B = (-1, 3)

C = (0, 2)

E = (-7, 0)   F = (-4, 0)

$f(x) = x^3 + 3x^2 + x + 2$

$f(x) = x^3 + 3x^2 + x + 2$
The slope of the secant that passes through the points E and B is:

23.

7E 1) If $f(x) = 2e^x - 7\ln(x)$; $then\ f'(x) =$

24.

7Fa 1) If $f(x) = 3x^2 - 7x + 89\ then\ f'(x) =$

25.

7Fb 1) If $f(x) = 2x^2 \cos(x)$; then $f'(x) =$

26.

7F.c. 1) If $f(x) = \tan(x)$; then $f'(x) = [\sec(x)]^2$

27.

7F.d. 1) If $f(x) = (x^2 - 4x + 5)^2$; then $f'(x) =$

28.

7G 1) If $f(x) = x^3 - 2x^2 + 3x - 4$; then $f''(x) =$

29.

7H 1) If $xy + x = 2$; $y' =$

30.

7Ia 1) If the first derivative is positive, the function is

_____.

31.

7I.b. 1) A differentiable function is a function whose _____ exists all over the domain.

Mark yourself

1	2	3	4	5	6	7	8	9	10
11	12	13	14	15	16	17	18	19	20
21	22	23	24	25	26	27	28	29	30
31									
# of Good Answers (NGA)=				NGA/total number of questions=Ratio				Percent=Ratio*100	
				YOUR PERCENT IS:					%

TEST 9

1.

4F.a. 5) The factored expression of $x^4 + 3x^2 + 2$ is:

2.

4F.b. 5) The term outside the square of expression $5x^2 - 15x + 3$ is:

3.

5A.a. 4) Determine if the relation below is a function.

X	Y
5	5
6	7
7	8
9	10

4.

5Ab 4) Determine if the following the expression represents a linear function.

$f(x) = 2x + 3$

5.

5A.c. 4) If $f(x) = y = \frac{3}{2x+4}$ the inverse function is: $f^{-1}(x) = y =$

6.

5B 2) Determine if the expression below represents a piecewise function.

$f(x) = \begin{cases} 2x \; for \; x < 0 \\ 3 \; for \; x \geq 0 \end{cases}$

7.

5C 4) In Quadrant 1 $\sin(\alpha)$ is _____.

8.

5D.b. 4) The tangent function is zero for _____

9.

5E 4) $Tan^{-1}(\sqrt{3})$ is:

10.

6A.b. 3) $\lim\limits_{x \to 2} \dfrac{x^2 + 2x - 8}{x - 2} =$

11.

6B 3) In the graph below $g(x) = -(0.5x - 3)^2 + 3 \; for \; x < 5$

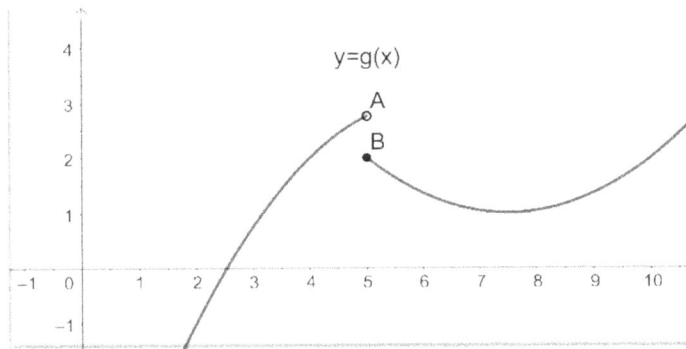

$g(x) = (0.4x - 3)^2 + 1 \; for \; x \geq 5$

$\lim\limits_{x \to 5} g(x) =$

12.

6C 3) $f(x) = -4x^3 + 6x^2 - x$: when x goes to -∞ it

goes up in the _____ quadrant

13.

6D 3) If f is a continuous function on the closed interval [-1,5] where
$f(-1) = 2; f(5) = 7$
$f(c) = -2$ for at least one value of c that is between _____

14.

6E 3) Using the graph below, Where: $f(x) = \begin{cases} 0.5x + 1, x \leq 2 \\ 2x - 3, x > 2 \end{cases}$

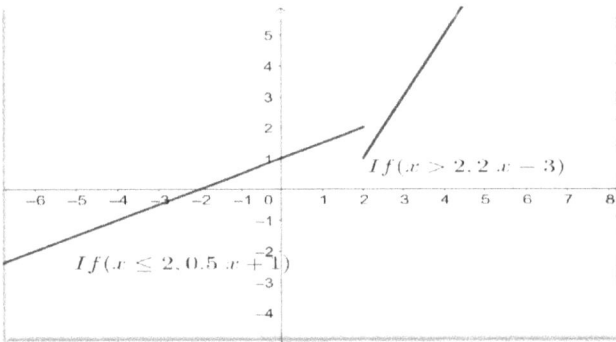

$$\lim_{x \to 2^-} f(x) =$$

15.

6F 3) $\lim\limits_{x \to \infty} \dfrac{3x^2+2x-45}{3x^3-5x^2+7x-3} =$

16.

6G 3) The following function is represented below: $f(x) =$

$$\begin{cases} 2x - 1, x \geq 2 \\ 3x - 3, 1.5 < x < 2 \\ 0.5x, x \leq 1.5 \end{cases}$$

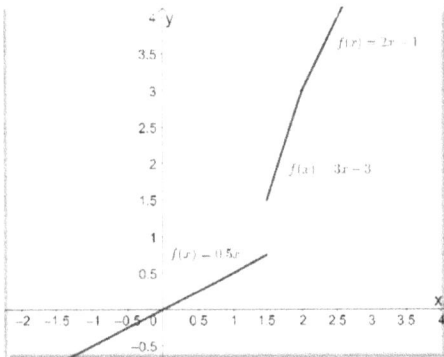

The function f(x) is _____ in x=1

17.

7B 2) The formula of the derivative is:

$$f'(x) = \lim_{h \to 0} \frac{f(x+h)-f(x)}{h}$$

Using this formula, calculate the following derivative.

$$f(x) = 3x^2 + 5x - 4 \text{ so } f'(x) =$$

18.

7C 2) Determine if this answer is correct: Another derivation notation is:

$$f(x) = y$$

19.

7Da 2) The average rate of change (speed) between 0 and 3 hours is:

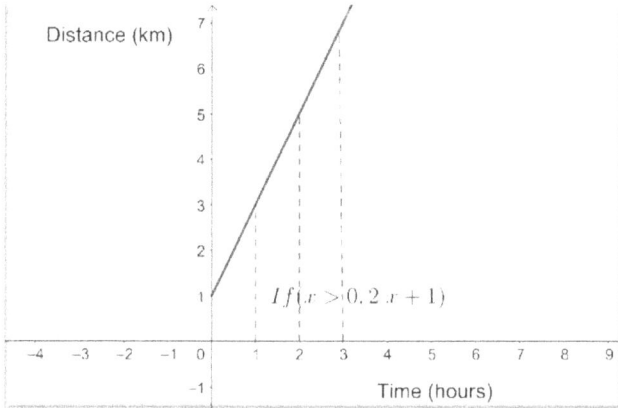

20.

7Db 2) The function represented below is: $f(x) = x^3 + 3x^2 + x + 2$

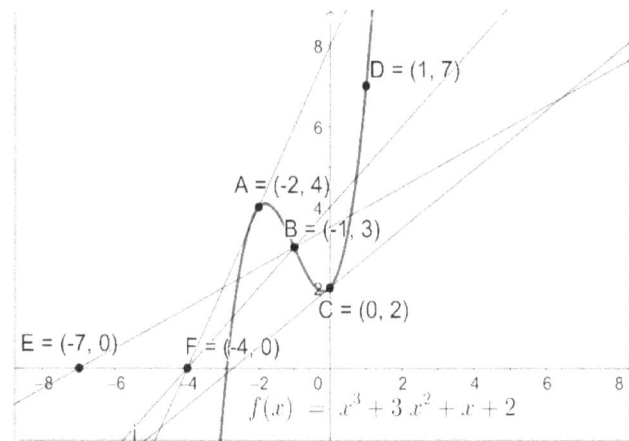

The slope of the secant that passes through the points F and B is:

21.

7E 2) If $f(x) = 3\ln(x) + 5e^x - ex$; then $f'(x) =$

22.

7Fa 2) If $f(x) = \frac{2}{5}x^2 + 27x - 63$ then $f'(x) =$

23.

7F.b. 2) If $f(x) = 3x^4\sqrt{x}$; then $f'(x) =$

24.

7F.c. 2) If $f(x) = \frac{\sin(x)}{x^2}$; then $f'(x) =$

25.

7Fd 2) If $f(x) = (3x^2 + 7x)^{\frac{3}{2}}$; then $f'(x) =$

26.

7G 2) If $f(x) = 7x^3 + 6x^2 + 5x - 4$; then $f'''(x) =$

27.

7H 2) If $x^2 + y^2 = 10$; $y' =$

28.

7I.a. 2) If the second derivative is positive, the function is concave

_____.

29.

7I.b. 2) In a two-dimensional curve between two points, there is at least one point at which the tangent to the curve is _____ to the secant through its two points.

30.

7I.c. 1) The square root of 135, using Newton's method is: 11.61

Mark yourself

1	2	3	4	5	6	7	8	9	10
11	12	13	14	15	16	17	18	19	20
21	22	23	24	25	26	27	28	29	30

# of Good Answers (NGA)=	NGA/total number of questions=Ratio	Percent=Ratio*100
	YOUR PERCENT IS:	%

TEST 10

1.

5A.a. 5) Determine if the values of the function is correct.

f(x) = 5x + 3   f(2) =15

2.

5A.b. 5) Determine if the expression represents a linear function.

$f(x) = x^2 - 3x + 4$

3.

5A.c. 5) If $f(x) = y = \frac{5}{2x^2+4}$ the inverse function is: $f^{-1}(x) = y =$

4.

5B 3) Determine if the expression below represents a piecewise function.

$f(x) = 4x + 3, x \in R$

5.

5C 5) In Quadrant 3 tan (α) is _____.

6.

5Da 1) Determine which statement is correct

For each of values of α, the values of sin (α) are:

α	0	$\frac{\pi}{6}$	$\frac{\pi}{2}$	$\frac{\pi}{3}$	π
sin (α)	0	$\frac{1}{2}$	1	$\frac{\sqrt{3}}{2}$	0

7.

5Db 5) For 60^0 the tangent is:

8.

5E 5) The domain of $f(x) = sin^{-1}(x)$ is:

9.

5F 1) The domain of $f(x) = cos^{-1}(x)$ is:

10.

6A.b. 4) $\lim\limits_{x \to 3} \left(\dfrac{1}{x-3} - \dfrac{6}{x^2-9} \right) =$

11.

6B 4) In the graph below $\lim\limits_{x \to 5^-} g(x) =$

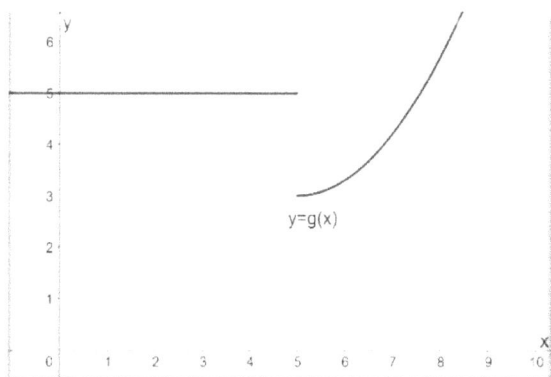

12.

6C 4) Determine what is the end behavior of this function using limits. $f(x) = x^3 - 36x^2 - x + 73$: when x goes to $-\infty$ f(x) it goes _____ in the third quadrant.

13.

6D 4) If f is a continuous function on the closed interval [-1,5] where
$f(-1) = 2; f(5) = 7$
$f(c) = 3$ for at least one value of c that is between -1 and 5

14.

6E 4) Using the graph below, determine:

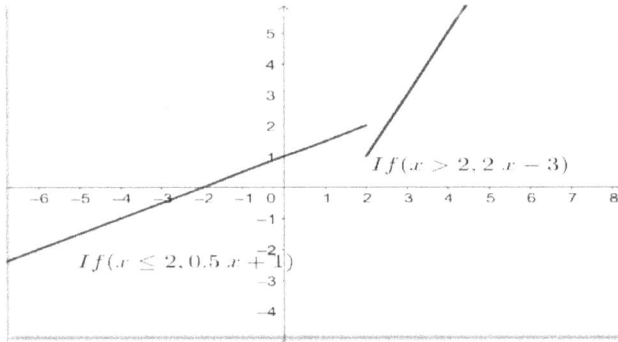

$$\lim_{x\to 2^+} f(x) =$$

15.

6F 4) $\lim_{x\to -\infty} \dfrac{x-4}{2x^2-8x-2} =$

16.

6G 4) The following function is represented below: $f(x) =$

$$\begin{cases} 2x - 1, x \geq 2 \\ 3x - 3, 1.5 < x < 2 \\ 0.5x, x \leq 1.5 \end{cases}$$

Is the function f(x) continuous in x=-7?

17.

7B 3) The formula of the derivative is: $f'(x) = \lim_{h\to 0} \dfrac{f(x+h)-f(x)}{h}$

Using this formula, calculate the following derivative

If $f(x) = Ax^2 + Bx - 37$ so $f'(x) =$

18.

7C 3) Joseph Louis Lagrange's notation is $f'(x)$

19.

7D.a. 3) A tire rolls by the relation between distance and time as follows: $s(t) = 2t^2 + 4t + 3$. The instantaneous velocity (first derivate) at t=2 seconds is:

20.

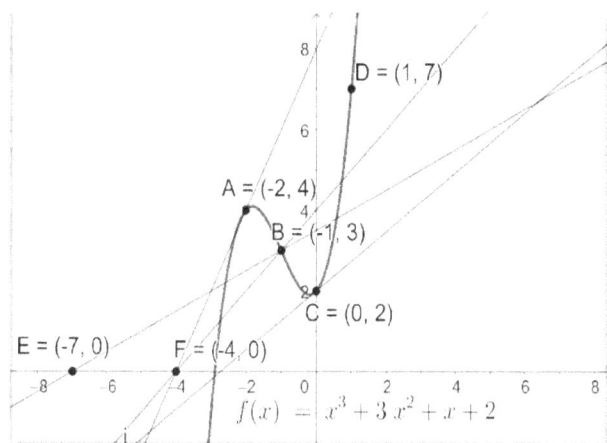

A = (-2, 4)
B = (-1, 3)
C = (0, 2)
D = (1, 7)
E = (-7, 0)
F = (-4, 0)
$f(x) = x^3 + 3x^2 + x + 2$

7D.b. 3) The slope of the tangent to the graph in point A equals:

21.

7E 3) If $f(x) = 37 - e^x$; then $f'(x) =$

22.

7Fa 3) If $f(x) = \frac{3}{7}x^4 - 2x^3 - 6x^2 + 81$ then $f'(x) =$

23.

7F.b. 3) If $f(x) = 4x^2 \tan(x)$; then $f'(x) =$

24.

7F.c. 3) If $f(x) = \frac{x}{e^x}$; then $f'(x) =$

25.

7F.d. 3) If $f(x) = \sqrt{x^3 - 5x}$; then $f'(x) =$

26.

7G 3) If $f(x) = 3x^5 + 4\sqrt{x} - 5x - 6$; then $f'''(x) =$

27.

7H 3) If $2x^2 + y^3 = 3$; $y'' =$

28.

7I.a. 3) The graph of the function $f(x) = \frac{2}{3}x^3 - 2x^2 - 6x + 7$ is increasing between -3 and 1

29.

7I.b. 3) If $f(x) = 3x^2 - 4x + 1$ and the points (-3,40) and, (1,0) the value of x where the tangent at the graph is parallel with the line that goes through (-3,40) and, (1,0) is:

30.

7Ic 2) The square root of 432, using Newton's method is:

Mark yourself

1	2	3	4	5	6	7	8	9	10
11	12	13	14	15	16	17	18	19	20
21	22	23	24	25	26	27			
# of Good Answers (NGA)=				NGA/total number of questions=Ratio				Percent=Ratio*100	
				YOUR PERCENT IS:					%

CHAPTER 8

INTEGRALS

8. A Definition of an integral and notation

Simplistically, an <u>integral</u> is a weighted sum of the values of the function multiplied with the infinitesimal widths dx.

There are two types of integrals; indefinite and definite.

The symbol for indefinite integral is $\int$.

The symbol for the definite integral is $\int_a^b$, where a and b are the ends of the interval for which the integral is calculated.

The mathematical process that helps us find an integral it is called <u>integration</u>.
If the variable of the function is x, the indefinite integral is symbolized $\int f(x)dx$.
The function f(x) under the integral sign is called <u>integrand</u>, and x is called the <u>integration variable</u>
Please note that dx is an infinitesimal quantity. To integrate something means as well that we add many products formed by the value of the function multiplied with the infinitesimal quantity dx.
But what do these pairs of f(x) times dx represent? If we try to visualize them on a system of axes, they represent the very small area of a rectangle with one side being dx and the height being f(x).

EXAMPLE

In the graph showed below, we want to calculate the area of the trapezoid ALQK. We

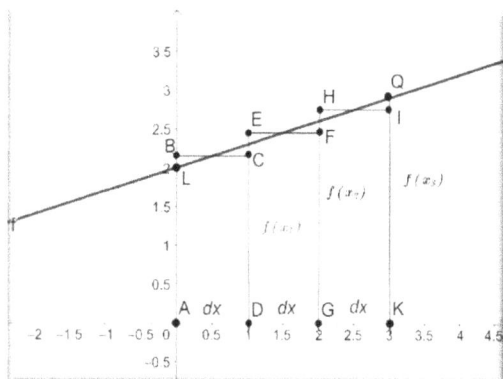

split the trapezoid into rectangles. In this case, the number is only 3 for clarity. The rectangles have the base dx=1 unit. The top side of the rectangles go through the value of f(x) where x is the middle of the base dx. The heights of the rectangles $DC = f(x_1); GF = f(x_2); KI = f(x_3)$

The area of ALQK, is approximately the sum of the areas of ABCD, DEFG, GHIK.

We write the sum as:

$Area = \sum_{i=1}^{3} f(x_i)dx = f(x_1)dx + f(x_2)dx + f(x_3)dx$

As we said before, dx is very small. So, to find the best approximation for area, we have to split the trapezoid into very small rectangles. We have then to add a lot of very small areas. We can approximate the area with:

$Area \cong \sum_{i=1}^{n} f(x_i)dx = f(x_1)dx + f(x_2)dx + \cdots f(x_n)dx$ where n is a very big number but finite.

When dx becomes infinitesimal, n approaches infinity so, instead of adding multiple

small areas we will use integration instead. $Area = \lim_{n \to \infty} \sum_{i=1}^{n} f(x_i)dx = \int_{0}^{n} f(x_i)dx$

In the next chapters we will discuss in more detail what this equation means, and how we should understand the many, many super-small intervals dx.

8.B Definite and indefinite integrals

If we have a function $f(x)$, an <u>anti-derivative</u> of this function is any function $F(x)$ such that

$F'(x) = f(x)$

The indefinite integral of $f(x)$ is:

$\int f(x)dx = F(x) + C$ where, $F(x)$ is the anti-derivative of $f(x)$, and C is any constant.

EXAMPLE

If $f(x) = 2x$, $\int f(x)dx = x^2 + C = F(x) + C$,

$F'(x) = (x^2)' = 2x = f(x)$

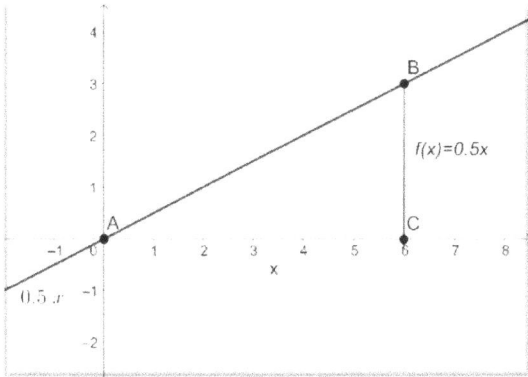

Finding Area Problem

We want to calculate the area of the triangle ABC. The function that constitutes the hypotenuse AB is:

$$f(x) = \frac{1}{2}x$$

The base of the triangle AC is x.
The height CB is $f(x) = 0.5x$

$$Area = \frac{AC*BC}{2} = \frac{1}{2}\left(\frac{1}{2}x\right)x = \frac{1}{4}x^2$$

What is the derivative of $\frac{1}{4}x^2$?

$(\frac{1}{4}x^2)' = \frac{1}{4} * 2x = \frac{1}{2}x = f(x)$, so, $\frac{1}{4}x^2 = F(x)$

Area of the surface that is below y=f(x) in between zero and 6 is the definite integral:

$\int_0^6 f(x)dx = F(x)$ between x=0 and x=6

In the future, we will discuss in more detail how to calculate the indefinite and definite integrals.

Some indefinite integrals:

$\int x^3 dx = \frac{x^4}{4} + C$

$\int \frac{dx}{x} = \ln|x| + C$

$\int e^x dx = e^x + C$

$\int \ln(x)\, dx = x\ln(x) - x + C$

$\int \sin(x)\, dx = -\cos(x) + C$

$\int a^x dx = \frac{a^x}{\ln(a)} + C$

$\int \cos(x)\, dx = \sin(x) + C$

$\sqrt{x} = x^{\frac{1}{2}}$ so,

$\int \sqrt{x}dx = \int x^{\frac{1}{2}}\, dx = \frac{2x^{\frac{3}{2}}}{3} + C$

8.C Approximations-Riemann sum, rectangle method, trapezoidal method

Riemann sum helps us approximate the definite integral $\int_a^b f(x)dx$ with the area beneath the graph of a function and for x values in a closed interval [a,b].

EXAMPLE

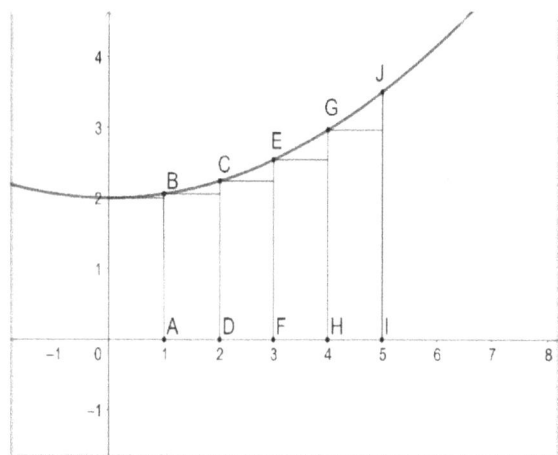

Let's have a=0 and b=**5**. The function is $f(x) = 0.06x^2 + 2$. We create rectangles that touch the graph at the top left corner. We add the areas of these rectangles. This is called <u>Riemann left sum</u>.

Coordinates of these top left corner points are shown below.

Point	X	F(x)
2	0	2
B	1	2.06
C	2	2.24
E	3	2.54
G	4	2.96

As it can be seen in the table, the distance between each consecutive x value is $\Delta x = 1$
The sum of these areas is:

$\int_0^4 f(x)dx \cong \sum_{i=0}^4 f(x_i) * \Delta x_i = f(0) * 1 + f(1) * 1 + f(2) * 1 + f(3) * 1 + f(4) * 1 = 2 + 2.06 + 2.24 + 2.54 + 2.96 = 11.8$

If we want a better approximation of the area beneath the graph, we need to choose much thinner rectangles with Δx much smaller.

<u>Trapezoidal method</u> helps us find the integral $\int_a^b f(x)dx$ by approximating it with the area beneath the graph of a function by partitioning the area in small trapezoids and calculate the total area of these trapezoids.

EXAMPLE

Let's use a=0 and b=**4**. We are using the same function as before: $f(x) = 0.06x^2 + 2$.
The formula for the area of a trapezoid is:

$Area = \frac{(Big\ base + small\ base)*height}{2}$

In this case height Δx is 1.
We approximate the integral with the sum of all the trapezoids the area is portioned in.
We consider four trapezoids 1 to 4.

$\int_1^4 f(x)dx \cong \sum_1^4 \frac{(Big\ base_i + small\ base_i)*\Delta x_i}{2}$

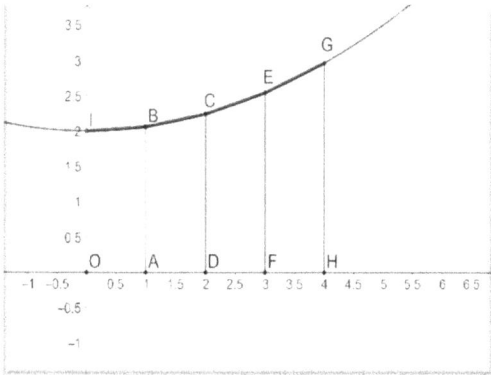

$$Area_{IBAO} = Area_1 = \frac{(OI+AB)*AO}{2} = \frac{(2+2.06)*1}{2} =$$

$$2.03$$

$$Area_{BADC} = Area_2 = \frac{(AB+DC)*AD}{2} =$$

$$\frac{(2.06+2.24)*1}{2} = 2.15$$

$$Area_{DCEF} = Area_3 = \frac{(DC+EF)*DF}{2} =$$

$$\frac{(2.24+2.54)*1}{2} = 2.39$$

$$Area_{DCEF} = Area_4 = \frac{(FEGH)*FH}{2} = \frac{(2.54+2.96)*1}{2} = 2.75$$

$$\int_1^4 f(x)dx \cong \sum_1^4 \frac{(Big\ base_i+small\ base_i)*\Delta x_i}{2} = \frac{(OI+AB)*AO}{2} + \frac{(AB+DC)*AD}{2} + \frac{(DC+EF)*DF}{2} +$$

$$\frac{(FEGH)*FH}{2} = 2.03 + 2.15 + 2.39 + 2.75 = 9.32$$

EXAMPLE

If $f(x) = 3x^3 - 2x^2 + 1$, using trapezoidal method, find the integral $\int_{-2}^2 f(x)dx$

The interval $\Delta x = 0.5$

$$\int_{-2}^2 f(x)dx \cong \sum_{i=-2}^{i=2} \frac{f(x_i)+f(x_{i+1})}{2} \Delta x_i = \frac{f(-2)+f(-1.5)}{2}*0.5 + \frac{f(-1.5)+f(-1)}{2}*0.5 + \frac{f(-1)+f(-0.5)}{2} *$$

$$0.5 + \frac{f(-0.5)+f(0)}{2}*0.5 + \frac{f(0)+f(0.5)}{2}*0.5 + \frac{f(0.5)+f(1)}{2}*0.5 + \frac{f(1)+f(1.5)}{2}*0.5 + \frac{f(1.5)+f(2)}{2}*0.5$$

$$f(-2) = 3(-2)^3 - 2(-2)^2 + 1 = -24 - 8 + 1 = -31$$

$$f(-1.5) = 3(-1.5)^3 - 2(-1.5)^2 + 1 = 3(-3.375) - 2(2.25) + 1 = -10.125 - 4.5 + 1 = -13.625$$

$$f(-1) = 3(-1)^3 - 2(-1)^2 + 1 = -3 - 2 + 1 = -4$$

$$f(-0.5) = 3(-0.5)^3 - 2(-0.5)^2 + 1 = 3(-0.125) - 2(0.25) + 1 = -0.375 - 0.5 + 1 = 0.13$$

$$f(0) = 3(0)^3 - 2(0)^2 + 1 = 1$$

$$f(0.5) = 3(0.5)^3 - 2(0.5)^2 + 1 = 3(0.125) - 2(0.25) + 1 = 0.375 - 0.5 + 1 = 0.875$$

$$f(1) = 3(1)^3 - 2(1)^2 + 1 = 3 - 2 + 1 = 2$$

$$f(1.5) = 3(1.5)^3 - 2(1.5)^2 + 1 = 3(3.375) - 2(2.25) + 1 = 10.125 - 4.5 + 1 = 6.625$$

$$f(2) = 3(2)^3 - 2(2)^2 + 1 = 24 - 8 + 1 = 17$$

$$\int_{-2}^2 f(x)dx \cong \sum_{i=-2}^{i=2} \frac{f(x_i)+f(x_{i+1})}{2} \Delta x_i = \frac{f(-2)+f(-1.5)}{2}*0.5 + \frac{f(-1.5)+f(-1)}{2}*0.5 + \frac{f(-1)+f(-0.5)}{2} *$$

$$0.5 + \frac{f(-0.5)+f(0)}{2}*0.5 + \frac{f(0)+f(0.5)}{2}*0.5 + \frac{f(0.5)+f(1)}{2}*0.5 + \frac{f(1)+f(1.5)}{2}*0.5 + \frac{f(1.5)+f(2)}{2} *$$

$$0.5 = \frac{-27-13.625}{2}*0.5 + \frac{-13.625-4}{2}*0.5 + \frac{-4-0.125}{2}*0.5 + \frac{-0.125+1}{2}*0.5 + \frac{1+0.875}{2}*0.5 +$$

$\frac{0.875+2}{2} * 0.5 + \frac{2+6.625}{2} * 0.5 + \frac{6.625+17}{2} * 0.5 = -11.16 - 4.40 - 0.969 + 0.218 + 0.468 +$

$0.718 + 2.156 + 5.9 = -7$

8.D Fundamental Theorem of Calculus

The fundamental theorem of calculus says that:

$\int_a^b f(x)dx = \int_0^b f(x)dx - \int_0^a f(x)dx = F(b) - F(a), where\ F'(x) = f(x)$

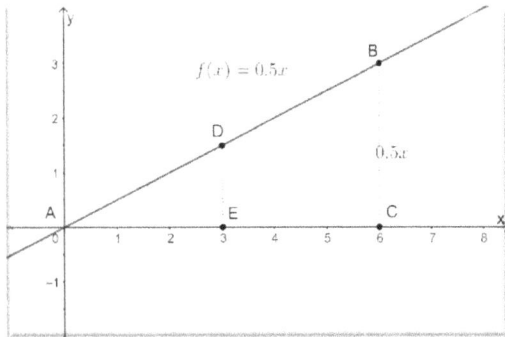

$f(x) = 0.5x$

Before we explain the formula, we go back to the area problem in 4.B.

Area Problem extended

We want to calculate the area of EDBC. First, we calculate the area of the triangle ABC. The function that constitutes the hypotenuse AB is:

$f(x) = \frac{1}{2}x$

The base of the triangle AC is x.

The height CB is $f(x) = 0.5x$

$Area = \frac{1}{2}AC * CB = \left(\frac{1}{2}x\right)x = \frac{1}{4}x^2$

What is the derivative of $\frac{1}{4}x^2$?

$(\frac{1}{4}x^2)' = \frac{1}{4} * 2x = \frac{1}{2}x = f(x)$, so,

$\frac{1}{4}x^2 = F(x) = f'(x)$

Area of the surface ABC that is below y=f(x) in between zero and 6 is the definite integral:

$\int_0^6 f(x)dx = F(x)$ between x=0 and x=6

$F(6) = \frac{1}{4}(6)^2 = \frac{36}{4} = 9$

$F(0) = 0$

Second, we calculate the area of the triangle ADE.

The function that constitutes the hypotenuse AD is:

$f(x) = \frac{1}{2}x$

Area of the surface ADE that is below y=f(x) in between zero and 3 is the definite integral:

$\int_0^3 f(x)dx = F(x)$ between x=0 and x=3

$F(3) = \frac{1}{4}(3)^2 = \frac{9}{4} = 2.25$

$Geometric\ Area_{EDBC} = \frac{(Big\ base+small\ base)*Height}{2} = \frac{(3+1.5)*3}{2} = 6.75$

What we notice is that if we subtract $F(6) - F(3) = 9 - 2.25 = 6.75$ equal with the area of EDBC

So, the area situated below the function $f(x) = \frac{1}{2}x$ between x=3 and x=6 is given by:

$\int_3^6 f(x)dx = \int_0^6 f(x)dx - \int_0^3 f(x)dx = F(6) - F(3) = 9 - 2.25 = 6.75, where\ F'(x) = f(x)$

EXAMPLE

Find the area under the graph of the function $f(x) = 0.3x^2 + 1$ between x=-1 and x=3 using the

Fundamental Theorem of Calculus.

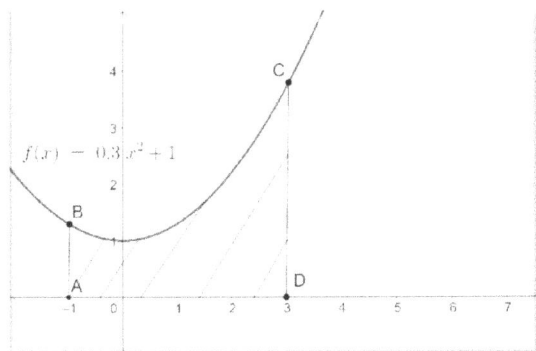

$\int_{-1}^3 f(x)dx = F(3) - F(-1)$

$F(x) = \int f(x)dx = \frac{0.3}{3}x^3 + x + C$

So,

$F(3) = \frac{0.3}{3}(3)^3 + (3) + C = 5.7 + C$

$F(-1) = \frac{0.3}{3}(-1)^3 + (-1) + C = -1.1 + C$

So,

$\int_{-1}^3 f(x)dx = F(3) - F(-1) = 5.7 + C -$

$(-1.1) - C = 6.8$

EXAMPLE

Find the area below the function $f(x) = \frac{2}{x} - 2x + 5$ for x=1 and x=4

First,

We calculate the indefinite integral of the function.

$F(x) = \int f(x)dx = \int \frac{2}{x}dx - \int 2xdx + \int 5dx = 2\ln(x) - \frac{2}{2}x^2 + 5x + C = 2\ln(x) - x^2 + 5x$

Second,

We calculate the values of the F(x) for x=4 and x=1

$F(4) = 2\ln(4) - (4)^2 + 5(4) = 2\ln(4) - 16 + 20 = 2\ln(4) + 4 = 2(1.38) + 4 = 2.77 + 4 = 6.77$

$F(1) = 2\ln(1) - (1)^2 + 5(1) = -1 + 5 = 4$

Third,

We apply the Fundamental Theorem of Calculus:

$\int_1^4 f(x)dx = F(4) - F(1) = 6.77 - 4 = 2.77$

8.E Integration

a. Antiderivatives of functions

The antiderivative F(x) is reverse of the derivative. We know that $F'(x) = original\ function\ f(x)$

Suppose that $f(x) = 3x + 1$. Then the derivate of the antiderivative $F'(x) = 3x + 1 = f(x)$ will be equal with the original function.

The indefinite integral of the function f(x) is $\int f(x)dx = \frac{3}{2}x^2 + x + C = F(x)$

We can see that, if we derivate F(x) we get f(x).

$F'(x) = (\frac{3}{2}x^2 + x + C)' = \frac{3*2}{2}x + 1 + 0 = 3x + 1 = f(x)$

The antiderivative is connected with the definite integral through the Fundamental Theorem of Calculus.

If we have an interval [a,b].

$\int_a^b f(x)dx = F(b) - F(a)\ where\ F(b)\ and\ F(a)$ are the antiderivatives of f(x) at x=b and x=b.

EXAMPLE

If the acceleration of a car is given by the function $f(t) = 5t + 3, where\ t\ is\ time,$ the velocity is given by the antiderivative $F(t) = \frac{5}{2}t^2 + 3t = \int f(t)dt = V(t)$

The derivative of F(t) is:

$F(t)' = V'(t) = (\frac{5}{2}t^2 + 3t)' = \frac{5*2}{2}t + 3 = 5t + 3 = f(t) = acceleration$

The distance is given by:

$s(t) = \int V(t)dt = \frac{5}{2*3}t^3 + \frac{3}{2}t^2 + C = \frac{5}{6}t^3 + \frac{3}{2}t^2 + C$

$s'(t) = (\frac{5}{6}t^3 + \frac{3}{2}t^2 + C)' = \frac{5*3}{6}t^2 + \frac{3*2}{2}t + 0 = \frac{5}{2}t^2 + 3t = V(t)$

8.E Integration

b. Methods of Integration - Substitution

Sometimes, we can differentiate one of the factors of a multiplication inside the integral and get part of the other term of the product.

If we have $\int 2x(x^2 + 1)dx$ we notice that if we derivate $x^2 + 1$, we get 2x.

We substitute $x^2 + 1$ with u:

$u' = (x^2 + 1)' = 2x = \frac{du}{dx}$

So,

$dx = \frac{du}{2x}$

We have:

$\int 2x(u) \frac{du}{2x} = \int u\,du = \frac{u^2}{2} = \frac{(x^2+1)^2}{2} + C$

So,

$\int 2x(x^2 + 1)dx = \frac{(x^2+1)^2}{2} + C$

EXAMPLE

If $f(x) = (4x - 3)(2x^2 - 3x + 5)^2$ then, find $F(x)$

We substitute $2x^2 - 3x + 5$ with u:

$u = 2x^2 - 3x + 5$ so,

Then we calculate the derivate of u:

$u' = \frac{du}{dx} = 4x - 3$

We substitute dx with du:

$dx = \frac{du}{4x-3}$

Then we calculate the integral in u:

$\int (4x - 3)u^2 \frac{du}{4x-3} = \int u^2 du = \frac{1}{3}u^3 + C$

We substitute back u with $2x^2 - 3x + 5$

$F(x) = \int f(x)dx = \frac{1}{3}(2x^2 - 3x + 5)^3 + C$

8. E Methods of integration

c. by parts

When we are using the integration by parts, we are using the product rule for differentiating, and substitution rule.

Remember the product rule for differentiating a product two functions f(x) and g(x) is given by the formula:

$[f(x) * g(x)]' = f'(x) * g(x) + f(x) * g'(x)$

If we integrate this formula we get:

$\int [f(x) * g(x)]'dx = \int f'(x) * g(x)dx + \int f(x) * g'(x)dx$

And then,

$f(x) * g(x) = \int f'(x) * g(x)dx + \int f(x) * g'(x)dx$

Now, the substitution rule is used.

Usually, u and v are used as the functions we substitute f(x) and g(x):

If $u = f(x)$, than $f'(x) = \frac{du}{dx}$

From here,

$du = f'(x)dx$

If $v = g(x)$, than $g'(x) = \frac{dv}{dx}$

From here,

$dv = g'(x)dx$

$f(x) * g(x) = \int f'(x) * g(x)dx + \int f(x) * g'(x)dx$ becomes:

$uv = \int vdu + \int udv$

From here:

$\int udv = uv - \int vdu$ the integration by parts formula.

EXAMPLE

If $f(x) = \frac{\ln(x)}{x^2}$ find $F(x)$

$u = \ln(x)$ so $u' = \frac{du}{dx} = \frac{1}{x}$

$du = \frac{1}{x}dx$

$dv = \frac{1}{x^2}dx$ so $v = \int \frac{1}{x^2}dx = \int x^{-2}dx = -x^{-1}$

By using the integration by parts formula, we have:

$\int udv = uv - \int vdu$

$\int \ln(x)\left(\frac{1}{x^2}dx\right) = \ln(x)(-x^{-1}) - \int(-x^{-1})\left(\frac{1}{x}dx\right) = -\frac{\ln(x)}{x} + \int(x^{-2}dx) = -\frac{\ln(x)}{x} - \frac{1}{x} =$

$-\frac{1}{x}[\ln(x) + 1]$

8.F Integration - Applications

a. Aria under a curve, volume of solids, average value of functions

Remember that the fundamental theorem of calculus says that:

$\int_a^b f(x)dx = \int_0^b f(x)dx - \int_0^a f(x)dx = F(b) - F(a)$, where $F'(x) = f(x)$

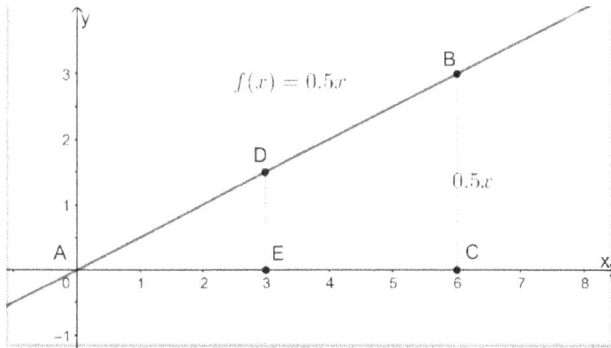

Before we explain the formula, we go back to the area problem in 4.B.

Area Problem extended

We want to calculate the area of EDBC. First, we calculate the area of the triangle ABC. The function that constitutes the hypotenuse is:

$$f(x) = \frac{1}{2}x$$

The base of the triangle AC is x.

The height CB is $f(x) = 0.5x$

$$Area = \frac{1}{2}AC * CB = \left(\frac{1}{2}x\right)x = \frac{1}{4}x^2$$

What is the derivative of $\frac{1}{4}x^2$?

$(\frac{1}{4}x^2)' = \frac{1}{4} * 2x = \frac{1}{2}x = f(x)$, so,

$$\frac{1}{4}x^2 = F(x) = f'(x)$$

Area of the surface ABC that is below y=f(x) in between zero and 6 is the definite integral

$\int_0^6 f(x)dx = F(x)$ between x=0 and x=6

$$F(6) = \frac{1}{4}(6)^2 = \frac{36}{4} = 9$$

$$F(0) = 0$$

Second, we calculate the area of the triangle ADE.

The function that constitutes the hypotenuse is:

$$f(x) = \frac{1}{2}x$$

Area of the surface ADE that is below y=f(x) in between zero and 3 is the definite integral

$\int_0^3 f(x)dx = F(x)$ between x=0 and x=3.

$$F(3) = \frac{1}{4}(3)^2 = \frac{9}{4} = 2.25$$

$$Geometric\ Area_{EDBC} = \frac{(Big\ base+small\ base)*Height}{2} = \frac{(3+1.5)*3}{2} = 6.75$$

What we notice is that if we subtract $F(6) - F(3) = 9 - 2.25 = 6.75$ equal with the area of EDBC

So, the area situated below the function $f(x) = \frac{1}{2}x$ between x=3 and x=6 is given by:

$\int_3^6 f(x)dx = \int_0^6 f(x)dx - \int_0^3 f(x)dx = F(6) - F(3) = 9 - 2.25 = 6.75, where\ F'(x) = f(x)$

EXAMPLE

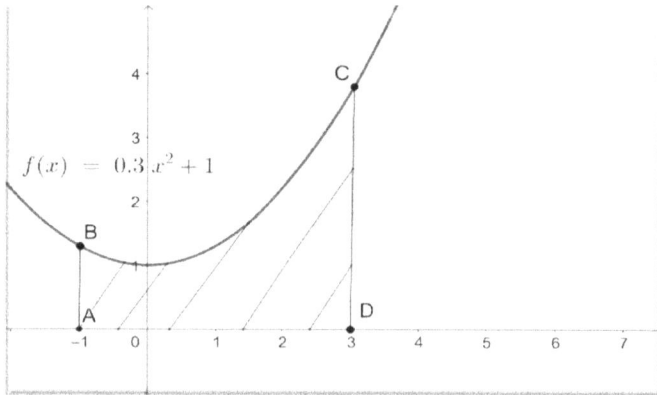

Find the area under the graph of the function $f(x) = 0.3x^2 + 1$ between x=-1 and x=3 using the Fundamental Theorem of Calculus.

$\int_{-1}^{3} f(x)dx = F(3) - F(-1)$

$F(x) = \int f(x)dx = \frac{0.3}{3}x^3 + x + C$

So,

$F(3) = \frac{0.3}{3}(3)^3 + (3) + C = 5.7 + C$

$F(-1) = \frac{0.3}{3}(-1)^3 + (-1) + C = -1.1 + C$

So,

$\int_{-1}^{3} f(x)dx = F(3) - F(-1) = 5.7 + C - (-1.1) - C = 6.8$

Volume for solids problem

Let us calculate the volume of a cone. The slant of the cone is the function y=f(x). This line will rotate around x axis. At any x value, y represents the radius of the circle at that x value.

Through the integral, we add all the surfaces with y=f(x) radius between the x initial and x final and obtain the volume of the cone.

EXAMPLE

Suppose we want to find the volume of a cone with the slant y=f(x)=x.

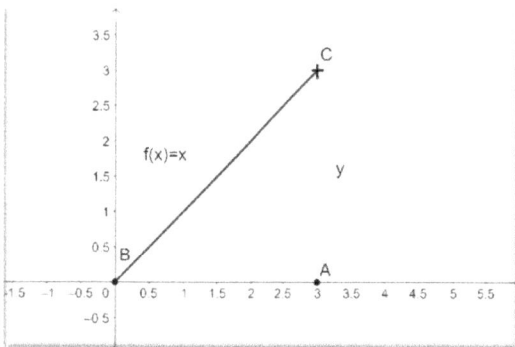

The slant rotates around x axis.

We consider y=f(x) as the radius of all the circles that constitute the cone.

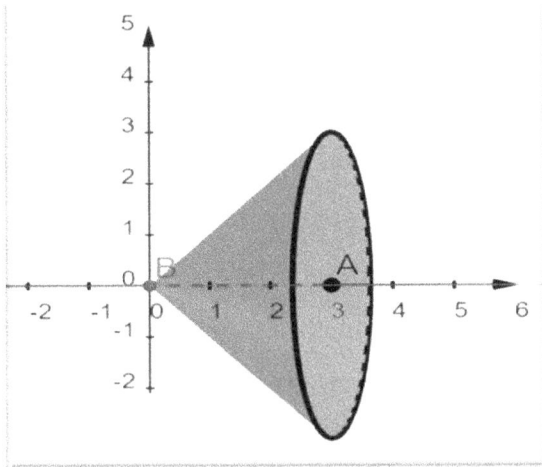

The area of the circle is given by the formula:

$$A = \pi Radius^2 = \pi[f(x)]^2 = \pi x^2$$

The sum of all the circles that is the volume of the cone is given by the definite integral:

$$Volume = \int_0^3 \pi x^2 dx$$

So,

$$Volume = \int_0^3 \pi x^2 dx = \pi(3)^2 - \pi(0)^2 = 9\pi$$

The average value of a function

The average value of a function on an interval [a,b] is calculated using the formula:

$$Average\ value = \frac{1}{b-a}\int_a^b f(x)dx$$

EXAMPLE

The average value of the function $f(x) = 2x^2 + \frac{1}{x}$ on the interval [2,6] is calculated:

$$\int_2^6 f(x)dx = F(6) - F(2)$$

First, we calculate the indefinite integral from f(x):

$$F(x) = \int f(x)dx = \int(2x^2 + \frac{1}{x})dx = 2\int x^2 dx + \int \frac{1}{x}dx = 2\frac{1}{3}x^3 + \ln(x) + C$$

Second, we calculate the value of F(x) for x=2 and x=6

$$F(6) = 2\frac{1}{3}(6)^3 + \ln(6) = 144 + 1.79 = 145.79$$

$$F(2) = 2\frac{1}{3}(2)^3 + \ln(2) = 5.33 + 0.69 = 6.02$$

We apply the Fundamental Theorem of Calculus.

$$\int_2^6 f(x)dx = F(6) - F(2) = 145.79 - 6.02 = 139.77$$

$$Average\ value = \frac{1}{6-2}\int_2^6 f(x)dx = \frac{139.77}{4} = 34.94$$

8.F Integration

b. Differential equations, Initial value problems, Slope fields

A differential equation is an equation where there are functions and their derivatives.

EXAMPLE

If y=f(x)=3x+1 solve the differential equation.

$y' + 2y - 3 = 0$

First, we calculate the derivative of y:

$(3x + 1)' = 3$

Second, we substitute y=3x+1 and the derivative of y in the equation.

$3 + 2(3x + 1) + 1 = 0$

Third, we solve the equation for x.

$3 + 6x + 2 + 1 = 0$

$6x + 6 = 0 \; so, x = -1$

<u>Initial value problems</u> are differential equations that have given an initial value. In physics models many times the model starts at an initial value.

EXAMPLE

If the differential equation is $y' = \frac{dy}{dx} = x^2 + 3x$, the initial condition is y(1)=2, find y:

$dy = (x^2 + 3x)dx$

We integrate both sides;

$\int dy = \int (x^2 + 3x)dx = \int x^2 dx + \int 3x dx = \frac{1}{3}x^3 + \frac{3}{2}x^2 + C$

$y = \frac{1}{3}x^3 + \frac{3}{2}x^2 + C$

We plug y=2 and x=1 in the formula to find C:

$2 = \frac{1}{3}(1)^3 + \frac{3}{2}(1)^2 + C$

$2 - \frac{1}{3} - \frac{3}{2} = C$

$\frac{12}{6} - \frac{2}{6} - \frac{9}{6} = \frac{1}{6} = 0.167 = C$

We write the final form of the function y with the constant C=0.167.

$y = \frac{1}{3}x^3 + \frac{3}{2}x^2 + 0.167$

<u>Slope fields</u> are all the antiderivative functions F(x) that we obtain for any constant C.

EXAMPLE

If one of the functions whose derivative is $f(x) = e^x$ is:

$F(x) = \int e^x dx = e^x + C$ So, one of the functions is $F(x) = e^x + 3$

TEST 11

Includes integrals

1.

5Ab 6) Determine if the expression $f(x) = 34x - 15$ represents a linear function.

2.

5A.c. 6) If $f(x) = y = \frac{\sqrt{x-1}}{3}$ the inverse function is: $f^{-1}(x) = y =$

3.

5B 4) Determine if the expression $f(x) = 7x - 3, x \in R$ represents piecewise function.

4.

5C 6) If $\cos(\alpha_1) = k, k \geq 0$ the other value of α that is a solution of the equation is:

5.

5D.a. 2) The graph of $\sin(\alpha)$ is:

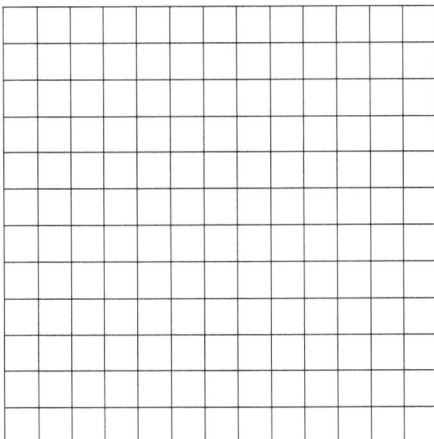

6.

5D.b. 6) Tangent of 45^0 is:

7.

5E 6) The range (principal value) of $f(x) = cos^{-1}(x)$ is _____.

8.

5F 2) Graph the function $f(x) = cos^{-1}(x)$

9.

6Aa 1) We can show limits through graphs or _____ of values.

10.

6Ab 5) $\lim\limits_{x \to \infty} \dfrac{4x+3}{x^2+4x-2} =$

11.

6B 5) In the graph below

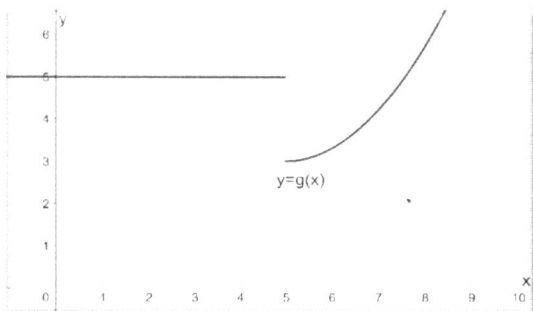

$$\lim_{x\to5^+} g(x) =$$

12.

6C 5) $f(x) = \dfrac{2x^3+3x^2-4x+5}{6x^3-5x+4}$: when x

goes to $+\infty$ it goes towards:

_____

13.

6D 5) Using the table below of a continuous function.

The function $f(x) = 3$ for at least one x that belong to $-1 \leq x \leq 3$. Yes or

x	-2	-1	0	3	5	7
$f(x)$	-1	0	3	4	5	9

now?

14.

6E 5) Using the graph below

Where: $f(x) = \begin{cases} 0.3x + 2 , x < 1 \\ 0.5x^2 + 0.5, x \geq 1 \end{cases}$

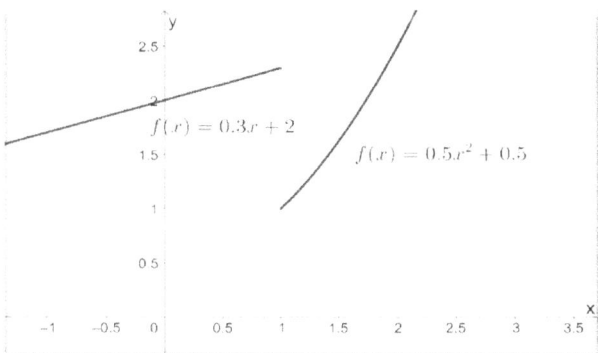

$$\lim_{x\to1^-} f(x) =$$

15.

6F 5) $\lim\limits_{x\to\infty} \dfrac{x^2+2x-45}{3x^2+37x-35} =$

16.

6G 5) The following function is represented below: $f(x) =$

$$\begin{cases} 2x - 1, x \geq 2 \\ 3x - 3, 1.5 < x < 2 \\ 0.5x, x \leq 1.5 \end{cases}$$

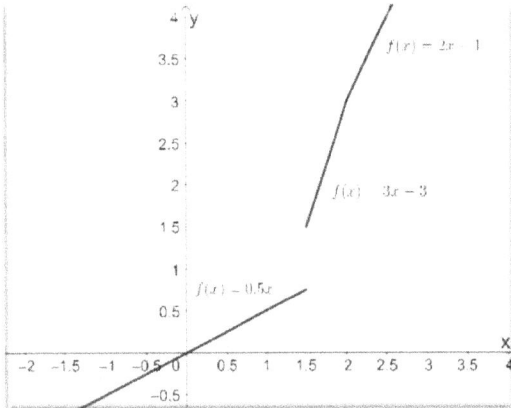

Is the function f(x) continuous in x=1.5?

17.

7B 4) Using Δx notation instead of h, find the derivative of the following function:

$$f(x) = 5x^2 - 4x + 3 \text{ so } f'(x) =$$

18.

7C 4) An interesting notation is d=V*t

19.

7D.a. 4) A tire rolls by the relation between distance and time: $s(t) = 2t^2 + 4t + 3$

The instantaneous velocity (first derivate) at t=5 seconds is:

20.

7D.b. 4) The function represented below is: $f(x) = x^3 + 3x^2 + x + 2$

The slope of the tangent to the graph in point C equals:

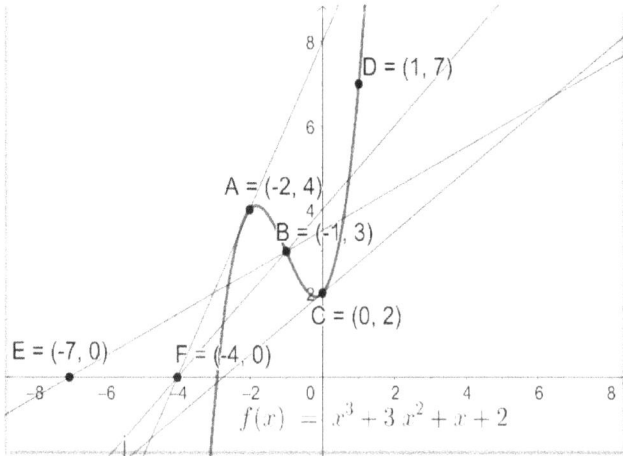

21.

7E 4) If $f(x) = 73 + \ln(124) - e^2$; then $f'(x) =$

22.

7F.a. 4) If $f(x) = x^4 - 2x^3 - 6x^2 + 81$ then $f'(x) =$

23.

7F.b. 4) If $f(x) = \frac{2}{5}x^2 e^x$; then $f'(x) =$

24.

7F.c. 4) If $f(x) = \frac{\sin(x)}{x^2+1}$; then $f'(x) =$

25.

7Fd 4) If $f(x) = \sqrt{e^x - 2x}$; then $f'(x) =$

26.

7G 4) If $f(x) = 3\sqrt{5x - 1} - 4x + 5$; then $f''(x) =$

27.

7H 4) If $4x^2 + 2y^2 = 9$; $y'' = 8x + 4y$

28.

7I.a. 4) The graph of the function $f(x) = 2x^3 - 7x^2 + 7$ is decreasing between _____

29.

7I.b. 4) If $f(x) = x^2 + 5x - 3$ and the points (-1,-7) and, (2,11) the value of x where the tangent at the graph is parallel with the line that goes through (-1,-7) and, (2,11) is: x=

30.

7Ic 3) Using Newton's method, the solution of $x^3 - 2x^2 = 4$ is:

Mark yourself

1	2	3	4	5	6	7	8	9	10
11	12	13	14	15	16	17	18	19	20
21	22	23	24	25	26	27	28	29	30
# of Good Answers (NGA)=				NGA/total number of questions=Ratio				Percent=Ratio*100	
				YOUR PERCENT IS:					%

TEST 12

1.

5Ab 7) Determine if the following graph represents a linear function

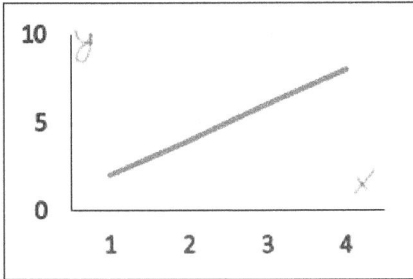

2.

5Db 7) Cotangent of 30^0 is:

3.

6Ab 6) $\lim\limits_{x \to \infty} \left(\sqrt{2x + 1} - \sqrt{2x}\right) =$

4.

6B 6) In the graph below

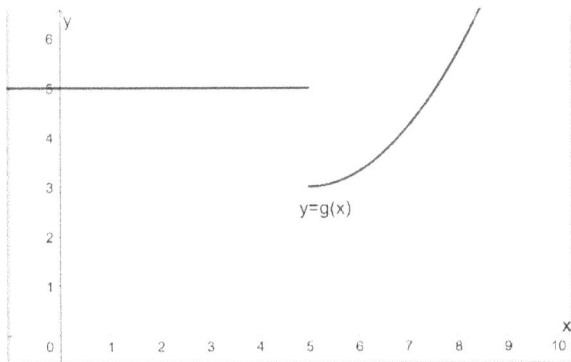

y=g(x)

$\lim\limits_{x \to 5} g(x) =$

5.

6C 6) $f(x) = \dfrac{x^3 + 2x^2 - 3x + 4}{6x^4 - 5x^3 + 4x^2 - 3x}$: when x goes to -∞ it goes towards:

6.

6D 6) Using the table below of a continuous function determine if the following question is correct.

x	-2	-1	0	3	5	7
$f(x)$	-1	0	3	4	5	9

The function $f(x) = 5$ for at least one x that belong to $-1 \leq x \leq 7$

7.

6E 6) Using the graph below,

Where: $f(x) = \begin{cases} 0.3x + 2 \, , x < 1 \\ 0.5x^2 + 0.5, x \geq 1 \end{cases}$

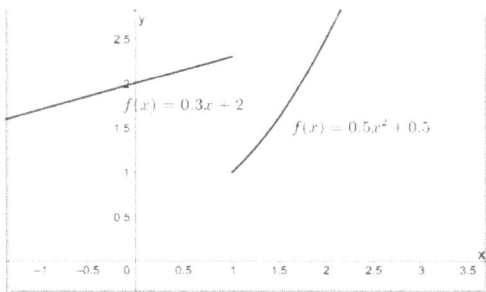

$$\lim_{x \to 1^+} f(x) =$$

8.

6F 6) $\lim_{x \to -\infty} \dfrac{3x^3 + 27x^2 - 4x + 5}{2x^3 - 54x^2 + 27x - 31} =$

9.

6G 6) The function g(x) is defined as follows: $g(x) = \begin{cases} \dfrac{9x^2 - 1}{3x - 1} & for \ x \neq 1/3 \\ 0 \ for \ x = 1/3 \end{cases}$

Does the function g(x) exists for x=1/3 ?

10.

7B 5) Using Δx notation instead of h, find the derivative of: $f(x) = 7x - 27$ so $f'(x) =$

11.

7C 5) Leonhard Euler's notation for the second derivative is: D_x^2

12.

7Da 5) If $R(t) = \frac{3000t^2 + 500t}{5} + 20{,}000$ represent the revenue a company earns in time.

After 10 days, the revenue is increasing with a speed of _____ per day.

13.

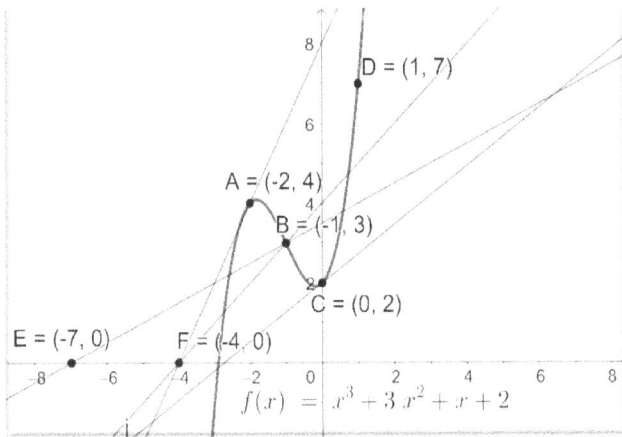

$$f(x) = x^3 + 3x^2 + x + 2$$

7Db 5) The slope of the tangent to the graph in point D equals:

14.

7Ea 5) If $f(x) = -4e^x + \ln(27)$; then $f'(x) =$

15.

7Fa 5) If $f(x) = \frac{1}{x}$ then $f'(x) =$

16.

7Fb 5) If $f(x) = 2^x \sin(x)$; then $f'(x) =$

17.

7Fc 5) If $f(x) = \frac{x^2 + 2x}{\sqrt{x}}$; then $f'(x) =$

18.

7Fd 5) If $f(x) = \sqrt{\sin(x)}$; then $f'(x) =$

19.

7G 5) If $f(x) = \sqrt{7x^2 + 6x}$; then $f''(x) =$

20.

7H 5) At point (2,3) the tangent slope to the curve $2x^2 + xy = 2$ *is*:

21.

7I.a. 5) The graph of the function $f(x) = \frac{2x-5}{x-1}$ is concave up for values of x less than 1

22.

7I.b. 5) If $f(x) = x^5 - 2x^3$ and the points (-1.6,-2.28. and, (1.6,2.28. the values of x where the tangents at the graph are parallel with the line that goes through (-1.6,-2.28. and, (1.6,2.28. are: x=

23.

7I.c. 4) Using Newton's method, the solution of $3x^4 - 4x^3 + 2x = 7$ is: x=

24.

7I.d. 1) If $x^3 - 2x + y = 1$ the first derivative with respect to z is:

25.

8B 1) If $f(x) = 3x^2 - e^x$, then $\int f(x)dx =$

26.

8C.a. 1) If $f(x) = 2x^2 - 3x$, using Riemann left sum, the integral $\int_{-2}^{3} f(x)dx =$
The interval $\Delta x = 1$

27.

8D 1) If $f(x) = 3x^2 + 3x - 2$ then, the $\int_{-4}^{5} f(x)dx =$

28.

8E.a. 1) If $f(x) = \sin(x) + 2x - \frac{3x^2+3x}{x}$, the antiderivative of $f(x)$ will be:

$F(x) =$

29.

8E.b. 1) If $f(x) = 4(2x + 1)$ then, $F(x) = \int f(x)dx =$

30.

8E.c. 1) If $f(x) = x^4 \ln(x)$ then $F(x) = \int f(x)dx =$

31.

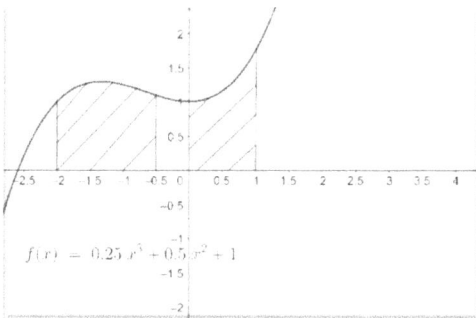

$f(r) = 0.25\,r^3 + 0.5\,r^2 + 1$

8Fa 1) The area under $f(x) = \frac{1}{4}x^3 + \frac{1}{2}x^2 + 1$ between x=-2 and x=-0.5 is:

32.

8F.b. 1) The trajectory of the Space Shuttle in the first minutes is represented by:

$h(t) = 2008 - 0.047t^3 + 18.3t^2 - 345t$

The velocity of the Space Shuttle at 20 seconds is:

Mark yourself

1	2	3	4	5	6	7	8	9	10
11	12	13	14	15	16	17	18	19	20
21	22	23	24	25	26	27	28	29	30
31	32								

# of Good Answers (NGA)=	NGA/total number of questions=Ratio	Percent=Ratio*100
	YOUR PERCENT IS:	%

TEST 13

1.

5A.b. 8. Determine if the following graph represents a linear function

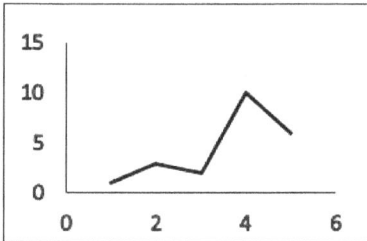

2.

5D.b. 8. The tangent function has an amplitude of:

3.

6Ab 7) $\lim\limits_{x \to 3}\left(\dfrac{16x^2-1}{4x-1}\right) =$

4.

6B 7) In the graph below

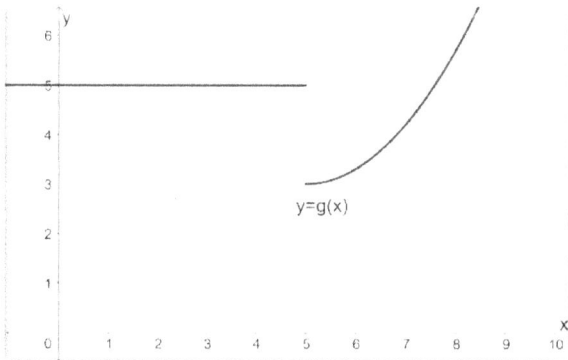

$\lim\limits_{x \to 5^+}\dfrac{5x}{x-5} =$

5.

6C 7) $f(x) = \dfrac{x^5+2x^2-3x+4}{6x^4-5x^3+4x^2-3x}$: when x goes to $+\infty$ it goes _____ in the first quadrant

6.

6D 7) Using the table below of a continuous function determine if the following question is correct. The function $f(x) = 0$ for at least one x that

x	-2	-1	0	3	5	7
$f(x)$	-1	0	3	4	5	9

belong to $0 \leq x \leq 3$

7.

6E 7) Using the graph below

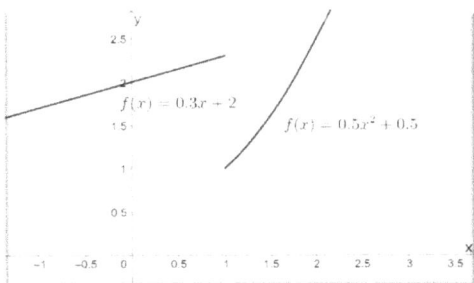

$f(x) = 0.3x + 2$

$f(x) = 0.5x^2 + 0.5$

$$\lim_{x \to 1} f(x) =$$

8.

6F 7) $$\lim_{x \to \infty} \frac{9x^2 - 4}{5x^2 + 7x - 3} =$$

9.

6G 7) The function g(x) is defined as follows: $g(x) = \begin{cases} \frac{9x^2-1}{3x-1} & for\ x \neq 1/3 \\ 0\ for\ x = 1/3 \end{cases}$

The limit $\lim_{x \to 1/3} g(x)$ _____.

10.

7B 6) Using Δx notation instead of h, find the derivative of the following function:

$f(x) = x^{-1}$ so $f'(x) =$

11.

7C 3) Joseph Louis Lagrange's notation is $f'(x)$. Yes or no?

12.

7D.a. 6) If $R(t) = \frac{3000t^2 + 500t}{5} + 20{,}000$ represent the revenue a company earns in time.

After 30 days, the revenue is increasing with a speed of _____ per day.

13.

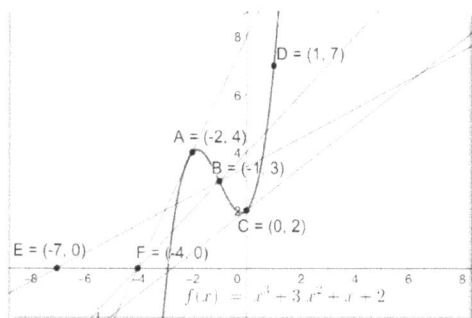

7D.b. 6) The slope of the tangent to the graph in point B equals:

14.

7E 6) If $f(x) = \cos(x) + 3x;\ then\ f'(x) =$

15.

7F.a. 6) If $f(x) = \frac{1}{x^4}$ then $f'(x) =$

16.

7F.b. 6) If $f(x) = 3\ln|x|x^2;\ then\ f'(x) =$

17.

7F.c. 6) If $f(x) = 2x\sec(x) = \frac{2x}{cox(x)};\ then\ f'(x) =$

18.

7F.d. 6) If $f(x) = 2[\ln(4x - 3)];\ then\ f'(x) =$

19.

7G 6) If $f(x) = \frac{2}{2x-3}\ ;\ then\ f'''(x) =$

20.

7H 6) At point (3,4) of the curve $2x^2 + 3xy - y^2 = 38$ the slope of the tangent line is:

21.

7I.a. 6) The function $f(x) = \frac{x^2}{x^2-4}$ is concave up between:

22.

7I.b. 6) If $f(x) = 4x^3 + 3x^2$ the values of x where the tangents of 1 at the graph are:

23.

7I.c. 5) Using Newton's method the solution of $2x^3 + 3x - 4 = \sin(x)$ is: x=

24.

7I.d. 2) If $2x^5 + 3x^4 - 4x + y^2 = 10$ the first derivative with respect to z is:

25.

8B 2) If $f(x) = 2x^4 + \ln(x)$, then $\int f(x)dx =$

26.

8C 2) If $f(x) = 4x^3 + 3x^2 - 2x + 1$, using Riemann left sum, the integral $\int_{-3}^{4} f(x)dx \cong$
The interval $\Delta x = 1$

27.

8D 2) If $f(x) = x^2 + 3\sqrt{x}$ then, the $\int_{1}^{5} f(x)dx =$

28.

8E.a. 2) If $f(x) = 5x + \frac{4x^3 + 3x^2 - x}{3}$ then, the antiderivative of $f(x)$ will be:

$F(x) =$

29.

8E.b. 2) If $f(x) = 30x(5x^2 - 3)^2$ then, $F(x) = \int f(x)dx =$

30.

8E.c. 2) If $f(x) = x^3 \cos(x)$ then $F(x) = \int f(x)dx =$

31.

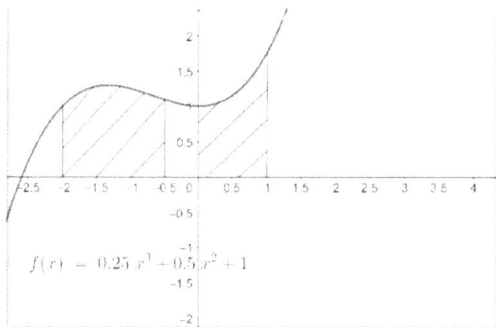

$f(x) = 0.25\,x^3 + 0.5\,x^2 + 1$

8F.a. 2) The area under $f(x) = \frac{1}{4}x^3 + \frac{1}{2}x^2 + 1$

between x=0 and x=1 is:

32.

8F.b. 2) The trajectory of the Space Shuttle in the first minutes is represented by:

$h(t) = 2008 - 0.047t^3 + 18.3t^2 - 345t$

The acceleration of the Space Shuttle at 20 seconds is:

Mark yourself

1	2	3	4	5	6	7	8	9	10
11	12	13	14	15	16	17	18	19	20
21	22	23	24	25	26	27	28	29	30
31	32								

# of Good Answers (NGA)=	NGA/total number of questions=Ratio		Percent=Ratio*100
	YOUR PERCENT IS:		%

TEST 14

1.

6B 8. In the graph below

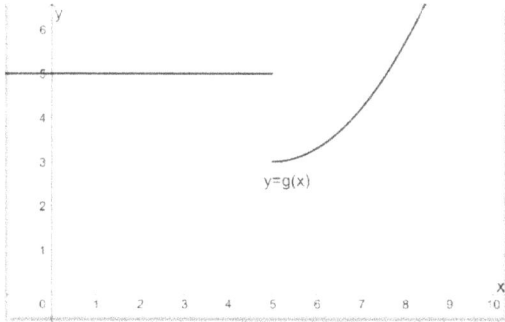

$$\lim_{x \to 5^-} \frac{5x}{x-4} =$$

2.

6C 8. $f(x) = \frac{2x^3+3x^2-4x+5}{6x^2-5x+4}$, when x goes to -∞ it goes _____ in the first quadrant

3.

6D 8. If $f(x) = \frac{2x+1}{-3x+7}$ determine if the question below is correct.

There will be a value c such that $f(c) = 1$ for $3 \leq c \leq 9$

4.

6E 8. Using the graph below

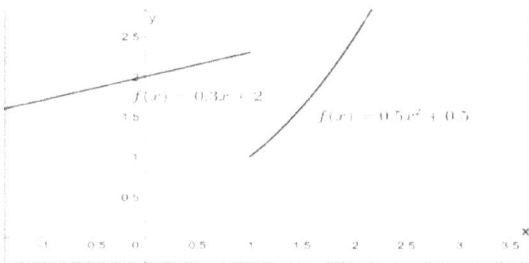

$$\lim_{x \to 3^-} f(x) =$$

5.

6F 8. $\lim_{x \to \infty} \frac{3x+7}{\sqrt{x^2-3x+4}} =$

6.

6G 8. The function g(x) is defined as follows: $g(x) = \begin{cases} \frac{9x^2-1}{3x-1} & for\ x \neq 1/3 \\ 0\ for\ x = 1/3 \end{cases}$

Is the function g(x) is continuous for x=0.33?

7.

7B 7) Using Δx notation instead of h, the derivative of the function:

$f(x) = 3x^{-1} + x$ so $f'(x) =$

8.

7C 1) Is Gottfried Leibniz's notation $\frac{dy}{dx}$?

9.

7D.a. 7) The following question follows the graph below. The graph represents the function: $f(x) = -0.2x^2 + 2x + 1$

The average rate of change between point A and B is:

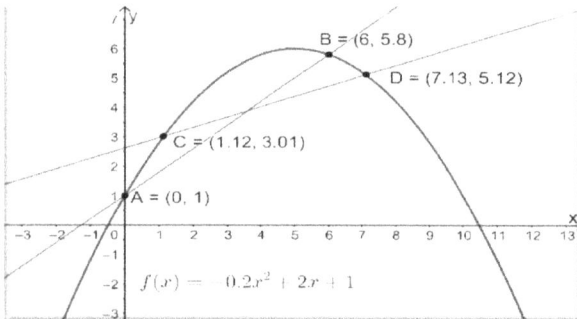

10.

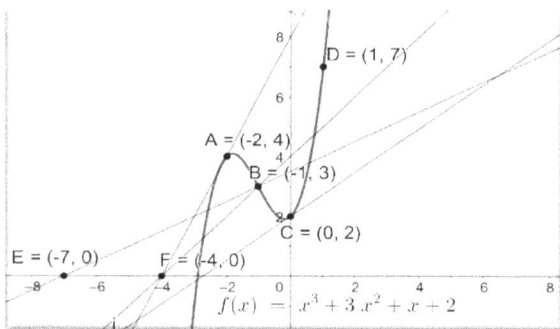

7D.b. 7) The slope of the secant that passes through the points A and B is:

11.

7E 7) If $(x) = 2\ln(x) + 3e^x + \tan(x)$; $then\ f'(x) =$

12.

7F.a. 7) If $f(x) = \sqrt{x^3}$ then $f'(x) =$

13.

7F.b. 7) If $f(x) = \ln|x|\sqrt{x};$ *then* $f'(x) =$

14.

7F.c. 7) If $f(x) = \frac{3x^2 - 4x + 2}{\sin(x)};$ *then* $f'(x) =$

15.

7F.d. 7) If $f(x) = [\ln(e^x - e^{-x})];$ *then* $f'(x) =$

16.

7G 7) If $(x) = e^x \ln(x);$ *then* $f'''(x) =$

17.

7H 7) The slope of the tangent line to the graph of $y = \frac{4}{\pi}x - \sin(xy)\, at\, \left(\frac{\pi}{2}, 1\right)$ *is:*

18.

7I.a. 7) The function $f(x) = \frac{2x}{x+3}$ is concave down

between_____

19.

7I.b. 7) If $f(x) = 6x^4 + 7$ the values of x where the tangents of -3 at the graph are:

20.

7I.c. 6) Using Newton method, the solution of $x^5 + 4x^3 - 5x = 3$ is: x=

21.

7I.d. 3) If $x^3 + 3y^2 - x^2 = \sin(x)$ the first derivative with respect to z is:

22.

8A 1) An integral is a _____ sum of the values of the function times the infinitesimal widths dx.

23.

8B 3) If $f(x) = \ln(x) - 2^x$, then $\int f(x)dx =$

24.

8C 3) If $f(x) = 3x^3 - 2x^2 + x - 1$, using Rectangle Method, the integral $\int_{-2}^{3} f(x)dx \cong$

The interval $\Delta x = 1$

25.

8D 3) If $f(x) = 4x - 3$ then, the $\int_{-4}^{5} f(x)dx =$

26.

8E.a. 3) If $f(x) = 7\tan(x) - sec^2(x)$ then, the antiderivative of $f(x)$ will be: $F(x) =$

27.

8E.b. 3) If $f(x) = 2x\cos(x^2)$ then, $F(x) = \int f(x)dx =$

28.

8E.c. 3) If $f(x) = \frac{\ln(x)}{x^3}$ then $F(x) = \int f(x)dx =$

29.

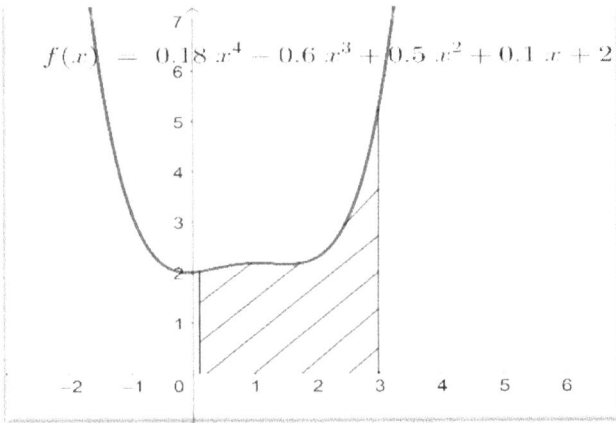

$$f(x) = 0.18x^4 - 0.6x^3 + 0.5x^2 + 0.1x + 2$$

8F.a. 3) The area under $f(x) = 0.18x^4 - 0.6x^3 + 0.5x^2 + x + 2$ between x=0.1 and x=3 is:

30.

8Fb 3) A formula one speed car has the acceleration formula: a(t)=t+4 m/s^2. The velocity of the car after 10 seconds is:

Mark yourself

1	2	3	4	5	6	7	8	9	10
11	12	13	14	15	16	17	18	19	20
21	22	23	24	25	26	27	28	29	30

# of Good Answers (NGA)=	NGA/total number of questions=Ratio	Percent=Ratio*100
	YOUR PERCENT IS:	%

TEST 15

1.

5A.a. 6) $f(x) = 3x^2 - 3x + 2$ so, $f(1) =$

2.

5A.b. 9. Determine if the following graph represents a linear function.

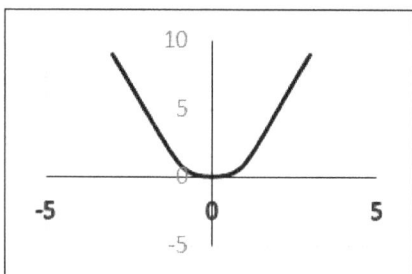

3.

5Ac 7) If $f(x) = y = \dfrac{3x}{x+3}$ the inverse function is: $f^{-1}(x) =$

4.

5B 5) Determine if the expression below represents a piecewise function

$$f(x) = \begin{cases} 0.75x^4 \ for \ x < 2 \\ 3x + 6 \ for \ x \geq 2 \end{cases}$$

5.

5C 7) By definition $\cos(\alpha) =$

6.

5D.a. 3) The minimum of $\cos(\alpha)$ is:

7.

5D.b. 9. Cotangent of 45^0 is:

8.

5E 7) The domain of $f(x) = tan^{-1}(x)$ is:

9.

5F 3) Graph the function $f(x) = sin^{-1}(x)$

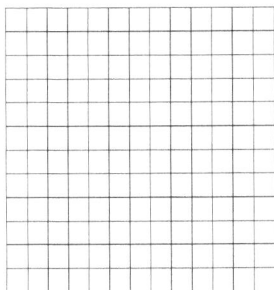

10.

6A.a. 2) If x=2 doesn't belong to the domain, the function _____ limit for f(2).

11.

7E 8. If $(x) = \frac{3}{7}\ln(x) + \frac{4}{9}e^x + \frac{5}{9}\tan(x)$; *then* $f'(x) =$

12.

7F.a. 8. If $f(x) = \frac{\sqrt{x}}{x^4}$ *then* $f'(x) =$

13.

7F.b. 8. If $f(x) = \cos(x) * e^x$; *then* $f'(x) =$

14.

7F.c. 8. If $f(x) = \frac{x^2+2x-3}{5x+4}$; *then* $f'(x) =$

15.

7F.d. 8. If $f(x) = \sin(x-1)$; *then* $f'(x) =$

16.

7G 8. If $f(x) = \frac{1}{2x} + \sqrt{x}$; *then* $f'''(x) =$

17.

7H 8. If $3y^2 + \ln(x) = 2y - \cos(x)$, $y' =$

18.

7I.a. 8. The function $f(x) = \frac{1}{4x^2-9}$ is _____ down between x=-1.5 and x=1.5

19.

7I.b. 8. If $f(x) = \frac{1}{x^2} - \frac{2}{5}$ the value of x where the tangent of 2 at the graph is:

X=

20.

7I.c. 7) Using Newton's method, the solution of $(x-3)^3 = \sin(x)$ is: x=

21.

7I.d. 4) If A=πR^2 the first derivative with respect to z is: $A =$

22.

8A 2) The number of the infinitesimal widths equals=

23.

8B 4) If $f(x) = 1 + \ln(x)$, then $\int f(x)dx =$

24.

8C 4) If $f(x) = x^2 + 2x + \sqrt[3]{x}$, using Rectangle Method, the integral $\int_{-3}^{3} f(x)dx \cong$
The interval $\Delta x = 1$

25.

8D 4) If $f(x) = 5x + 6\sqrt{x}$ then, the $\int_{2}^{7} f(x)dx =$

26.

8E.a. 4) If $f(x) = e^x - \frac{x^2-3x-10}{x+2} + 7$, the antiderivative of $f(x)$ will be: $F(x) =$

27.

8E.b. 4) If $f(x) = \frac{5}{3x-73}$ then, $F(x) = \int f(x)dx =$

28.

8E.c. 4) If $f(x) = x^2 sin(x)$ then $F(x) = \int f(x)dx =$

29.

8F.a. 4) The volume of the object that is made by the function $f(x) = x^2 + 2$

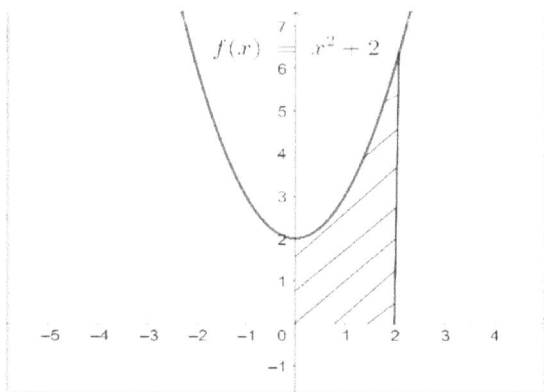

that rotate around x axis between x=0 and x=2 is: V=

30.

8F.b. 4) A formula one speed car has the acceleration formula: a(t)=t+4 m/s^2

The distance traveled by the car after 10 seconds is:

Mark yourself

1	2	3	4	5	6	7	8	9	10
11	12	13	14	15	16	17	18	19	20
21	22	23	24	25	26	27	28	29	30

# of Good Answers (NGA)=				NGA/total number of questions=Ratio				Percent=Ratio*100	
					YOUR PERCENT IS:				%

TEST 16

1.

6A.a. 3) The table below represent the value of $f(x) = 0.3x^2 - 2$ around -0.8

for x around _____

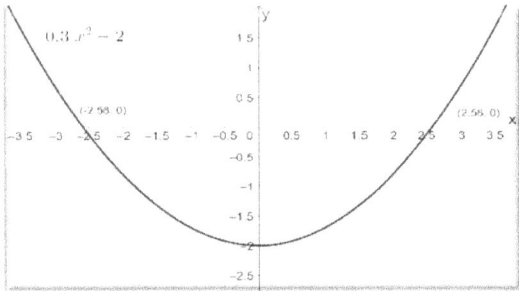

X	$f(x) = 0.3x^2 - 2$
1.8	-1.028
1.9	-0.917
1.999	-0.8012
2.001	-0.7988
2.01	-0.7879
2.1	-0.677

2.

6A.b. **8.** $\lim\limits_{x \to 1} \left(\dfrac{x^4 - 1}{x - 1} \right) =$

3.

6B **9.** $\lim\limits_{x \to 3^-} \dfrac{5x}{x - 5} =$

4.

6C **9.** Graph the following function $f(x) = \dfrac{x^3 - 2x^2 + 3x}{x^4 + 1}$ and determine the end behavior for x going towards $+\infty$

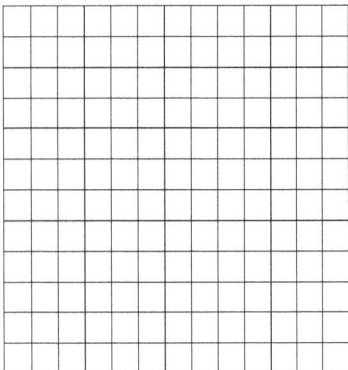

5.

6D 9. If $f(x) = \dfrac{2x+1}{-3x+7}$ determine if the question below is correct.

There will be a value c such that $f(c) = -2$ for $3 \leq c \leq 7$

6.

6E 9. Using the graph below;

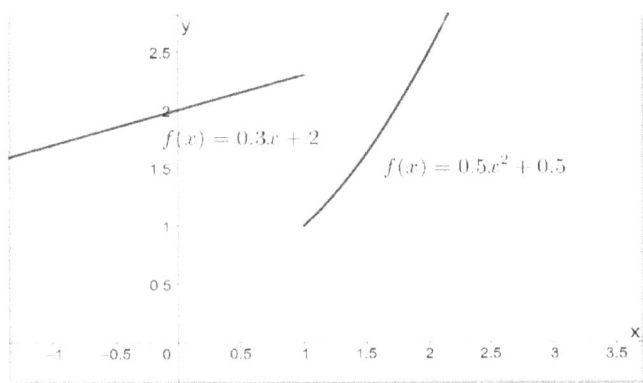

$f(x) = 0.3x + 2$

$f(x) = 0.5x^2 + 0.5$

Where: $f(x) =$
$$\begin{cases} 0.3x + 2, x < 1 \\ 0.5x^2 + 0.5, x \geq 1 \end{cases}$$

Calculate $\lim\limits_{x \to 3^+} f(x) = 5$

7.

6F 9. $\lim\limits_{x \to -\infty} \dfrac{5x-6}{\sqrt{x^2-5}} =$

8.

6G 9. The function $f(x) = \begin{cases} 3x \ if \ -1 \leq x \leq 2 \\ x + 2 \ if \ x > 2 \end{cases}$ is continuous over the interval

_____

9.

7B 8. Using the formula $f'(x) = \lim\limits_{h \to 0} \dfrac{f(x+h)-f(x)}{h}$ for $f(x) = x^{-1} + x^2$ find the derivative

10.

7C 7) One of Isaac Newton's notation is: $\dot{y}$ for first derivative

11.

7D.a. 8.

The following question follow the graph below. The graph represents the function:

$$f(x) = -0.2x^2 + 2x + 1$$

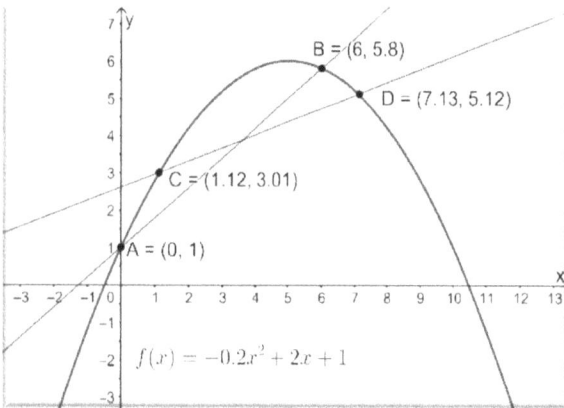

The average rate of change between point C and D is:

12.

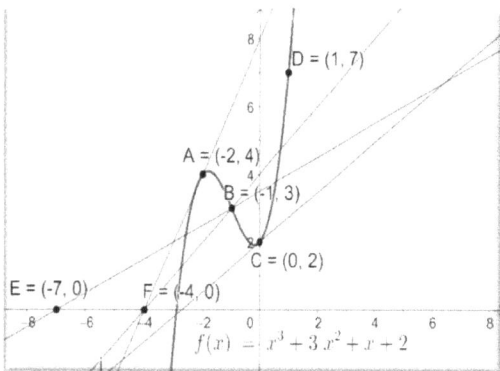

7D.b. 8. The slope of the secant that passes through the points B and C is:

13.

7I.b. 9. If $f(x) = \ln(x) + 10$ the value of x where the tangent of 5 at the graph is: x=

14.

7I.c. 8. The solution of $(x - 6)^3 = \ln(x)$ is: x=

15.

7I.d. 5) We have a rectangular yard with perimeter of 300 m. It has to be fenced. The dimensions of the yard that will give us the greatest area are:

16.

7I.d. 6) We have to build a box with the length of the base four times the width. The height is the 3 times length minus 10 cm. The dimensions that will minimize the volume are: Length= _____ , Width=_____ , height= _____

17.

8A 3) The notation of definite integral in the interval [a,b] is:

18.

8A 4) The sign $\int$ represents the _____ between $f(x)$ and dx.

19.

8B 5) If $f(x) = 2 - cos\ (x)$, then $\int f(x)dx =$

20.

8B 6) If $f(x) = x^2 - 2x$, then $\int_0^x f(x)dx = \frac{1}{3}x^3 - x^2 + C$

21.

8C 5) If $f(x) = 4x^3 + sin(x)$, using Riemann left sum, the integral
$\int_{-3}^3 f(x)dx \cong$
The interval $\Delta x = 1$

22.

8C 6) If $f(x) = 3x^3 - 2x^2 + 1$, using left side Riemann sum, the integral
$\int_{-2}^2 f(x)dx \cong$
The interval $\Delta x = 1$

23.

8D 5) If $f(x) = \frac{7}{x} + x - 3$ then, the $\int_1^5 f(x)dx =$

24.

8D 6) If $f(x) = \frac{4+x^2}{x} - 3x + 7$ then, the $\int_2^8 f(x)dx = 7$

25.

8E.a. 5) If $f(x) = \cos(x) - \frac{2x^2 + 3x}{x} + 3$ then, the antiderivative of $f(x)$ will be:

$F(x) =$

26.

8E.a. 6) If $f(x) = \frac{2x^2 + 3x}{x^2} - 5\sec^2(x)$ then, the antiderivative of $f(x)$ will be:

$F(x) =$

27.

8E.b. 5) If $f(x) = \frac{7\ln(x)}{5x}$ then, $F(x) = \int f(x)dx =$

28.

8E.b. 6) If $f(x) = 3x^3 e^{x^4}$ then, $F(x) = \int f(x)dx =$

29.

8E.c. 5) If $f(x) = e^x \cos(x)$ then $F(x) = \int f(x)dx =$

30.

8E.c. 6) If $f(x) = x^2 e^x$ then $F(x) = \int f(x)dx =$

31.

8F.a. 5) The volume of the object that is made by the function $f(x) = -0.3x^2 + 4$ that rotate around x axis between x=0.5 and x=3 is: V=

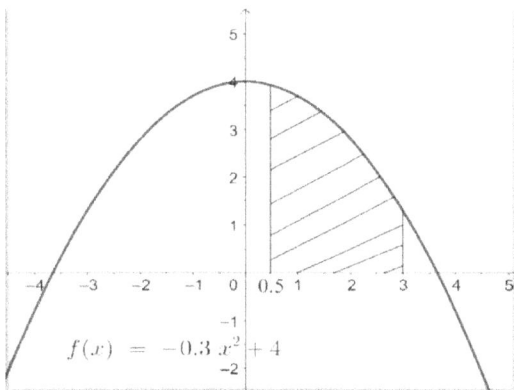

$f(x) = -0.3\,x^2 + 4$

32.

8F.a. 6) The volume of the sphere that is made by the function $4 = x^2 + y^2$ that rotate around x axis between x=-2 and x=2 is: V=

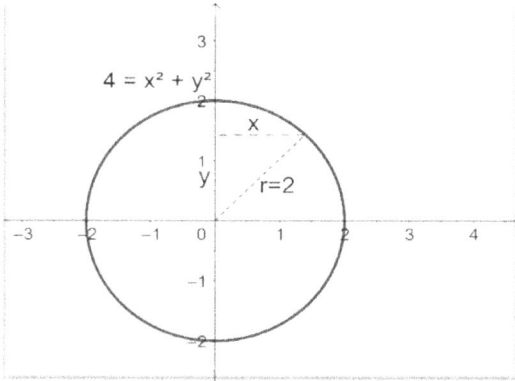

33.

8Fb 5) If the differential equation is $y' = \dfrac{dy}{dx} = 15 + x$, the initial condition is y(0)=3, then $y =$

34.

8Fb 6) If the differential equation is $y' = \dfrac{dy}{dx} = 2x^2 - 3x$, the initial condition is y(1)=2, then $y =$

Mark yourself

1	2	3	4	5	6	7	8	9	10
11	12	13	14	15	16	17	18	19	20
21	22	23	24	25	26	27	28	29	30
31	32	33	34						

| # of Good Answers (NGA)= | | | | NGA/total number of questions=Ratio | | | | Percent=Ratio*100 | |
| YOUR PERCENT IS: | | | | | | | | | % |

TEST 17

1.

6F 10) $\lim\limits_{x\to-\infty} \dfrac{3x^2-2x+7}{3x^3-6x^2+8x-1} =$

2.

7B 9. Using the formula $f'(x) = \lim\limits_{h\to0} \dfrac{f(x+h)-f(x)}{h}$ find the derivative for $f(x) = x^2 + 3x$.

$for\ x = 2\ ,f'(2) =$

3.

7C 3) Is this correct? Joseph Louis Lagrange's notation is $f'(x)$

4.

7D.a. 9. The following questions follow the graph below. The graph represents the function: $f(x) = -0.2x^2 + 2x + 1$

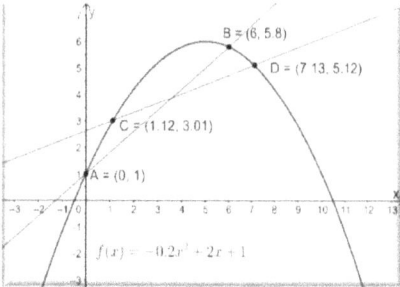

The instantaneous rate of change (first derivate) at point C equals:

5.

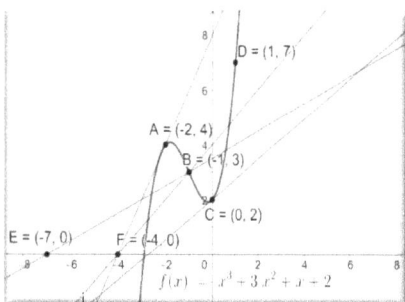

7D.b. 9. The slope of the secant that passes through the points B and D is:

6.

7E 9. If $(x) = \dfrac{5}{9}\ln(2) + \dfrac{1}{2}e^x + \cos(x)$;

$then\ f'(x) =$

7.

7F.a. 9. If $f(x) = \frac{\sqrt{x}}{x^2} + 3x$ then $f'(x) =$

8.

7F.b. 9. If $f(x) = e^x 3^x$; then $f'(x) =$

9.

7F.c. 9. If $f(x) = \frac{x^2}{lnx}$; then $f'(x) =$

10.

7F.d. 9. If $f(x) = e^x + 2\ln(x + 7)$; then $f'(x) =$

11.

7G 9. If $f(x) = \sin(x) + x^3$; then $f''(x) =$

12.

7H 9. If $\cot(y) + 2x = 5y - y^2$, $y' =$

13.

7I.a. 9. The function $f(x) = \frac{x}{x^2-4}$ is concave up between _____

14.

7I.b. 10) If $f(x) = x^3 + 2$ and the points (-0.5,1.875) and, (0.5,2.125) the values of x where the tangents at the graph are parallel with the line that goes through (-0.5,1.875) and, (0.5,2.125) are: x=_____

15.

7I.c. 9. Using Newton's method, the solution of $2(x - 9.^3 = \ln(x) + 2x$ is: x=

16.

7I.d. 7) We have to build a box with a base that have length two times the width, and we have 25 m square of material. The dimensions for the maximum volume are:

Length= _____ m , Width= _____ m   and Height= _____ m

17.

7I.d. 8. We have to build a tunnel that has a cylinder shape that has 50 Liters in volume. The dimensions of the tunnel in order to have the smallest surface area are: R= ____cm

And h = _____cm

18.

8A 5) The notation of indefinite integral is:

19.

8A 6) The function $f(x)$ under the integral sign it is called _____.

20.

8B 7) If $f(x) = 2x^2 + 3x - 4$, then $\int_0^x f(x)dx =$

21.

8B 8. If $f(x) = x^2 - 4\cos(x)$, then $\int_0^6 f(x)dx =$

22.

8C.a. 7) If $f(x) = 3x^3 - 2x^2 + 1$, using left side Riemann sum, the integral $\int_{-2}^2 f(x)dx \cong$

The interval $\Delta x = 0.5$

23.

8C.a. 8. If $f(x) = 3x^3 - 2x^2 + 1$, using trapezoidal method, the integral $\int_{-2}^{2} f(x)dx \cong$

The interval $\Delta x = 1$

24.

8D 7) If $f(x) = 4x + e^x$ then, the $\int_{-2}^{6} f(x)dx =$

25.

8E.a. 7) If $f(x) = \frac{sin^2(x) + cos^2(x)}{x} - e^x$ then, the antiderivative of $f(x)$ will be:

$F(x) =$

26.

8E.a. 8. If $f(x) = \frac{sin^2(x) - cos^2(x)}{sin(x) + cos(x)} + 1 + \ln(x)$ then, the antiderivative of $f(x)$ will be: $F(x) =$

27.

8E.b. 7) If $f(x) = (6x - 4)(3x^2 - 4x + 5)^2$ then, $F(x) = \int f(x)dx = 6x^2 - 10x + C$

28.

8E.b. 8. If $f(x) = \frac{8x^3 + 3}{\sqrt{2x^4 + 3x - 5}}$ then, $F(x) = \int f(x)dx =$

29.

8E.c. 7) If $f(x) = \frac{\ln(x)}{x}$ then $F(x) = \int f(x)dx = \frac{1}{2}\ln^2(x) + C$

30.

8E.c. 8. If $f(x) = x\sqrt{1 - sin^2(x)}$ then $F(x) = \int f(x)dx =$

31.

8F.a. 7) The volume of the object that is made by the function $f(x) = x^2$ that rotate around x axis between x=0 and x=3 is:

32.

8F.a. 8. The volume of the object that is made by the function $f(x) = x$ that rotate around x axis between x=1 and x=5 is:

33.

8F.b. 7) If the differential equation is $y' = \frac{dy}{dx} = e^x + 4x$, the initial condition is y(0)=5, then $y =$

34.

8F.b. 8. One of the functions whose derivative is $f(x) = 4x$ is $F(x) = 2x^2 + 37$

Mark yourself

1	2	3	4	5	6	7	8	9	10
11	12	13	14	15	16	17	18	19	20
21	22	23	24	25	26	27	28	29	30
31	32	33	34						
# of Good Answers (NGA)=				NGA/total number of questions=Ratio				Percent=Ratio*100	

TEST 18

1.

A = (-6, 3) B = (-2, 3) C = (4, 3)

K = (-5, 1) G = (-3, 1) H = (2, 1) I = (6, 1)

D = (-6, -2) E = (1, -2) F = (5, -2)

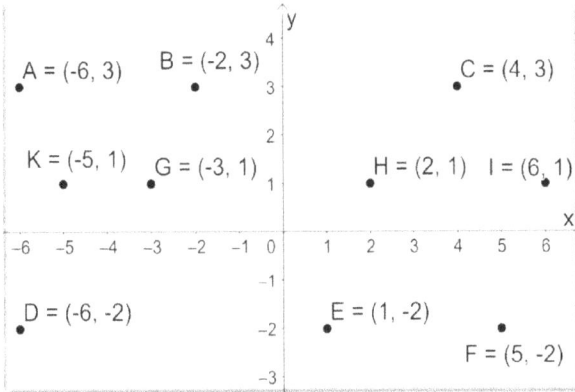

1B.a. **8.** The distance between point G and point H is:

2.

1B.c. **8.** The distance between point B and point D is:

(same graph)

3.

B = (-2, 3) C = (2, 3)

A = (-5, 3)

G = (-2, 1) I = (5, 1)

K = (-5, 0) H = (2, 1)

D = (-5, -2) E = (2, -2) F = (5, -2)

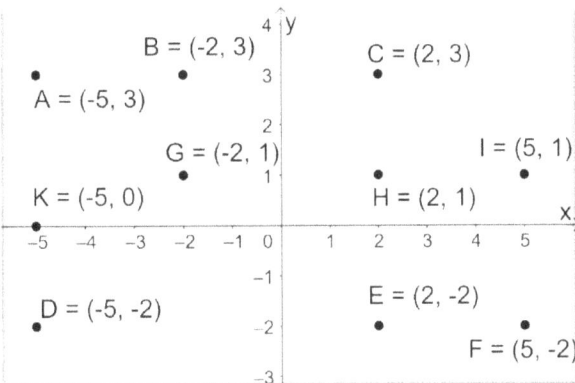

1B.b. 8) The distance between point G and point E is:

4.

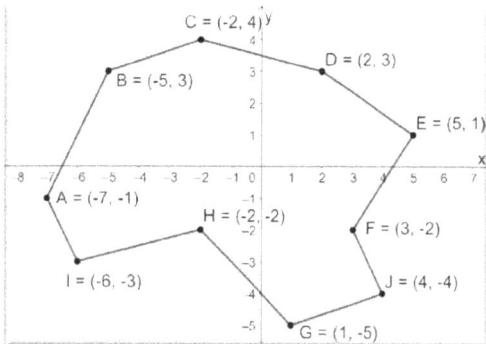

1B.d. 8) The mid-point coordinates of segment GH are:

5.

1C 8) The slope of segment GH is:

6.

4A.b. 4) The perimeter of a rectangle with length $L = 3\sqrt{20} + 5\sqrt{5}$ and width $W = 5\sqrt{20}$ is

7.

4A.b. 5) The area of the above rectangle is:

8.

4A.c. 5) The mixed radical of $\sqrt{288}$ is:

9.

4B.a. 5) $2 \times 3 \times (9 \times 5)$ is equal with $(2 \times 9 \times 3) \times 5$

10.

4B.b. 5) $(x^2 - 5)(2x^3 + 7x - 3)$ is equal with:

11.

4C.a. 5) The single logarithm of expression $3(\log_5 a + \log_5 b) - 2\log_5 b$; $a, b > 0$ is:

12.

4C.b. 5) The solution of the equation $\log_4 36 = \log_4(x + 3) + \log_4(x - 3)$; $x > 3$ is $x =$

13.

4D.a. 5) The area of a triangle with the base $=5x + 4$ and the height $= 3x + 1$ is:

14.

4D.b. 5) The area of a rectangle is $\sqrt{7} + 2\sqrt{5}$ and length $\sqrt{3} - 1$. The width is:

15.

4E.a. 3) The slope of the line $-3x + 5y = -1$ is:

16.

6F 11) $\lim\limits_{x \to \infty} \dfrac{2x+3}{\sqrt{x^3-27}} =$

17.

7I.d. 9) The profit relation for a company is $P = Revenue - expenses =$ $\dfrac{1230}{p} - \dfrac{550}{(p)^2}$ The price per unit (p) for maximum profit P is: $p =$

18.

8A 7) The _____documented technique that tried to calculate the integral was used by the ancient Greek astronomer Euxodus around 370 BC.

19.

8B 9) If $f(x) = \ln(x) + 3x$, then $\int_0^4 f(x)dx =$

20.

8C 9) If $f(x) = 3x^3 - 2x^2 + 1$, using trapezoidal method, the integral $\int_{-2}^{2} f(x)dx \cong$
The interval $\Delta x = 0.5$

21.

8D 8) If $f(x) = e^x + \sin(x)$ then, the $\int_1^6 f(x)dx =$

22.

8D 9) If $f(x) = 2e^x + 3x^2 - 4x + 5$ then, the $\int_{-3}^8 f(x)dx =$

23.

8E.a. 9) If $f(x) = \sin^2(x) + \cos^2(x) + 1 - \ln(x)$, the antiderivative of $f(x)$ will be: $F(x) =$

24.

8E.a. 10) If $f(x) = \frac{2}{\sqrt{x}} - \sqrt{x}$ then, the antiderivative of $f(x)$ will be: $F(x) =$

25.

8E.b. 9) If $f(x) = \frac{9x^2 - 8x}{\sqrt{3x^3 - 4x^2 + 5}}$ then, $\int_1^3 f(x) = F(3) - F(1) =$

26.

8E.b. 10) If $f(x) = \frac{1}{\sqrt{x^2 + 9}} = \frac{1}{\sqrt{x^2 + 3^2}}$ then, $F(x) = \int f(x)dx =$

27.

8E.c. 9) If $f(x) = x\ln(x)$ then $F(x) = \int f(x)dx =$

28.

8E.c. 10) If $f(x) = e^x \sin(2x)$ then $F(x) = \int f(x)dx =$

29.

8F.a. 9) The average value of the function $f(x) = 2x + \sin(x)$ on the interval [1,3] is:

30.

8F.a. 10) The average value of the function $f(x) = x^2 + \frac{1}{x}$ on the interval [1,5] is:

31.

8F.b. 9) One of the functions whose derivative is $f(x) = 3x^{-1}$ is $F(x) =$

32.

8F.b. 10) One of the functions whose derivative is $f(x) = e^x$ is $F(x) = \frac{5e^{2x}}{2} +$ 3

Mark yourself

1	2	3	4	5	6	7	8	9	10
11	12	13	14	15	16	17	18	19	20
21	22	23	24	25	26	27	28	29	30
31	32								
# of Good Answers (NGA)=				NGA/total number of questions=Ratio				Percent=Ratio*100	

TEST 19

1.

1D.a. 7) The intersection to x axis of the line $y = -3x + 1$ is $P(\frac{1}{3}, 0)$

2.

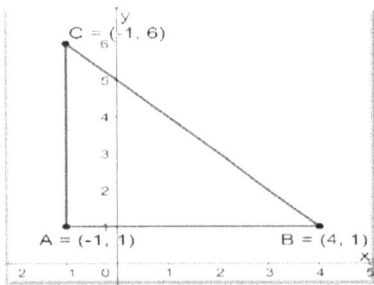

1D.b. 7) The equation of the line that passes through the points A and B in the figure below is $x =$

3.

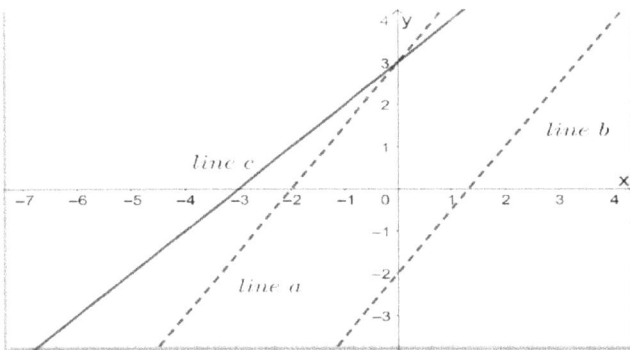

1E.a. 8) The equation of the line b is $y =$

4.

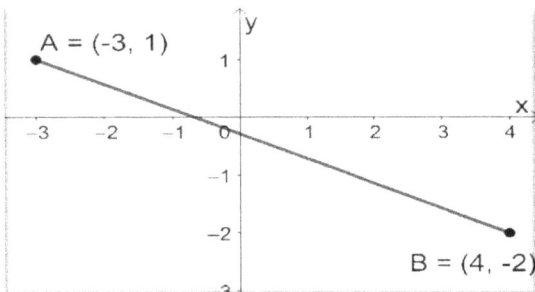

1E.b. 8) The equation of the line perpendicular to AB in problem 7 will intersect x axis in $x =$

5.

1F.a. 8) The point symmetric to M (2,6) about the vertical x=-2 has the x=

6.

1F.b. 8) The symmetric point to P (-3,-1) by the point S (2,4) is L (   )

7.

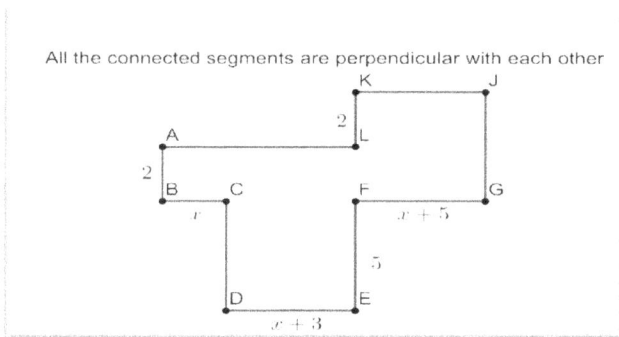

All the connected segments are perpendicular with each other

2A.a. 6) The area of the rectangle FKJG is:

8.

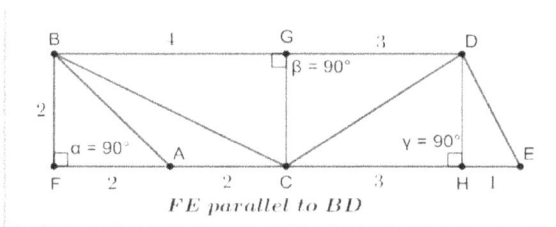

FE parallel to BD

2A.b. 6) The area of the triangle GCD is:

9.

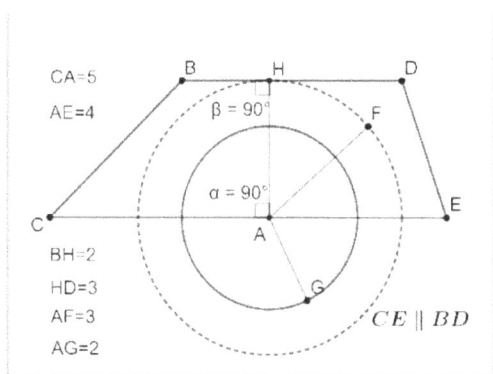

CA=5
AE=4
BH=2
HD=3
AF=3
AG=2
$CE \parallel BD$

2A.c. 6) The area of the Circle of radius AF=3 is:

10.

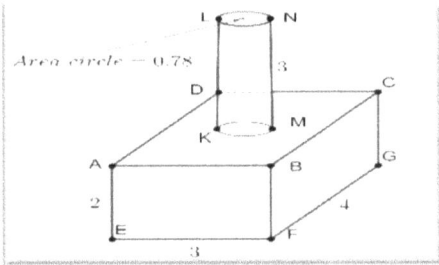

2B.a. 6) The volume of the prism is:

11.

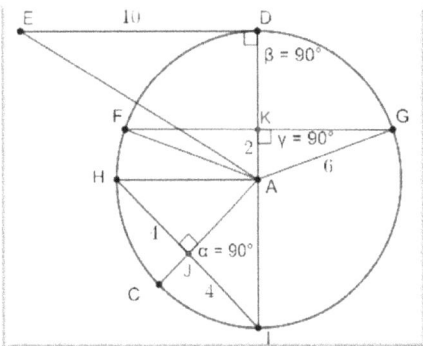

2C.a. 6) The diameter of a circle is the line that unites _____ opposite points on the circle, and passes through the center.

12.

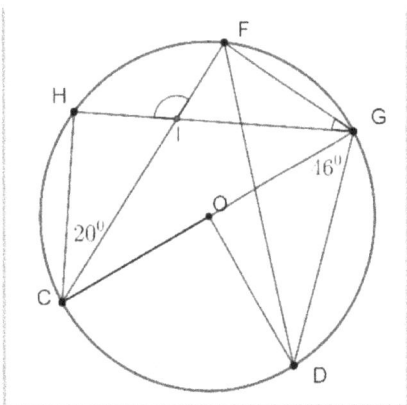

2C.b. 6) The measure of angle CFD is:

13.

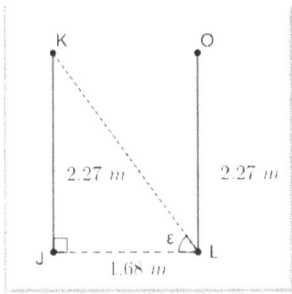

3A.a. 6) $\angle \varepsilon = 53.49^0$. The distance between the top (K) and the bottom is:

14.

4A.a. 6) The result of the expression $\left(\frac{1}{7}\right)^{-2}$ is:

15.

4A.b. 6) The simplified expression of $\dfrac{3\sqrt[3]{x^5 y}}{27\sqrt{x^2 y^4}}$; $x, y \neq 0$ is:

16.

4A.c. 6) The mixed radical of $\sqrt{375}$ is:

17.

4Ba

6) Is $\dfrac{3 \times 5 \times (7 \times 9.}{2 \times (4 \times 3)}$ equal with $\dfrac{3 \times (5 \times 77) \times 9}{2 \times 4 \times 3}$?

18.

4B.b. 6) The value of x from $\dfrac{x}{3} = \dfrac{7}{5}$ is

19.

4C.a. 6) The single logarithm of expression $4 \log_3 6 - 3$ is:

20.

4C.b. 6) The solution of the equation $\log_5(3x + 9) - \log_5(x + 3) = \log_5(x - 4)$; $x > 4$ is:

21.

4D.a. 6) The area of the figure shown below is:

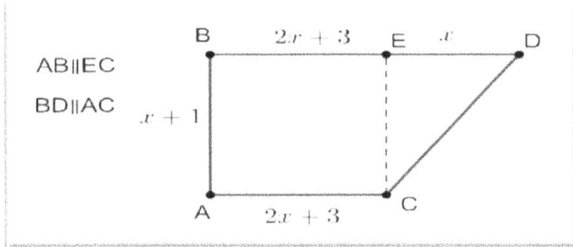

AB∥EC

BD∥AC

$x + 1$

B $2x + 3$ E x D

A $2x + 3$ C

22.

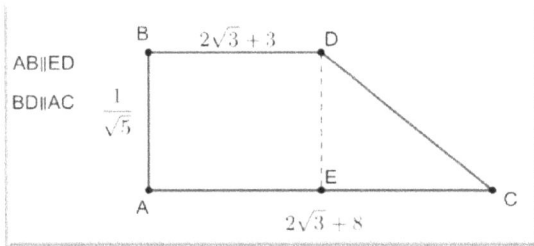

AB∥ED

BD∥AC

$\dfrac{1}{\sqrt{5}}$

B $2\sqrt{3} + 3$ D

A $2\sqrt{3} + 8$ E   C

4D.b. 6) The area of the rectangle ABDE in the figure shown below is:

23.

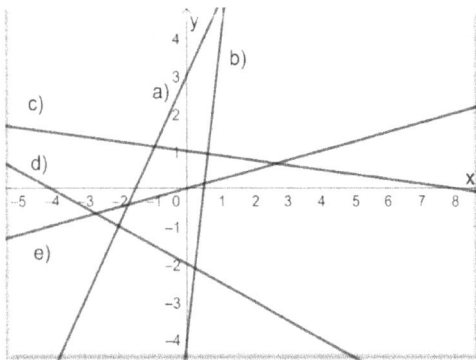

4E.a. 6) The line a) has the equation:

24.

4F.a. 6) The factored expression of $16x^4 - 1$ is:

25.

4F.b. 6) The term outside the square of expression $x^2 - 5x + a$ is:

26.

5A.a. 7) If G(s) = $\dfrac{s^2 - 3s + 7}{s + 3}$ G(2) = 1

27.

5A.b. 10) Determine if the following expression represents a quadratic function.

$f(x) = 2x^2 - 3x + 4$

28.

5A.c. 8) If $f(x) = y = \frac{1}{5x+3}$ the inverse function is: $f^{-1}(x) =$

29.

5B 6) Is this a piecewise function? $f(x) = 3\sin(x - 2) + 3, x \in R$

30.

7I.d. 10) We need to fence two adjacent rectangular lots of land. We have 240 m of fence. The dimensions x and y to maximize the area are: x = ____ m and y = _____ m

Mark yourself

1	2	3	4	5	6	7	8	9	10
11	12	13	14	15	16	17	18	19	20
21	22	23	24	25	26	27	28	29	30
# of Good Answers (NGA)=				NGA/total number of questions=Ratio				Percent=Ratio*100	

TEST 20

1.

A = (-6, 3) B = (-2, 3) C = (4, 3)
K = (-5, 1) G = (-3, 1) H = (2, 1) I = (6, 1)
D = (-6, -2) E = (1, -2) F = (5, -2)

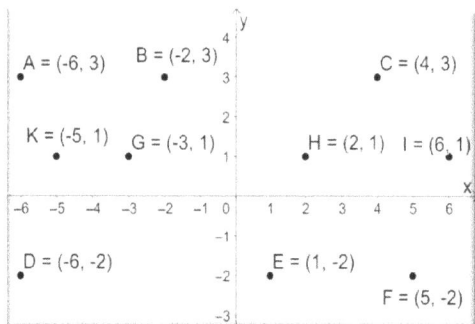

1B.a. 9) The horizontal distance between point G and point I is:

2.

B = (-2, 3) C = (2, 3)
A = (-5, 3)
G = (-2, 1) I = (5, 1)
K = (-5, 0) H = (2, 1)
D = (-5, -2) E = (2, -2) F = (5, -2)

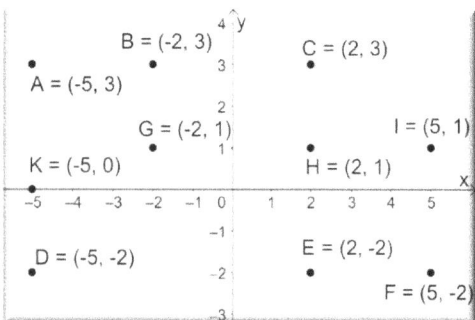

1Bb 9) The vertical distance between point A and point K is:

3.

C = (-2, 4)
B = (-5, 3) D = (2, 3)
E = (5, 1)
A = (-7, -1) H = (-2, -2) F = (3, -2)
I = (-6, -3) J = (4, -4)
G = (1, -5)

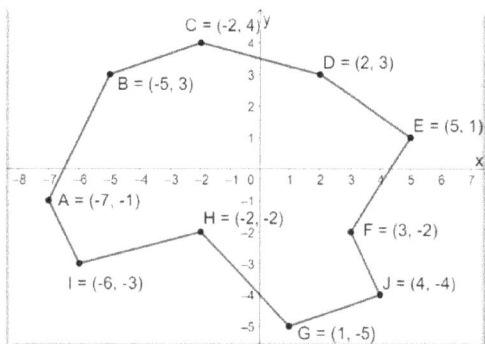

1B.c. 9) In the graph above, the distance between point B and point E is:

4.

1B.d. 8) The mid-point coordinates of segment GH are:

5.

1Da 8) The slope of the line with x intercept = 4 and passing through M(2,3) is $m =$

6.

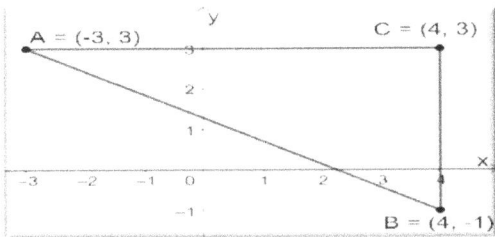

1D.b. 8) The equation of the segment AC is $y =$

7.

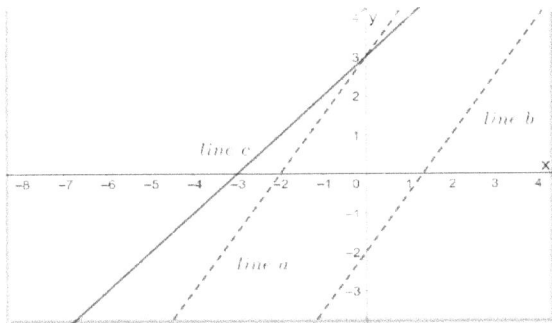

1E.a. 9) Line c is _____ with line a and line b

8.

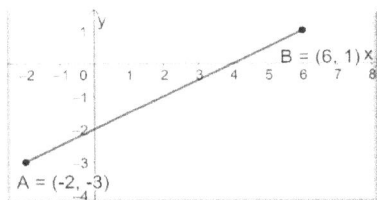

1E.b. 9) The perpendicular to AB through the point B (6,1) will intersect y axis in $b =$

9.

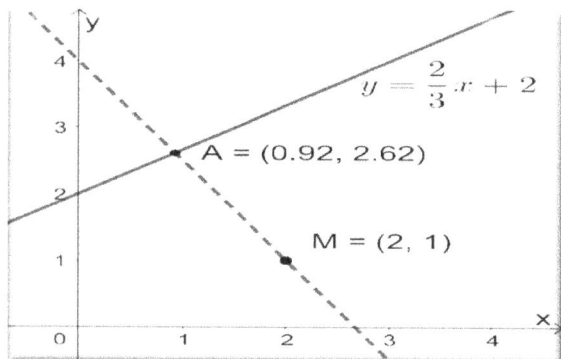

1F.a. 9) The point symmetric to M (2,1) about the line $y = \frac{2}{3}x + 2$ has

x=

10.

1F.b. 9) We have the point P (-3,-1., the point L (3,6). The point S (1,4) is _____the middle point of segment PL.

11.

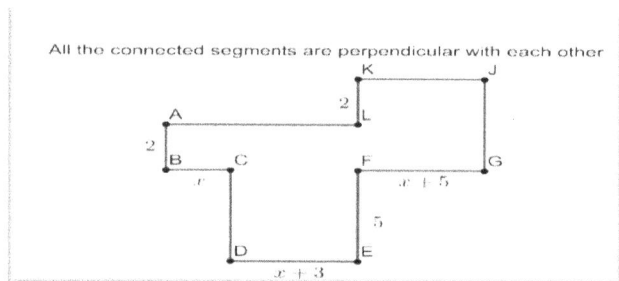

2A.a.7) The area of the whole figure is:

12.

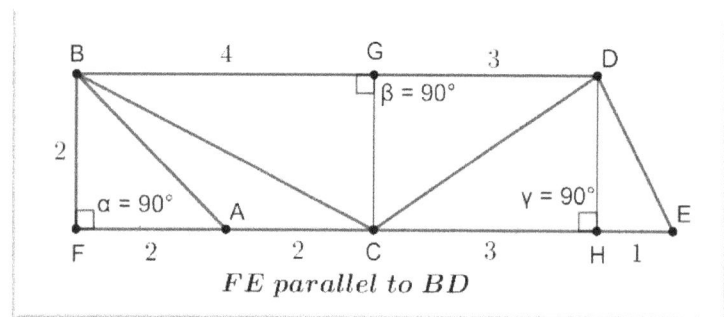

2A.b. 7) The area of the triangle BCD is:

13.

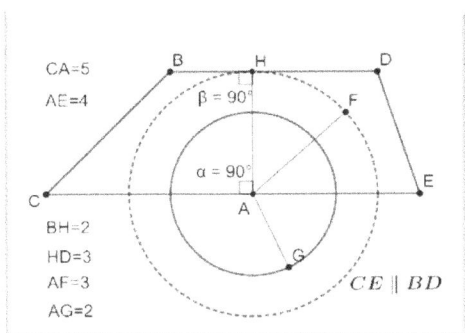

CA=5
AE=4
β = 90°
α = 90°
BH=2
HD=3
AF=3
AG=2
CE ∥ BD

2A.c. 7) The area of the Circle of radius AG=2 is 48π

14.

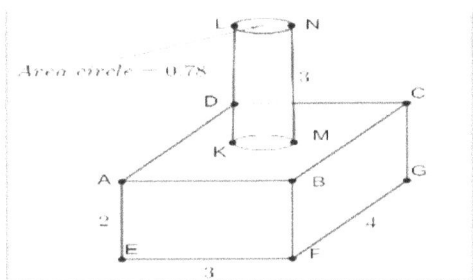

Area circle = 0.78

2B.a. 7) The volume of the whole figure is:

15.

2C.a. 7) A tangent to a circle is _____ to the radius at the point where the tangent touches the circle.

16.

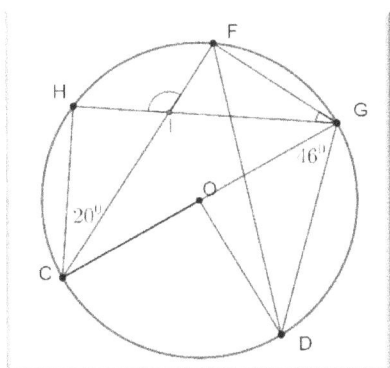

2C.b. 7) The measure of angle COD is:

17.

3A.a. 7) The height of a pole whose shade of 4 meters and makes an angle of 30^0 with the ground is:

18.

4A.a. 7) The result of the expression $(\frac{2}{5})^{-1}$ is:

19.

4A.b. 7) The simplified expression of $\sqrt[4]{\dfrac{625a^6b}{a^2b^5}}$ is:_____ ; $b \neq 0$

20.

4A.c. 7) The entire radical of $8\sqrt{7}$ is:

21.

4B.a. 7) Is $\dfrac{3\times(5+7)}{2\times(4-3)}$ equal with $\dfrac{3\times5+3\times7}{2\times4-2\times3}$?

22.

4B.b. 7) The value of x from $\dfrac{x-2}{4} = \dfrac{2x}{7}$ is: x=

23.

4C.a. 7) The single logarithm of expression $2\ln x + \ln(3x - 4)$ is

_____ ; $x > \dfrac{4}{3}$

24.

4C.b. 7) The solution of the equation $3^{x+3} = 7^{2x+5}$ is $x =$

25.

4D.a. 7) The result of $x^2 + 3x + 4$ divided by $x + 1$ is:

26.

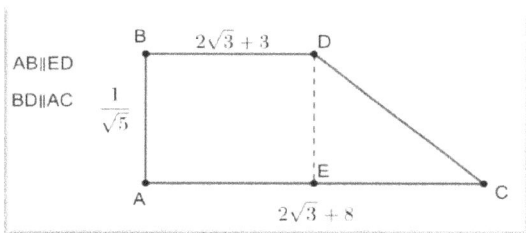

4D.b. 7) The area of the right-angle triangle CDE in the figure shown is:

AB∥ED

BD∥AC

$\dfrac{1}{\sqrt{5}}$

B $2\sqrt{3}+3$ D

E

A

$2\sqrt{3}+8$

C

27.

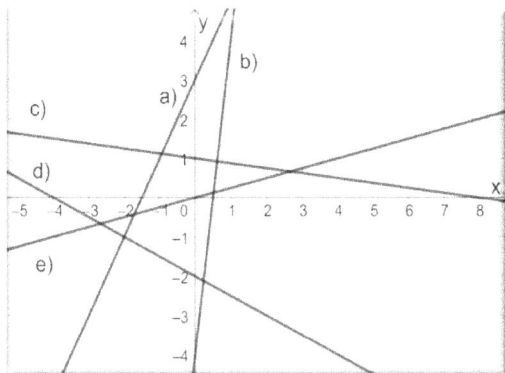

4E.a. 7) The line b) has the equation:

28.

4F.a. 7) The solution of $x^2 + 2x - 3 = 0$ by factoring is $x =$ ____ or $x =$ _____

29.

4F.b. 7) The completed square form of $x^2 + 6x - 4$ is:

30.

8A 4) The sign $\int$ represents the _____ between $f(x)$ and dx.

Mark yourself

1	2	3	4	5	6	7	8	9	10
11	12	13	14	15	16	17	18	19	20
21	22	23	24	25	26	27	28	29	30
# of Good Answers (NGA)=				NGA/total number of questions=Ratio				Percent=Ratio*100	

TEST 21

1.

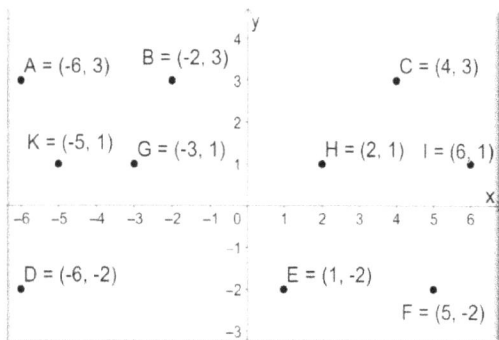

1B.a.10) The horizontal distance between point H and point I is:

2.

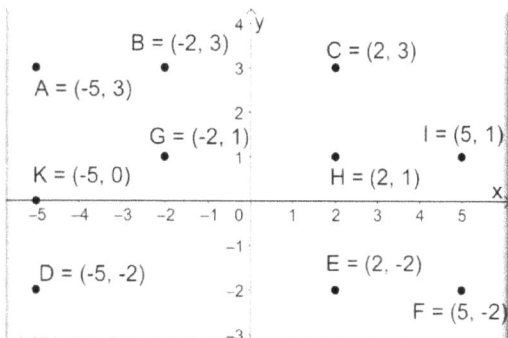

1B.b.10) The vertical distance between point I and point H is:

3.

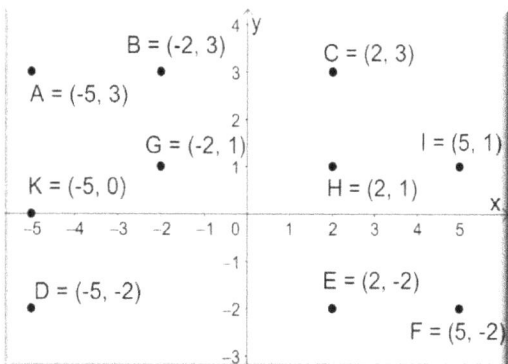

1B.c. 10) The distance between point B and point F is:

4.

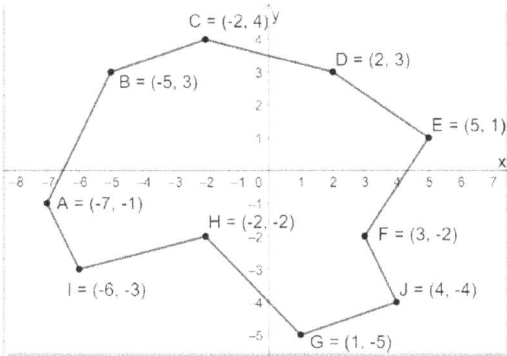

1B.d. 10) The mid-point coordinates of segment IA are: (   )

5.

1C 9) The slope of segment HI is:

6.

1Da 9) The slope of the line that has x intercept the point P(2,0) and passes through the point M(6,1) is m =

7.

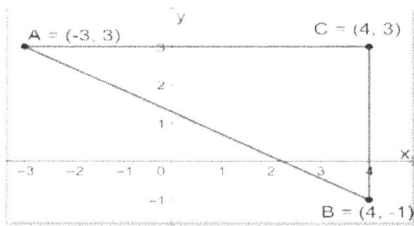

1D.b. 9) The equation of the segment CB is:

8.

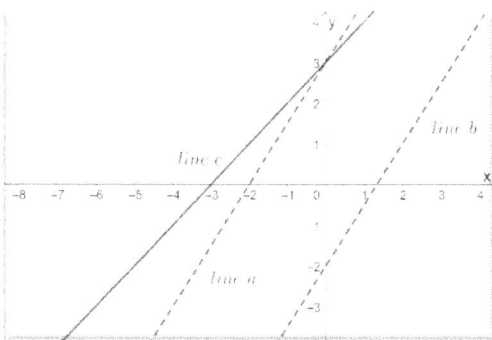

1E.a. 10) The equation of the line c is:

9.

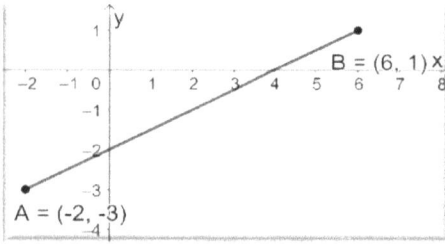

1E.b. 10) **The perpendicular to AB through the point B (6,1) will intersect x axis at** $x =$

10.

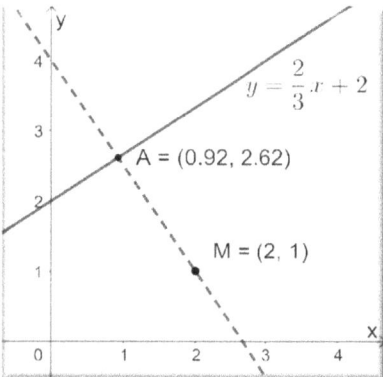

1F.a. 10) **The point symmetric to M (2,1) about the line** $y = \frac{2}{3}x + 2$ **has the y coordinate y=**

11.

1F.b. 10) **The symmetric point to L (3,6) by the point K (3,0) is Q (   )**

12.

2A.a. 8) **The height of a rectangle is half of the base which is 10 units. The area is:**

13.

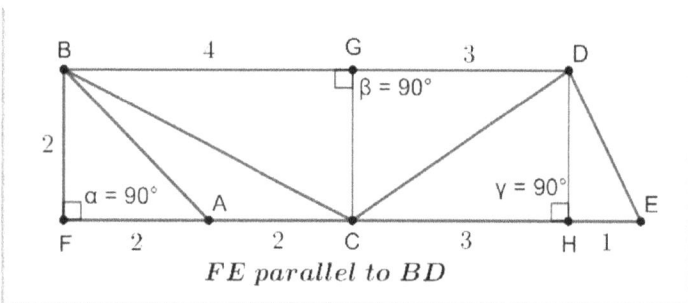

FE parallel to BD

2A.b. 8) **The area of the triangle CDE is:**

14.

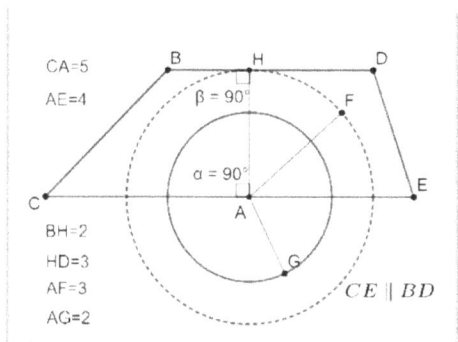

CA=5
AE=4
β = 90°
α = 90°
BH=2
HD=3
AF=3
AG=2
CE ∥ BD

2Ac 8) The difference between the two areas of circle with radius AF and circle with radius AG is:

15.

The next problem will use the figure shown below. The approximate dimensions of the Apollo spacecraft are shown below.

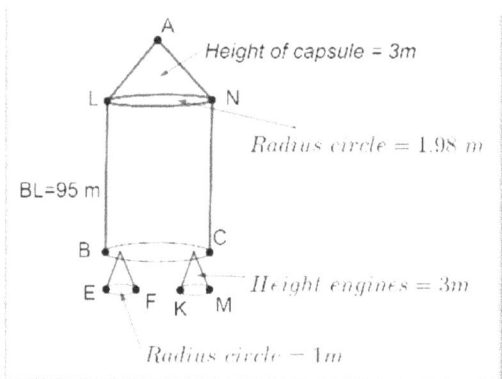

Height of capsule = 3m
Radius circle = 1.98 m
BL=95 m
Height engines = 3m
Radius circle − 1m

2B.a. 8) The volume of the capsule is:

_____m^3

16.

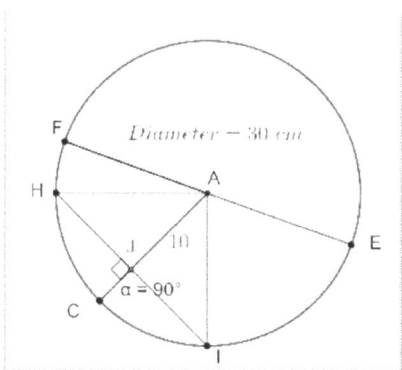

Diameter − 30 cm
10
α = 90°

2C.a. 8) The length of the radius of the circle is:

17.

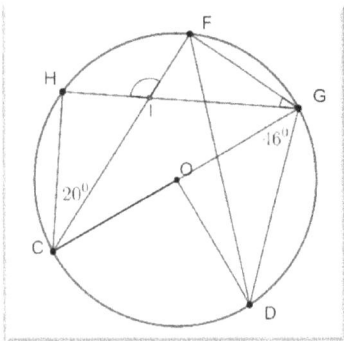

2C.b. 8) The measure of angle HIF is:

18.

3A 8) A hiker climbs a 30 m long ramp that makes 60^0 with the ground. The vertical height is:

19.

4A.a. 8) The result of the expression $15x^5 \div 5x^8$ is:

20.

4A.b. 8) The numerical value of $\sqrt[3]{-81} \div \sqrt[3]{-3}$ is:

21.

4A.c. 8) The entire radical of $7\sqrt{8}$ is:

22.

4Ba 8) is $4\times 5 \times (3 \times 2)$ equal with $(44\times 5) \times (3 \times 2)$?

23.

4B.b. 8) The ratio between the width (x+4) and length (2x-6) of a rectangle is ¾. The length is:

24.

4C.a. 8) The expression $\log_3(\frac{a^2}{b^4})$ in terms of $\log_3 a$ and $\log_3 b$ is:_____; $a, b > 0$

25.

4C.b. 8) The solution of the equation $5 * 3^x = 2^{x-2}$ is $x =$

26.

4D.a. 8) The result of $2x^2 - 5x + 6$ divided by $x - 2$ is:

27.

4D.b. 8) The rationalized expression of $\dfrac{7}{\sqrt{3}+\sqrt{2}}$ is:

28.

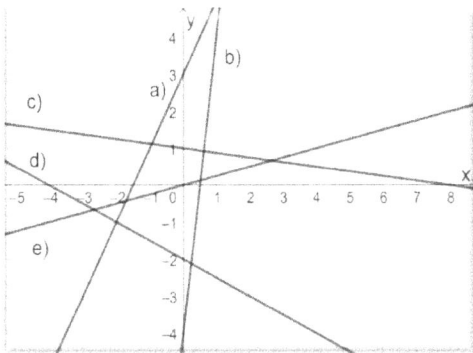

4E.a. 8) The line c) has the equation:

29.

4F.a. 8) The solution of $6x^2 - x - 12 = 0$ by factoring is $x = 1.5 \; or \; x = -1.3$

30.

8B10) If $f(x) = 2x + \sqrt{x}$, then $\int_0^3 f(x)dx =$

Mark yourself

1	2	3	4	5	6	7	8	9	10
11	12	13	14	15	16	17	18	19	20
21	22	23	24	25	26	27	28	29	30

# of Good Answers (NGA)=	NGA/total number of questions=Ratio	Percent=Ratio*100

TEST 22

1.

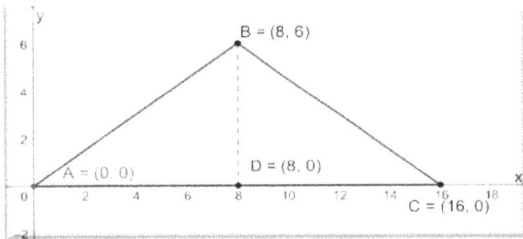

1D.a. 10) **The slope of line BC is** $m =$

2.

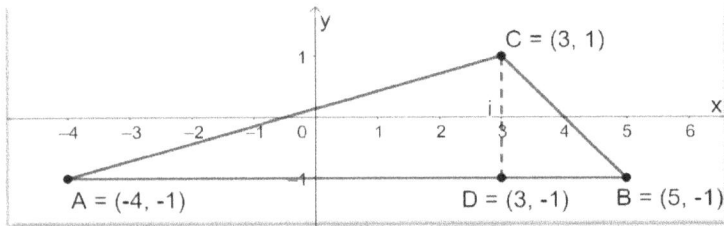

1D.b. 10) **The equation of the segment CD is:**

3.

2A.a. 9) **The base of a rectangle is** $2x$. **The height is 5 units. The area is:**

4.

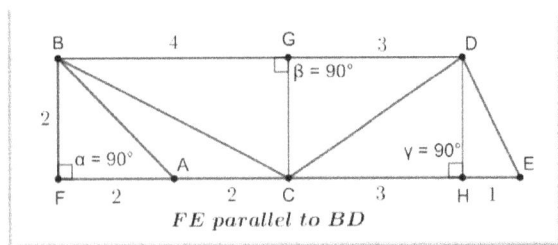

2A.b. 9) **The area of the BAC is:**

5.

2A.c. 9) The area of the difference of the two circles inside the trapezoid CBDE is:

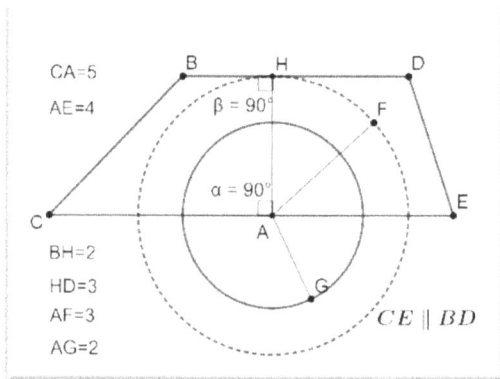

CA=5
AE=4
B
H
D
β = 90°
F
α = 90°
E
C
A
BH=2
HD=3
AF=3
AG=2
G
$CE \parallel BD$

6.

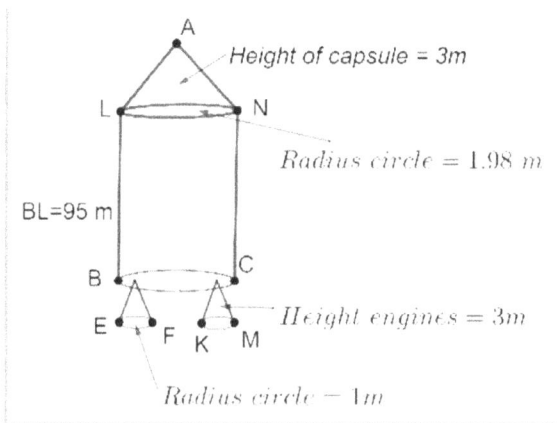

A
Height of capsule = 3m
L
N
Radius circle = 1.98 m
BL=95 m
B
C
E
F
K
M
Height engines = 3m
Radius circle = 1m

2B.a. 9) The volume of the vertical cylinder is:

7.

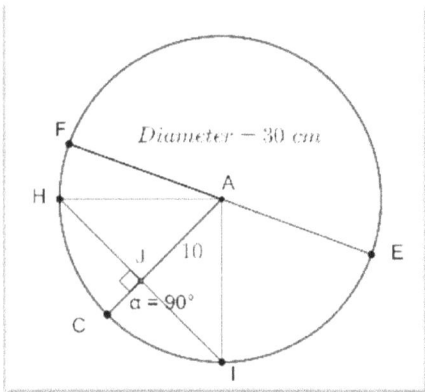

2C.a. 9) The length of HJ is:

8.

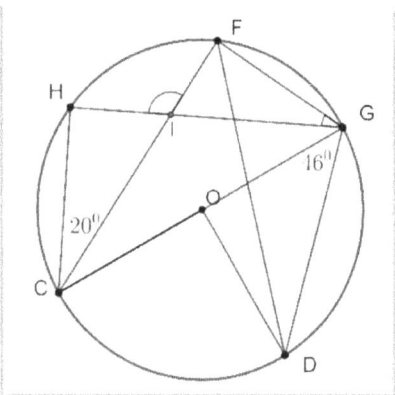

2C.b. 9) The measure of the angle FGI is:

9.

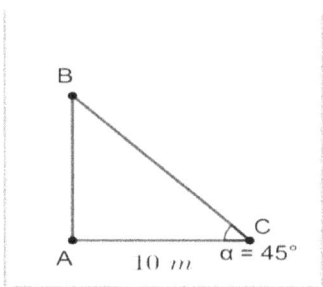

3A 9) An electric pole breaks due to storm. The broken part bends so that the top of the pole touches the ground making an angle 45° with it. The distance between the foot of the tree to the point where the top touches the ground is 10 m. The vertical height of the pole is:

10.

4Aa **9.** The simplified expression of $\frac{27x^3y^{-5}}{2x^{-5}y^{-3}} * \frac{4x^{-5}}{-9xy^{-4}} * \frac{1}{yz^{-2}}$; $x, y, z \neq 0$ is:

11.

4A.b. 9) The area of a right-angle triangle with the base $B = \sqrt{5} \, m$ and height $H = \sqrt{125} \, m$ is:

12.

4Ac 9) The entire radical of $6\sqrt{3}$ is:

13.

4Ba 9) Is $3 \times (7 - 4)$ is equal with $3 \times 7 - 3 \times 4$?

14.

4B.b. 9) The value of x from the expression $\frac{x+7}{x} = \frac{3x-2}{5}$ $(x \neq 0)$ is

$x =$ _____ $or \; x =$ _____

15.

4C.a. 9) The result of $\log_4 8$ in base 2 is:

16.

4C.b. 9) The solution of the equation $\log x + \log(x - 3) = \log 5 + \log(x + 5)$; $x > 3$ is $x =$

17.

4D.a. 9) The result of $6x^3 - 3x - 4$ divided by $3x^2 + x - 4$ is:

18.

4D.b. 9) The rationalized expression of $\frac{8 - 2\sqrt{15}}{\sqrt{5} - \sqrt{3}}$ is:

19.

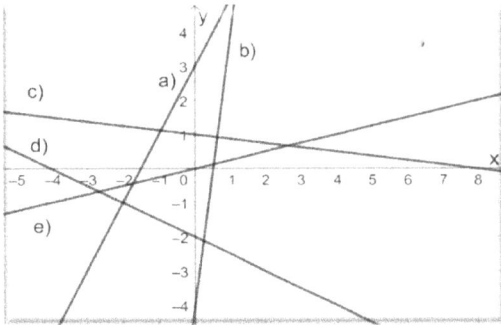

4E.a. 9) The line d) in the graph has the equation:

20.

4F.a. 9) The solution of $3x^4 - 3x^2 - 60 = 0$ is:

21.

4F.b. 8) The completed square form of $x^2 + bx - 4$ is:

22.

5A.a. 8) Determine if the range of the following function is correct. The domain is given.

G(t) = 3 – 2t   D = {-1, -2, 3}   R= {5, 7, -3}

23.

5A.b. 11) Determine if the following expression represents a quadratic function.

$f(x) = x^2 - 3x$

24.

5C 8) $\tan(\alpha) =$

25.

5D.a. 4) The maximum of $\sin(\alpha)$ is:

26.

5D.b. 10) The formula of tangent in terms of sine and cosine is $\tan(\alpha) =$

27.

5E 8) **The angle** $\propto$ **in the figure bellow is:**

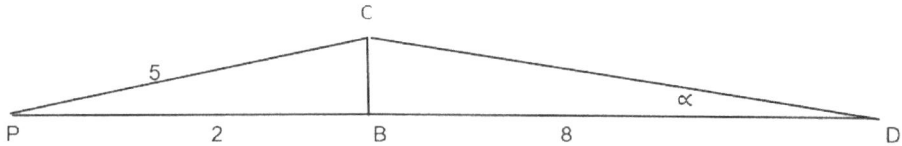

28.

5F 1) **The domain of** $f(x) = cos^{-1}(x)$ **is:**

29.

6A.a. 4) **The table below represents the value of** $f(x) = x^2 + 3$ **around 7.6 for** x around:

X	$f(x) = x^2 + 3$
4.8	27
4.99	27.5
4.999	27.6
5.001	27.9
5.01	28
5.1	29

30.

8C.a. 10) **If** $f(x) = x^2 - x - 6$, **using trapezoidal method, the integral**
$\int_{-2}^{2} f(x)dx \cong$
The interval $\Delta x = 0.5$

Mark yourself

1	2	3	4	5	6	7	8	9	10
11	12	13	14	15	16	17	18	19	20
21	22	23	24	25	26	27	28	29	30
# of Good Answers (NGA)=				NGA/total number of questions=Ratio				Percent=Ratio*100	

TEST 23

1.

2A.a. 10) The length of a rectangle is $x + 3$ and the height is 5 units. If the perimeter of this rectangle is 20 units, the area will be:

2.

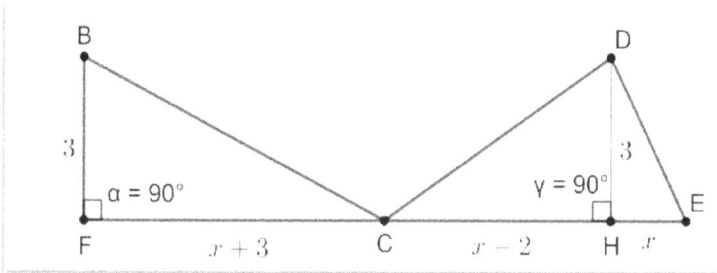

2A.b. 10) The areas of the two triangles in the figure are the same. The area of the triangle BFC is:

3.

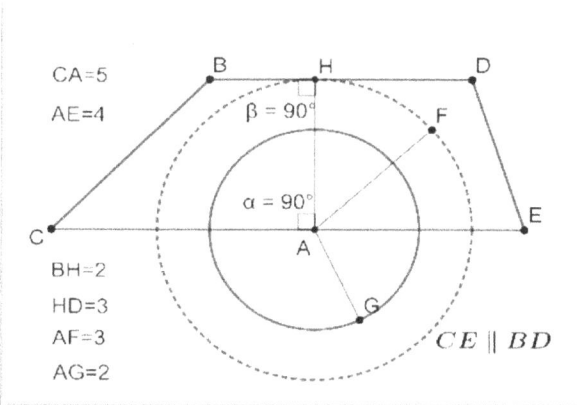

2A.c. 10) The area inside the trapezoid CBDE but exterior to the circle with the radius AF is:

4.

Height of capsule = 3m

Radius circle = 1.98 m

BL=95 m

Height engines = 3m

Radius circle – 1m

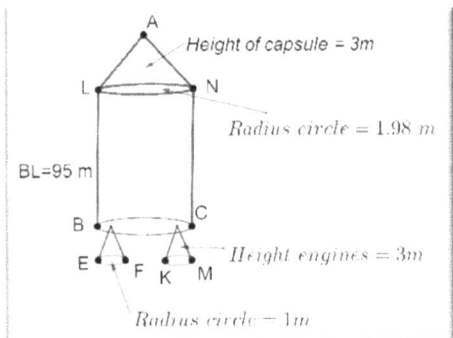

2B.a. 10) The approximate dimensions of the Apollo spacecraft are shown below. The total volume of 5 engines in the form of cones is:

5.

2C.a. 10) The length of segment JC is:

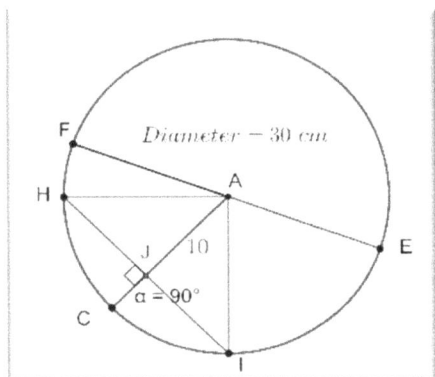

Diameter – 30 cm

10

α = 90°

6.

2C.b. 10) All inscribed angles subtended by a semicircle are _____ angles.

7.

B

A   10 m α = 45°   C

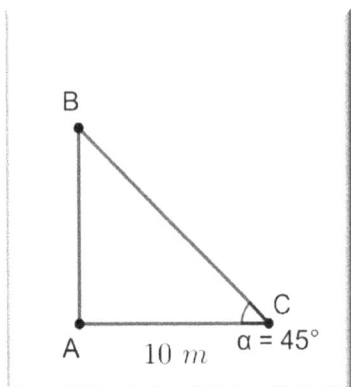

3A 10) An electric pole breaks due to storm. The broken part bends so that the top of the pole touches the ground making an angle 45° with it. The distance between the foot of the tree to the point where the top touches the ground is 10 m. The total height of the above pole is:

8.

4A.a. 10) The simplified expression of $(\frac{8x^{-5}}{24xy^{-3}})^{-1}$ is: $; x, y \neq 0$

9.

4A.b. 10) The simplified form of $\dfrac{\sqrt{\sqrt{16}}}{\sqrt{\sqrt{625}}}$ is:

10.

4A.c. 10) The entire radical of $12\sqrt{5}$ is:

11.

4B.a. 10) Is (5-7)× 3 equal with $5 \times 3 - 4 + 7 \times 3$?

12.

4B.b. 10) Area of a triangle with the $base = x + 3$, $height = 2x + 4$ and, the ratio base/height $\frac{3}{5}$ is $Area =$

13.

4C.a. 10) The result of $\log_{125} 25$ in base 5 is:

14.

4C.b. 10) The solution of the equation $\log(x^2 + x - 6) - \log(x + 3) = 1$ is $x =$

15.

4D.a. 10) The length of a rectangle with area $= 4x^2 + 2x$ and width $= 2x$ is:

16.

4D.b. 10) The expression $\dfrac{15\sqrt{2}}{\sqrt{10}} - \dfrac{5\sqrt{25}}{\sqrt{125}}$ can be written as $k * \sqrt{5}$. The value of k is:

17.

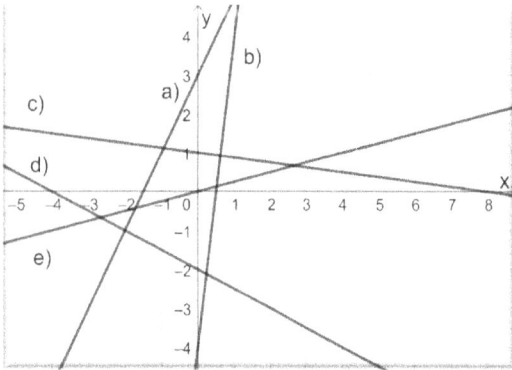

4Ea10) The line e) has the equation

18.

4F.a. 10) The solution of $(x^4 - 625) = 0$ is $x =$

19.

4F.b. 9) The completed square form of $ax^2 + 2x - 4, a \neq 0$ is:

20.

5A.a. 9) Determine if the range of the following function is correct. The domain is given. F(x) = $x^2 - 5x + 1$ D = {-2, 0, 3}   R= {15, 1, -3}

21.

5A.c. 9) Graph the following functions and their inverse functions on the same graph.

$f(x) = y = \frac{2}{3x+2}$ the inverse is: $f^{-1}(x) = \frac{2-2x}{3x}$

22.

5B 7) Determine if the expression below represents a piecewise function.

7) $f(x) = \begin{cases} x^2 \ for \ x \leq -2 \\ 4 \ for -2 < x < 2 \\ 2x \ for \ x \geq 2 \end{cases}$

23.

5C 9) $\sin \left(\frac{\pi}{2}\right) =$

24.

5D.a. 5) The graph of $\cos(\alpha)$ is:

25.

5D.b. 11) Tangent of 30^0 is:

26.

5E 9) The $cos^{-1}\left(\frac{\sqrt{2}}{2}\right)$ is:

27.

5F 2) Graph the function $f(x) = cos^{-1}(x)$

28.

6A.a. 5) The table below represents the value of $f(x) = x^2 + 3$ around _____ for x around 5

X	$f(x) = x^2 + 3$
4.8	27
4.99	27.5
4.999	27.6
5.001	27.9
5.01	28
5.1	29

29.

6Ab 9) $\lim\limits_{x \to 1}\left(\dfrac{x^4-1}{x^2-1}\right) =$

30.

8D 10) If $f(x) = x^2 - 3x + 7$ then, the $\int_{-3}^{3} f(x)dx = 60$

Mark yourself

1	2	3	4	5	6	7	8	9	10
11	12	13	14	15	16	17	18	19	20
21	22	23	24	25	26	27	28	29	30
# of Good Answers (NGA)=				NGA/total number of questions=Ratio				Percent=Ratio*100	

TEST 24

1.

4F.b. 10) To form a perfect square k in $x^2 + kx + 3$, $k =$

2.

5A.a. 10) Determine if the range of the following function is correct. The domain is given.

$H(c) = \frac{c^2 - 2c}{c+2}$ $D = \{-3, 0, 3\}$ $R = \{-15, 0, \frac{3}{5}\}$

3.

5A.b. 12) Determine if the following expressions represent a quadratic function.

$f(x) = x^4 - 3x + 4$

4.

5A.c. 10) Graph the following function and its inverse on the same graph.

$f(x) = y = \frac{2x-1}{x+2}$ the inverse is: $f^{-1}(x) = \frac{-2x-1}{x-2}$

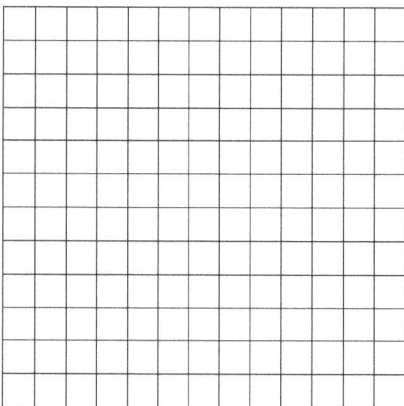

5.

5B 5) Determine if the expression below represents a piecewise function.

$f(x) = \begin{cases} 0.75x^4 \ for \ x < 2 \\ 3x + 6 \ for \ x \geq 2 \end{cases}$

6.

5C 10) $\cos\left(\frac{\pi}{4}\right) =$

7.

5D.a. 6) For each of values of α, the values of $\cos(\alpha)$ are:

α	0	$\dfrac{\pi}{6}$	$\dfrac{\pi}{2}$	$\dfrac{\pi}{3}$	π
$\cos(\alpha)$	1	$\dfrac{1}{2}$	2	$\dfrac{1}{2}$	0

8.

5D.b. 12) Cotangent of 60^0 is:

9.

5E 10) The $sin^{-1}(0.3)$ is:

10.

5F 3) Graph the function $f(x) = sin^{-1}(x)$

11.

6A.a. 6) The table below represents the value of $f(x) = x^2 + 2x$ around 10 for x around:

X	$f(x) = x^2 + 3$
1.89	6.5
1.999	6.59
2.001	6.61
2.01	6.62
2.1	6.9

12.

6Ab 10) $\lim\limits_{h \to 0} \dfrac{f(x+h) - f(x)}{h}$ *when* $f(x) = 3x^2 + 4x$ *is:*

13.

6B 10) $\lim\limits_{x\to3^-}\dfrac{x}{x-5} =$

14.

6C 10) Graph the following function and determine the end behavior for x going towards +∞

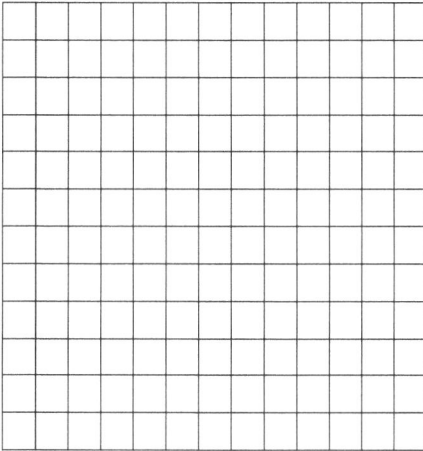

$$f(x) = \frac{2x^2+3x-2}{x^3+1}$$

15.

6D 10) If $f(x) = \dfrac{2x+1}{-3x+7}$, Will there be a value

c such that $f(c) = 3$ for $-2 \le c \le 2$?

16.

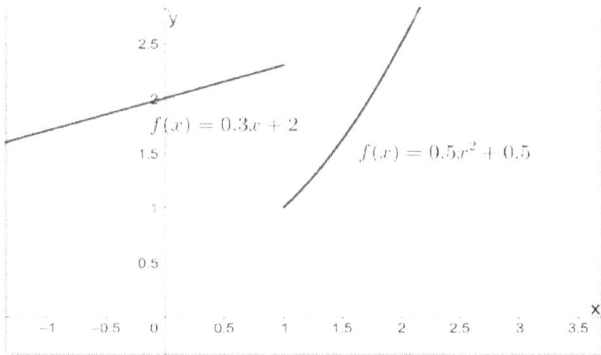

$f(x) = 0.3x + 2$

$f(x) = 0.5x^2 + 0.5$

6E 10) In the graph $\lim\limits_{x\to7^-} f(x) =$

17.

6F

12) $\lim\limits_{x\to\infty}\dfrac{x^2+2x-45}{4x^2-3x^2+7x-3} = \dfrac{1}{4}$

18.

6G 10) Is the function $f(x) = \begin{cases} 5x \text{ if } -1 \le x \le 3 \\ 2x + 2 \text{ if } x > 3 \end{cases}$ is continuous over [-5,5]?

19.

7B 10) Using the formula $f'(x) = \lim\limits_{h\to0}\dfrac{f(x+h)-f(x)}{h}$ find the following derivative.

$f(x) = x^2 + 33 \; so for \; x = 3, f'(3) =$

20.

7C 10) Did Isaac Newton also used this notation for first derivative: $\square \dot{y}$?

21.

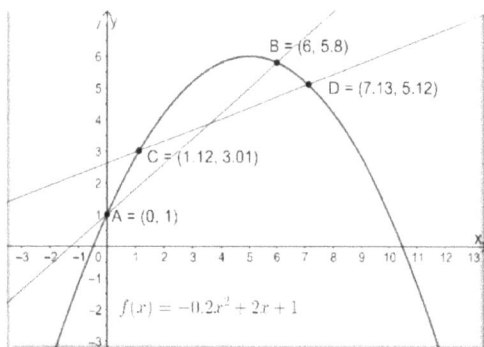

7D.a. 10) The instantaneous rate of change (first derivate) at point D equals:

22.

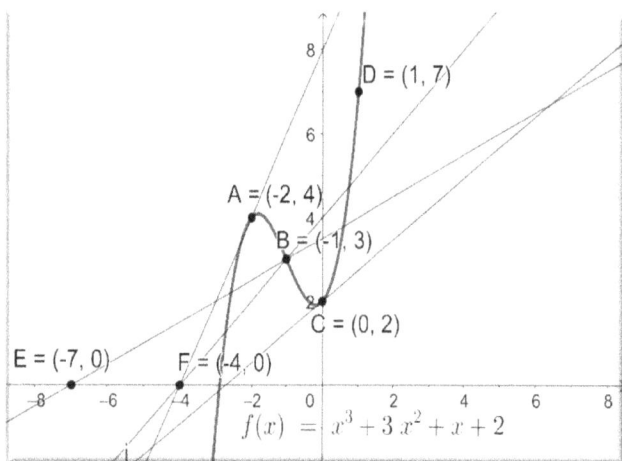

7D.b. 10) The slope of the secant that passes through the points C and D is:

23.

7E.a. 10) If $(x) = 4e^x + \dfrac{5}{6}\ln(x) - 6\cot(x)$;

$then \; f'(x) =$

24.

7F.a. 10) If $f(x) = \dfrac{x^3}{x^{-2}}$ $then \; f'(x) =$

25.

7F.b. 10) If $f(x) = e^x x^5; then \; f'(x) =$

26.

7F.c. 10) $f(x) = \dfrac{2x^2 - 4x + 6}{2x}$; *then* $f'(x) =$

27.

7F.d. 10) $f(x) = \ln(x^2 - 4x)$; *then* $f'(x) =$

28.

7G 10) If $f(x) = \cos(x) + e^x$; *then* $f''(x) =$

29.

7H 10) If $(y - 1)^2 = 6y + x^3 + 2x,\ y' =$

30.

7I.a. 10) The function $f(x) = x^5 - 2x^3$ is concave up for $ \leq x$ *and*

____ $\leq x \leq$ ____.

Mark yourself

1	2	3	4	5	6	7	8	9	10
11	12	13	14	15	16	17	18	19	20
21	22	23	24	25	26	27	28	29	30

# of Good Answers (NGA)=	NGA/total number of questions=Ratio	Percent=Ratio*100

STEP BY STEP
SOLUTIONS

CHAPTER 1

Chapter 1. B. a. Distance between points-horizontal lines

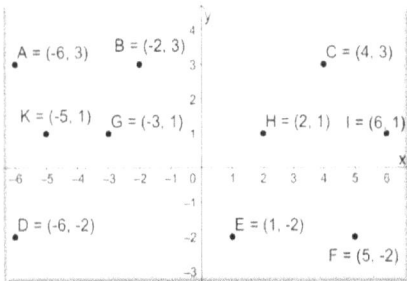

1. The distance between point A and point C is 10

$|4 - (-6)| = |10| = 10$

2. The distance between point A and point B is 5

$|-2 - (-6)| = |4| = 4$

3. The distance between point D and point E is 7

$|1 - (-6)| = |7| = 7$

4. The distance between point E and point F is 7

$|5 - (1.| = |4| = 4$

5. The distance between point K and point G is 2

$|-3 - (-5)| = |2| = 2$

6. The distance between point K and point H is 7

$|2 - (-5)| = |7| = 7$

7. The distance between point K and point I is 9

$|6 - (-5)| = |11| = 11$

8. The distance between point G and point H is 5

$|2 - (-3)| = |5| = 5$

9. The distance between point G and point I is 8

$|6 - (-3)| = |9| = 9$

10. The distance between point H and point I is 4

$|6 - 2| = |4| = 4$

Chapter 1. B. b. Distance between points-Vertical lines

1. The distance between point A and point K is 10

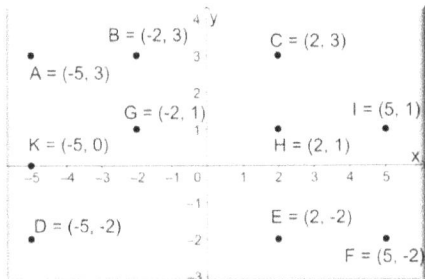

2. The distance between point A and point D is 5

$|-2 - 3| = |-5| = 5$

3. The distance between point B and point G is 3

$|1 - 3| = |-2| = 2$

4. The distance between point C and point H is 7

$|1 - 3| = |-2| = 2$

5. The distance between point C and point E is 5

6. The distance between point K and point D is 7

$|-2 - (0)| = |-2| = 2$

7. The distance between point H and point E is 3

$|-2 - 1| = |-3| = 3$

8. The distance between point I and point F is 7

$|-2 - (-6)| = |4| = 4$

9. The distance between point A and point K is 9

$|0 - 3| = |-3| = 3$

10. The distance between point I and point H is 4

$|5 - 2| = |3| = 3$ (*Horizontal distance*)

Chapter 1.B. c.   Distance between points – non-horizontal and, non-vertical distance

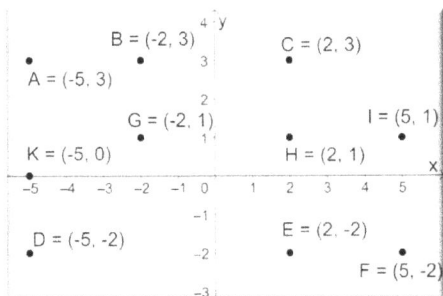

1. The distance between point B and point H is $2\sqrt{5} = 4.47$

BH=$\sqrt{(x_2 - x_1)^2 + (y_2 - y_1)^2} =$

$\sqrt{[(2 - (-2)]^2 + (1 - 3)^2} = \sqrt{16 + 4} = \sqrt{20} = \sqrt{4 \times 5} = 2\sqrt{5}$

2. The distance between point K and point C is 7.61

GC=$\sqrt{(x_2 - x_1)^2 + (y_2 - y_1)^2} =$

$\sqrt{(2 - (-5))^2 + (3 - 0)^2} = \sqrt{49 + 9} = \sqrt{58} = 7.61$

3. The distance between point A and point G is

AG=$\sqrt{(x_2 - x_1)^2 + (y_2 - y_1)^2} = \sqrt{(-2 - (-5))^2 + (1 - 3)^2} = \sqrt{9 + 4} = \sqrt{13} = 3.6$

4. The distance between point A and point H is

AH=$\sqrt{(x_2 - x_1)^2 + (y_2 - y_1)^2} = \sqrt{(2 - (-5))^2 + (1 - 3)^2} = \sqrt{49 + 4} = \sqrt{53} = 7.28$

5. The distance between point A and point I is

AI=$\sqrt{(x_2 - x_1)^2 + (y_2 - y_1)^2} = \sqrt{(5 - (-5))^2 + (1 - 3)^2} = \sqrt{100 + 4} = \sqrt{104} = 10.19$

6. The distance between point A and point F is

AF=$\sqrt{(x_2 - x_1)^2 + (y_2 - y_1)^2} = \sqrt{(5 - (-5))^2 + (-2 - 3)^2} = \sqrt{100 + 25} = \sqrt{125} = 11.18$

7. The distance between point A and point E is

$AE=\sqrt{(x_2 - x_1)^2 + (y_2 - y_1)^2} = \sqrt{(2 - (-5))^2 + (-2 - 3)^2} = \sqrt{49 + 25} = \sqrt{74} = 8.6$

8. The distance between point B and point D is

$BD=\sqrt{(x_2 - x_1)^2 + (y_2 - y_1)^2} = \sqrt{(-5 - (-2))^2 + (-2 - 3)^2} = \sqrt{9 + 25} = \sqrt{34} = 5.83$

9. The distance between point B and point E is

$BE=\sqrt{(x_2 - x_1)^2 + (y_2 - y_1)^2} = \sqrt{(2 - (-2))^2 + (-2 - 3)^2} = \sqrt{16 + 25} = \sqrt{41} = 6.4$

10. The distance between point B and point F is

$BF=\sqrt{(x_2 - x_1)^2 + (y_2 - y_1)^2} = \sqrt{(5 - (-2))^2 + (-2 - 3)^2} = \sqrt{49 + 25} = \sqrt{74} = 8.6$

Chapter 1. B. d.　Midpoint Coordinates

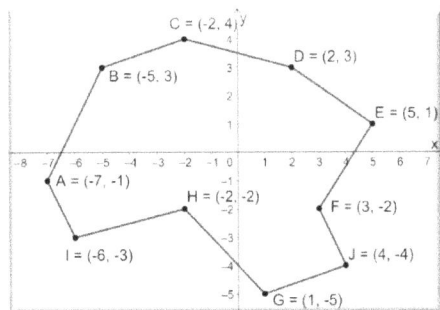

1. The mid-point coordinates of segment AB are:

$x = \frac{x_2+x_1}{2} = \frac{-5+(-7)}{2} = \frac{-12}{2} = -6$

$y = \frac{y_2+y_1}{2} = \frac{3+(-1.}{2} = \frac{2}{2} = 1$

2. The mid-point coordinates of segment BC are:

$x = \frac{x_2+x_1}{2} = \frac{-2+(-5)}{2} = \frac{-7}{2} = -3.5$

$y = \frac{y_2+y_1}{2} = \frac{4+3}{2} = \frac{7}{2} = 3.5$

3. The mid-point coordinates of segment CD are:

$x = \frac{x_2+x_1}{2} = \frac{2+(-2)}{2} = \frac{0}{2} = 0$

$y = \frac{y_2+y_1}{2} = \frac{4+3}{2} = \frac{7}{2} = 3.5$

4. The mid-point coordinates of segment DE are:

$x = \frac{x_2+x_1}{2} = \frac{2+5}{2} = \frac{7}{2} = 3.5$

$y = \frac{y_2+y_1}{2} = \frac{1+3}{2} = \frac{4}{2} = 2$

5. The mid-point coordinates of segment EF are:

$x = \frac{x_2+x_1}{2} = \frac{3+5}{2} = \frac{8}{2} = 4$

$y = \frac{y_2+y_1}{2} = \frac{-2+1}{2} = \frac{-1}{2} = -0.5$

6. The mid-point coordinates of segment FJ are:

$x = \frac{x_2+x_1}{2} = \frac{3+4}{2} = \frac{-7}{2} = -3.5$

$y = \frac{y_2+y_1}{2} = \frac{-2-4}{2} = \frac{-6}{2} = -3$

7. The mid-point coordinates of segment JG are:

$x = \frac{x_2+x_1}{2} = \frac{1+4}{2} = \frac{5}{2} = 2.5$

$y = \frac{y_2+y_1}{2} = \frac{-5+(-4)}{2} = \frac{-9}{2} = -4.5$

8. The mid-point coordinates of segment GH are:

$x = \frac{x_2+x_1}{2} = \frac{-2+1}{2} = \frac{-1}{2} = -0.5$

$$y = \frac{y_2+y_1}{2} = \frac{-2+(-5)}{2} = \frac{-7}{2} = -3.5$$

9. The mid-point coordinates of segment HI are:

$$x = \frac{x_2+x_1}{2} = \frac{-6+(-2)}{2} = \frac{-8}{2} = -4$$

$$y = \frac{y_2+y_1}{2} = \frac{-3+(-2)}{2} = \frac{-5}{2} = -2.5$$

10. The mid-point coordinates of segment IA are:

$$x = \frac{x_2+x_1}{2} = \frac{-7+(-6)}{2} = \frac{-13}{2} = -6.5 \quad y = \frac{y_2+y_1}{2} = \frac{-1+(-3)}{2} = \frac{-4}{2} = -2$$

Chapter 1. C   Slope of a line

1. The slope of segment AB is:

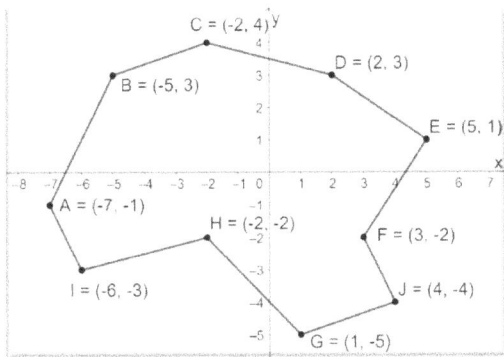

$$m = \frac{y_2-y_1}{x_2-x_1} = \frac{3-(-1.)}{-5-(-7)} = \frac{4}{2} = 2$$

2. The slope of segment BC is

$$m = \frac{y_2-y_1}{x_2-x_1} = \frac{4-3}{-2-(-5)} = \frac{1}{3}$$

3. The slope of segment CD is

$$m = \frac{y_2-y_1}{x_2-x_1} = \frac{3-4}{2-(-2)} = \frac{-1}{4}$$

4. The slope of segment DE is

$$m = \frac{y_2-y_1}{x_2-x_1} = \frac{1-3}{5-2} = \frac{-2}{3}$$

5. The slope of segment EF is

$$m = \frac{y_2-y_1}{x_2-x_1} = \frac{-2-1}{3-5} = \frac{-3}{-2} = 1.5$$

6. The slope of segment FJ is:

$$m = \frac{y_2-y_1}{x_2-x_1} = \frac{-4-(-2)}{4-3} = \frac{-2}{1} = -2$$

7. The slope of segment JG is

$$m = \frac{y_2-y_1}{x_2-x_1} = \frac{-5-(-4)}{1-4} = \frac{-1}{-3} = \frac{1}{3}$$

8. The slope of segment GH is

$$m = \frac{y_2-y_1}{x_2-x_1} = \frac{-2-(-5)}{-2-1} = \frac{3}{-3} = -1$$

9. The slope of segment HI is

$$m = \frac{y_2-y_1}{x_2-x_1} = \frac{-3-(-2)}{-6-(-2)} = \frac{-1}{-4} = \frac{1}{4}$$

10. The slope of segment IA is

$$m = \frac{y_2-y_1}{x_2-x_1} = \frac{-1-(-3)}{-7-(-6)} = \frac{2}{-1} = -2$$

Chapter 1. D. a.   Equation of a straight line – Non-vertical and non-horizontal line

1. The equation of the line through M (-3,**1.** and slope -2 is $y = -2x - 5$

$$m = \frac{y - y_1}{x - x_1}$$

$$-2 = \frac{y - 1}{x - (-3)}$$

$$-2x - 6 = y - 1$$

$$y = -2x - 5$$

2. The y intercept of the line $y = -2x - 5$ is y=-5

3. In the slope relation, $m = \frac{y - 5}{x + 4}$, the y intercept in terms of the slope m, is $b = 4m + 5$

$$m = \frac{y - 5}{x + 4}$$

$$m(x + 4) = y - 5$$

$$mx + 4m = y - 5$$

$$mx + 5 + 4m = y$$

$$y = mx + 4m + 5$$

$$y = mx + b$$

$$b = 4m + 5$$

4. The equation of the parallel line with $y = 3x + 1$ that passes through the point M (5,6) is $y = 3x - 9$

$$m = \frac{y - 6}{x - 5} = 3$$

$$3(x - 5) = y - 6$$

$$3x - 15 = y - 6$$

$$y = 3x - 9$$

5. Y intercept of the parallel line with $y = 3x + 1$ in problem 4 is b=-9

6. The intersection to x axis of $y = 4x - 8$ is P (2,0)

$$y = 0$$

So,

$$0 = 4x - 8$$

$$8 = 4x$$

$$x = \frac{8}{4} = 2$$

7. The intersection to x axis of the line $y = -3x + 1$ is $P(\frac{1}{3}, 0)$

$$0 = -3x + 1$$

$$-1 = -3x$$

$$x = \frac{1}{3}$$

8. The slope of the line with x intercept = 4 and passing through M(2,3) is $m = \frac{-3}{2}$

Slope is $m = \frac{y - y_1}{x - x_1}$

$$m = \frac{y - 3}{x - 2}$$

We know that, the x intercept has y=0. In our case, x intercept has the coordinates (4,0)

$$m = \frac{0-3}{4-2} = \frac{-3}{2}$$

9. The slope of the line that has x intercept the point P(2,0) and passes through the point M(6,**1.** is m = -1/4

Slope is $m = \frac{y-y_1}{x-x_1}$

$$m = \frac{y-1}{x-6}$$

We know that, the x intercept has y=0. In our case, x intercept has the coordinates (2,0)

$$m = \frac{0-1}{2-6} = \frac{-1}{4}$$

10. The slope of line BC is

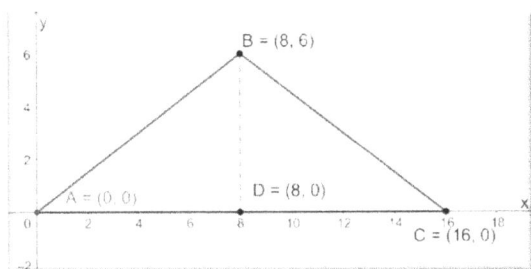

$$m = \frac{y_2-y_1}{x_2-x_1} = \frac{6-0}{8-16} = \frac{6}{-8} = \frac{-3}{4}$$

Chapter 1. D. b.   Equation of a straight line – Vertical and Horizontal lines

1. The equation of horizontal line through M (3,4) is $y = 4$

Problems 2,3,4,5 will be based on the figure shown below.

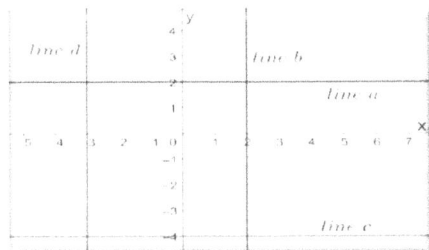

2. The equation of line a is $y = 2$

3. e equation of line b is $x = 2$

4. e equation of line c is $y = -4$

5. e equation of line d is $x = -3$

Problems 6 and 7 will be based on the figure shown below.

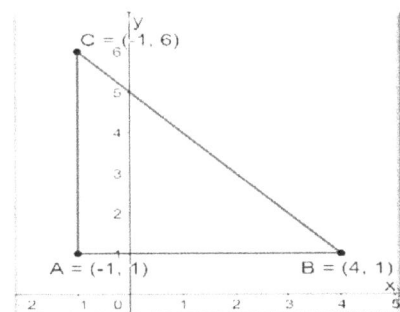

6. e equation of the line that passes through the points C and A in the figure below is $x = -1$

7. e equation of the line that passes through the points A and B in the figure below is $y = -1$

Problems 8 and 9 will be based on the figure shown below.

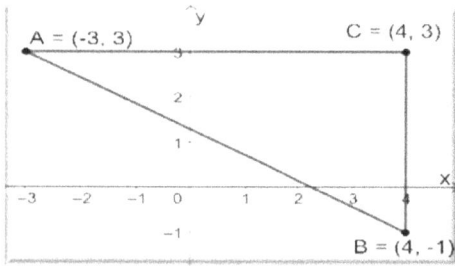

8. The equation of the segment AC is $y = 3$

9. The equation of the segment CB is $x = 4$

Problem 10 will be based on the figure shown below.

10. The equation of the segment CD is $x = 3$

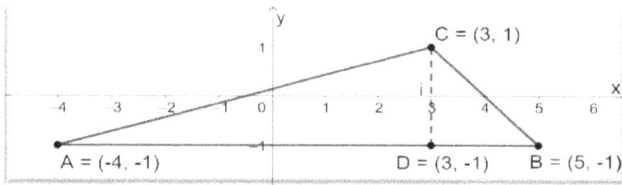

Chapter 1. E. a.   Parallel lines

1. The equation of the parallel line with $y = x - 1$ that intersects y axis at point M (0,5) is $y = x + 5$

2. The equation of the parallel line with $y = -3x + 2$ that intersects y axis at point M (0,-3) is $y = -3x - 3$

3. e line $y = -5x + 3$ is not parallel with $y = -4x + 3$

4. The $y = 3x - 1$ is the same with $y = 3x - 1$

5. The line $y = 4x + 3$ is parallel with $y = 4x - 25$

Problems 6 and 7 will be based on the figure shown below.

6. The line a is parallel to line b

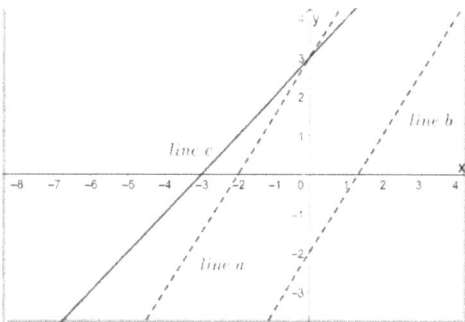

7. The equation of the line a is $y = \frac{3}{2}x + 3$

8. The equation of the line b is $y = \frac{3}{2}x - 2$

9. Line c is not parallel with line a and line b

10. The equation of the line c is $= x + 3$

Chapter 1. E. b.   Perpendicular lines

1. The lines that have the equations $y = 3x + 4$ and $y = -\frac{1}{3}x - 5$ are perpendicular.

2. The lines that have the equations $y = 3x - 3$ and $y = 3x + 3$ are not perpendicular.

3. The line perpendicular to the line $y = -2x + 7$ has the slope $m = \frac{1}{2}$.

4. The equation of the line perpendicular to $y = 5x - 1$ that passes through M (3,4) is

$$y = -\frac{1}{5}x + 4.6$$

$$m = -\frac{1}{5}$$

The perpendicular line goes through M (3,4)

$$4 = -\frac{1}{5}(3) + b$$

$$4 + \frac{1}{5}(3) = b$$

$$b = \frac{20+3}{5} = \frac{23}{5} = 4.6$$

$$y = -\frac{1}{5}x + 4.6$$

5. The equation of the line perpendicular to $y = -2x + 3$ through M (-2,-3) is

$$y = \frac{1}{2}x - 2$$

The slope of the perpendicular line is $m = \frac{1}{2}$

We substitute the coordinates of M into the formula $y = mx + b$

$$-3 = \frac{1}{2}(-2) + b$$

$$-3 = -1 + b$$

$$-3 + 1 = +b$$

$$b = -2$$

So, the equation of the line perpendicular to $y = -2x + 3$ through M (-2,-3) is

$$y = \frac{1}{2}x - 2$$

6. The equation of the line perpendicular to $y = x + 2$ through K (1,2) intersects the y axis in M (0,3)

The slope of the perpendicular to $y = x + 2$ will be $m = -1$

We substitute the coordinates of K into the formula $y = mx + b$

$$2 = (-1)(1) + b$$

$$2 + 1 = b$$

$$b = 3$$

So, the equation of the line perpendicular to $y = x + 2$ through K (1,2) is $y = -x + 3$

Y intercept will be 3

7. The equation of the line perpendicular to AB in point B, is $y = \frac{7}{3}x - 9$

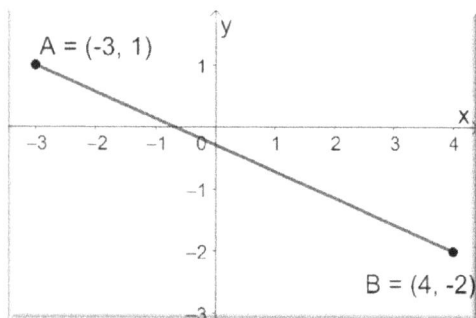

Slope of AB is $m = \frac{y_2-y_1}{x_2-x_1} = \frac{-2-1}{4-(-3)} = \frac{-3}{7}$

So,

The slope of the perpendicular to AB is $m = \frac{7}{3}$

The perpendicular goes through B so, we substitute point B coordinates into $y = mx + b$

$-2 = \frac{7}{3}(4) + b$

$-2 - \frac{28}{3} = b$

$b = -11.33$

So, the equation of the line perpendicular to AB through B (4,-2) is $y = \frac{7}{3}x - 11.33$

8. The equation of the line perpendicular to AB in problem 7 will intersect x axis in $x =$

The equation of the line perpendicular to AB through B (4,-2) is $y = \frac{7}{3}x - 11.33$.

The x intercept will have the y coordinate equal zero.

$0 = \frac{7}{3}x - 11.33$

$\frac{7}{3}x = 11.33$

$x = \frac{11.33*3}{7} = \frac{33.99}{7} = 4.84$

9. The perpendicular to AB through the point B (6,**1.** will intersect y axis in $b = 13$

The slope of AB is $m = \frac{y_2-y_1}{x_2-x_1} = \frac{1-(-3)}{6-(-2)} = \frac{4}{8} = \frac{1}{2}$

So, the slope of the perpendicular to AB will be $m = -2$

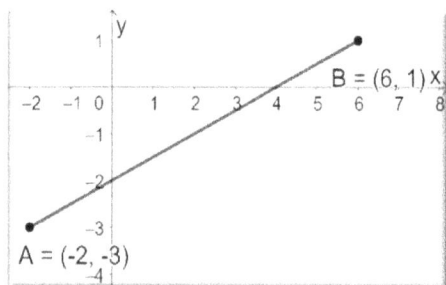

The perpendicular goes through B so, we substitute point B coordinates into $y = mx + b$

$1 = -2(6) + b$

$1 + 12 = b$

$b = 13$

10. The perpendicular to AB through the point B (6,**1.** will intersect x axis at $x = 6.5$

We know that for x intercept y coordinate will be zero.

$0 = -2x + 13$

$2x = 13$

$x = \frac{13}{2} = 6.5$

Chapter 1. F. a.   Symmetry about a line

1. The point symmetric to M (3,4) about y axis has x coordinate equal -3

2. The point symmetric to M (3,4) about x axis has y coordinate equal -4

3. The point symmetric to M (5,4) about the vertical x=1 has the x coordinate equal -3

4. The point symmetric to M (4,3) about the vertical x=2 has the x coordinate equal 0

5. The point symmetric to M (-1,2) about the vertical x=3 has the x coordinate equal 7

6. The point symmetric to M (-1,**1.** about the horizontal y=2 has the y coordinate equal 3

7. The point symmetric to M (-3,-2) about the vertical x=3 has the y coordinate equal 9

8. The point symmetric to M (2,6) about the vertical x=-2 has the x coordinate equal -6

9. The point symmetric to M (2,**1.** about the line $y = \frac{2}{3}x + 2$ has the x coordinate equal to x=-0.16

The perpendicular to $y = \frac{2}{3}x + 2$ through M (2,**1.** has the slope $m = -\frac{3}{2}$.

To find the y intercept of the perpendicular we substitute the coordinates of M into $y = mx + b$

$1 = -\frac{3}{2}(2) + b$

$1 + 3 = b = 4$

So, the equation of the perpendicular to $y = \frac{2}{3}x + 2$ is $y = -\frac{3}{2}x + 4$

These two perpendicular lines intersect at point A

To find the intersection of the two perpendiculars we equalize both equations

$y = \frac{2}{3}x + 2 = -\frac{3}{2}x + 4$

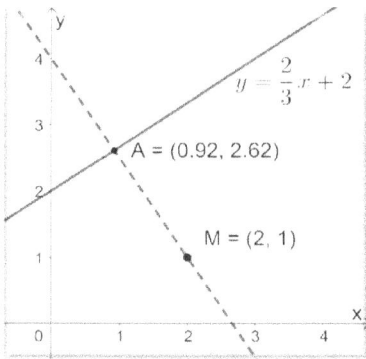

A = (0.92, 2.62)

M = (2, 1)

So, $\frac{2}{3}x + 2 = -\frac{3}{2}x + 4$

$\frac{2}{3}x + \frac{3}{2}x = 4 - 2$

$\frac{4+9}{6}x = 2$

$\frac{13}{6}x = 2$

$x = \frac{6*2}{13} = \frac{12}{13} = 0.92$

$y = \frac{2}{3}(0.92) + 2$

$y = 2.62$

The symmetric point to M by the line $y = \frac{2}{3}x + 2$ (K), will be at the equal distance from this line, but situated on the perpendicular line $y = -\frac{3}{2}x + 4$

Point A (0.92, 2.62) will be the middle point of the segment MK, where M (2, 1).

We use the formulas for middle point

To find x coordinate of K

$0.92 = \frac{2+x_k}{2}$

$1.84 = 2 + x_k$

$x_k = -0.16$

10. The point symmetric to M (2,**1.** about the line $y = \frac{2}{3}x + 2$ has the y coordinate equal to y=4.24

To find y coordinate of K

$2.62 = \frac{1+y_k}{2}$

$5.24 = 1 + y_k$

$y_k = 4.24$

Chapter 1. F. b.   Symmetry about a point

1. The symmetric point to B (2,3) by the origin is C (-2,-3)

Point 1 is B (2,3)

Point 2 is C (x_2, y_2)

The middle point of the segment BC is the origin O (0,0)

We are using the middle point formula to find x and y coordinates of point C.

$x_{midle} = \frac{x_1+x_2}{2}$

$0 = \frac{2+x_2}{2}$

$x_2 = -2$

$y_{midle} = \frac{y_1+y_2}{2}$

$0 = \frac{3+y_2}{2}$

$y_2 = -3$

2. The symmetric point to R (0,2) by the origin is B (0,-2)

Point 1 is R (0,2)

Point 2 is B (x_2, y_2)

The middle point of the segment RB is the origin O (0,0)

We are using the middle point formula to find x and y coordinates of point B.

$x_{midle} = \frac{x_1+x_2}{2}$

$0 = \frac{0+x_2}{2}$

$x_2 = 0$

$y_{midle} = \frac{y_1+y_2}{2}$

$0 = \frac{2+y_2}{2}$

$y_2 = -2$

3. The symmetric point to R (3,0) by the point S (-1,3) is K (-5,6)

Point 1 is R (3,0)

Point 2 is K (x_2, y_2)

The middle point of the segment RK is the point S (-1,3)

We are using the middle point formula to find x and y coordinates of point K.

$x_{midle} = \frac{x_1+x_2}{2}$

$-1 = \frac{3+x_2}{2}$

$-2 = 3 + x_2$

$x_2 = -5$

$y_{midle} = \frac{y_1 + y_2}{2}$

$3 = \frac{0 + y_2}{2}$

$y_2 = 6$

4. The symmetric point to P (-2,-3) by the point S (-1,2) is L (0,7)

Point 1 is P (0,2)

Point 2 is L (x_2, y_2)

The middle point of the segment PL is the point S (-1,2)

We are using the middle point formula to find x and y coordinates of point L.

$x_{midle} = \frac{x_1 + x_2}{2}$

$-1 = \frac{-2 + x_2}{2}$

$-2 = -2 + x_2$

$x_2 = 0$

$y_{midle} = \frac{y_1 + y_2}{2}$

$2 = \frac{-3 + y_2}{2}$

$4 = -3 + y_2$

$y_2 = 7$

5. The symmetric point to P (-3,4) by the point S (0,0) is L (3,-4)

Point 1 is P (-3,4)

Point 2 is L (x_2, y_2)

The middle point of the segment PL is the point S (0,0)

We are using the middle point formula to find x and y coordinates of point L.

$x_{midle} = \frac{x_1 + x_2}{2}$

$0 = \frac{-3 + x_2}{2}$

$x_2 = 3$

$y_{midle} = \frac{y_1 + y_2}{2}$

$0 = \frac{4 + y_2}{2}$

$y_2 = -4$

6. The symmetric point to P (3,4) by the point S (0,0) is L (-3,-4)

Point 1 is P (3,4)

Point 2 is L (x_2, y_2)

The middle point of the segment PL is the point S (0,0)

We are using the middle point formula to find x and y coordinates of point L.

$x_{midle} = \frac{x_1 + x_2}{2}$

$0 = \frac{3 + x_2}{2}$

$x_2 = -3$

$y_{midle} = \frac{y_1 + y_2}{2}$

$0 = \frac{4 + y_2}{2}$

$y_2 = -4$

7. The point L (-1, 1.5) is not symmetric to P (2,-3) by the point S (0,0)

Point 1 is P (2,-3)

Symmetric Point 2 is K (x_2, y_2)

We have to find the middle point of the segment PL.

We are using the middle point formula to find x and y coordinates of point K.

$x_{midle} = \frac{x_1 + x_2}{2}$

$x_{midle} = \frac{2 - 1}{2} = \frac{1}{2} = 0.5$

$y_{midle} = \frac{y_1 + y_2}{2}$

$y_{midle} = \frac{-3 + 1.5}{2} = -\frac{1.5}{2} = -0.75$

We can see that S (0,0) is not the middle point, so L is not symmetric to P by S (0,0)

8. The symmetric point to P (-3,-1. by the point S (2,4) is L (7,9.

Point 1 is P (-3,-1.

Point 2 is L (7,9.

We have to show that the middle point of the segment PL is the point S (2,4)

We are using the middle point formula to find x and y coordinates of point S.

$x_{midle} = \frac{x_1 + x_2}{2}$

$x_{midle} = \frac{-3 + 7}{2} = \frac{4}{2} = 2$

$y_{midle} = \frac{y_1 + y_2}{2}$

$y_{midle} = \frac{-1 + 9}{2} = \frac{8}{2} = 4$

9. We have the point P (-3,-1., the point L (3,6). The point S (1,4) is not the middle point of segment PL.

We are using the middle point formula to find x and y coordinates of point S.

$x_{midle} = \frac{x_1 + x_2}{2}$

$x_{midle} = \frac{-3 + 3}{2} = \frac{0}{2} = 0$

$y_{midle} = \frac{y_1 + y_2}{2}$

$y_{midle} = \frac{-1 + 6}{2} = \frac{5}{2} = 2.5$

10. The symmetric point to L (3,6) by the point K (3,0) is Q (3,-6)

Point 1 is L (3,6)

Point 2 is Q (x_2, y_2)

We are using the middle point formula to find x and y coordinates of point Q.

$$x_{midle} = \frac{x_1 + x_2}{2}$$

$$3 = \frac{3 + x_2}{2}$$

$$6 = 3 + x_2$$

$$x_2 = 3$$

$$y_{midle} = \frac{y_1 + y_2}{2}$$

$$0 = \frac{6 + y_2}{2}$$

$$y_2 = -6$$

CHAPTER 2

Chapter 2.A. a.   Area of rectangles and squares

The next problems will use the figure shown below

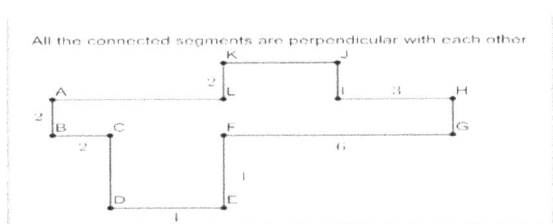

All the connected segments are perpendicular with each other

1. The area of the rectangle ABFL is 12

2. The area of the square CDEF is 16

3. The area of FKJIHGF is 18

4. The area of the whole figure is 46

The next problems will use the figure shown below

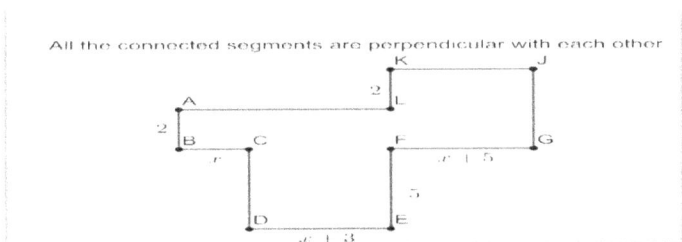

All the connected segments are perpendicular with each other

5. The area of the square CDEF is $5x + 15$

6. The area of the rectangle FKJG is $4x + 20$

7. The area of the whole is $13x + 41$

$$5x + 15 + 4x + 20 + 2(2x + 3)$$
$$= 9x + 35 + 4x + 6$$
$$= 13x + 41$$

8. The height of a rectangle is half of the base which is 10 units. The area is 50

9. The base of a rectangle is $2x$. The height is 5 units. The area is $10x$

10. The length of a rectangle is $x + 3$ and the height is 5 units. If the perimeter of this rectangle is 20 units, the area will be: 25

$$P = 2(x + 3) + 2 * 5 = 20$$

$$2x + 6 + 10 = 20$$

$$2x = 4$$

$$x = 2$$

So,

Area $= (x + 3) * 5 = 5 * 5 = 25$

Chapter 2.A. b.   Area of triangles

The fastest car in 2018 with 261 mph was Bugatti ....Chiron

The next problems will use the figure shown below.

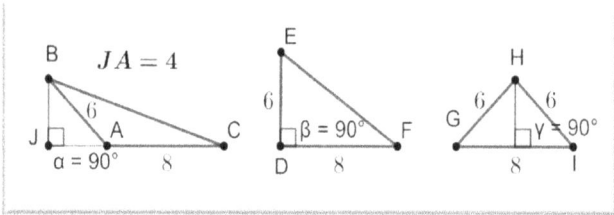

1. The area of the triangle BAC is

$$BJ = \sqrt{6^2 - 4^2} = \sqrt{36 - 16} = \sqrt{20}$$
$$= 2\sqrt{5}$$
$$Area = \frac{BJ * AC}{2} = \frac{2\sqrt{5} * 8}{2} = 8\sqrt{5}$$

2. The area of the triangle DEF is

$$Area = \frac{DE*DF}{2} = \frac{6*8}{2} = 24$$

3. The area of triangle GHI is

$$height = \sqrt{6^2 - 4^2} = \sqrt{36 - 16} = \sqrt{20} = 2\sqrt{5}$$
$$Area = \frac{height*GI}{2} = \frac{2\sqrt{5}*8}{2} = 8\sqrt{5}$$

The next problems will use the figure shown below.

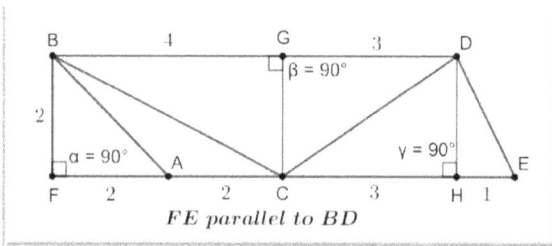

4. The area of triangle BFA is 2

5. The area of triangle BFC is 4

6. The area of the triangle GCD is 3

7. The area of the triangle BCD is 7

8. The area of the triangle CDE is 4

9. The area of the BAC is 2

10. The areas of the two triangles in the figure below are the same. The area of the triangle BFC is 12

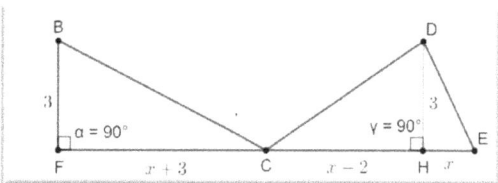

$$\frac{3(x+3)}{2} = \frac{3(x-2+x)}{2}$$
$$x + 3 = 2x - 2$$
$$x = 5$$

So, $Area\ of\ \Delta BFC = \frac{8*3}{2} = 12$

Chapter 2. A. c.   Area of trapezoids and circles

The next problems will use the figure shown below

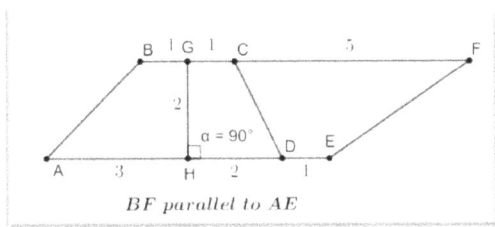

BF parallel to AE

1. The area of the trapezoid ABGH is 4

$$Area = \frac{(big\ base + small\ base) \times heigth}{2}$$

$$= \frac{(3+1)*2}{2} = 4$$

2. The area of the trapezoid ABCD is 7

$$Area = \frac{(big\ base+small\ base)\times heigth}{2} = \frac{(5+2)*2}{2} = 7$$

3. The area of the trapezoid CDEF is 6

$$Area = \frac{(big\ base+small\ base)\times heigth}{2} = \frac{(5+1)*2}{2} = 6$$

4. The area of the trapezoid GHEF is 9

$$Area = \frac{(big\ base+small\ base)\times heigth}{2} = \frac{(6+3)*2}{2} = 9$$

5. The area of the trapezoid GHCD is 3

$$Area = \frac{(big\ base+small\ base)\times heigth}{2} = \frac{(2+1)*2}{2} = 3$$

The next problems will use the figure shown below.

6. The area of the Circle of radius AF=3 is 9π

$Area = \pi R^2 = \pi * (3)^2 = 9\pi$

7. The area of the Circle of radius AG=2 is 9π

$Area = \pi R^2 = \pi * (2)^2 = 4\pi$

8. The difference between the two above areas is 5π

$Diff = 9\pi - 4\pi = 5\pi$

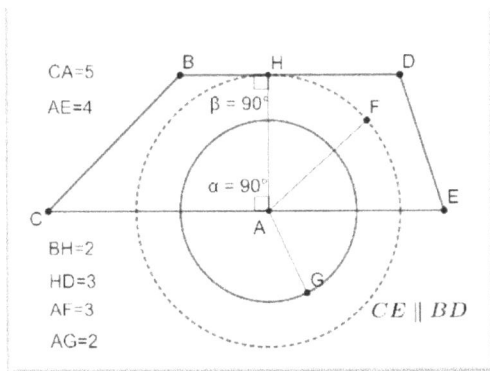

CA=5
AE=4
β = 90°
α = 90°
BH=2
HD=3
AF=3
AG=2
CE ∥ BD

9. The area of the difference of the two circles inside the trapezoid CBDE is 2.5π

$$Area = \frac{5\pi}{2} = 2.5\pi$$

10. The area inside the trapezoid CBDE but exterior to the circle with the radius AF is 6.87

The area of the Circle of radius AF=3 is $9\pi = 9 * 3.14 = 28.26$

The area of the half of Circle of radius AF=3 is

$\frac{28.26}{2} = 14.13$

The area of the trapezoid is $Area = \frac{(big\ base+small\ base)\times heigth}{2} = \frac{(9+5)*3}{2} = \frac{14*3}{2} = 21$

So, the area inside the trapezoid CBDE but exterior to the circle with the radius AF is $21 - 14.13 = 6.87$

Chapter 2. B. a.   Volume of Prisms, cubes, cylinders and cones

1. The volume of the prism with a rectangular base where Length =5, Width = 4 and Height = 3 units is 60

2. The volume of a right triangular prism with the base a triangle with area = 20 and height = 5 is 100 units

3. The volume of a cube with the side of 5 is 125

4. The volume of a cylinder with radius = 3 and height = 6 is 54π

$Volume = \pi R^2 * h = \pi(3)^2 * 6 = 54\pi$

The next three problems will use the figure shown below

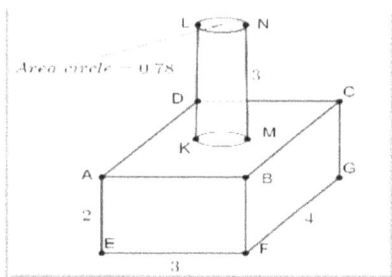

5. The volume of the cylinder is 2.34

$Volume = Area\ base * height = 0.78 * 3 = 2.34$

6. The volume of the prism is 24

7. The volume of the whole figure is 26.34

The next three problems will use the figure shown below. The approximate dimensions of the Apollo spacecraft are shown below.

8. The volume of the capsule is

$Volume = \dfrac{area\ base*height}{3} = \dfrac{\pi(1.98.^2*3)}{3} = 12.31m^3$

9. The volume of the vertical cylinder is

$Volume = Area\ base * height = \pi(1.98.^2 * 95 = 1,169.45m^3$

10. The total volume of 5 engines in the form of cones is

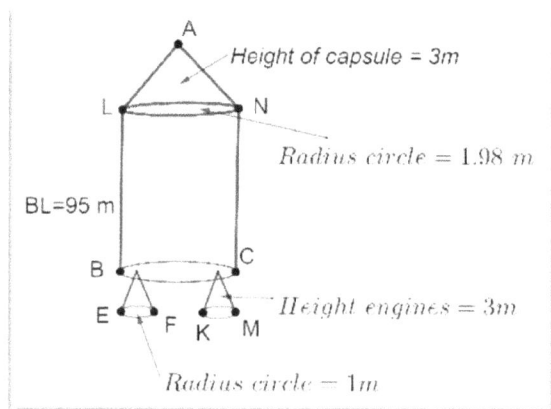

$Volume = 5 * \dfrac{area\ base*height}{3} = 5 * \dfrac{\pi(1)^2*3}{3} =$

$5 * 3.14 = 15.7m^3$

Chapter 2.C. a. Circle – Diameter, secant, tangent, properties of tangents to a circle, unit circle

1. The secant is the line that intersects the circle in two points.

The following problems use the figure shown below

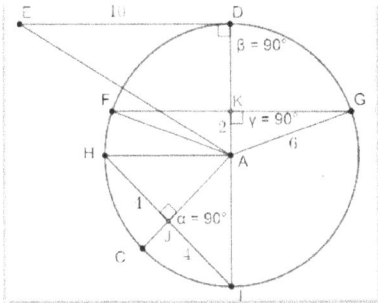

2. The length of the line EA is 11.66

In the right-angle triangle EAD $EA^2 = AD^2 + ED^2 = 6^2 + 10^2 = 36 + 100 = 136$

$EA = \sqrt{136} = 11.66$

3. The length of the segment KD is 4

4. The length of the segment KG is 5.65

In the right-angle triangle AKG, $KG = \sqrt{AG^2 - AK^2} = \sqrt{36 - 4} = \sqrt{32} = 5.65$

5. The length of the segment CJ is

In the right-angle triangle AJH, $AJ = \sqrt{AH^2 - HJ^2} = \sqrt{6^2 - 4^2} = \sqrt{36 - 16} = \sqrt{20} = 4.47$

So, $CJ = AC - AJ = 6 - 4.47 = 1.52$

6. The diameter of a circle is the line that unites two opposite points on the circle, and passes through the center.

7. A tangent to a circle is perpendicular to the radius at the point where the tangent touches the circle.

The following problems use the figure shown below

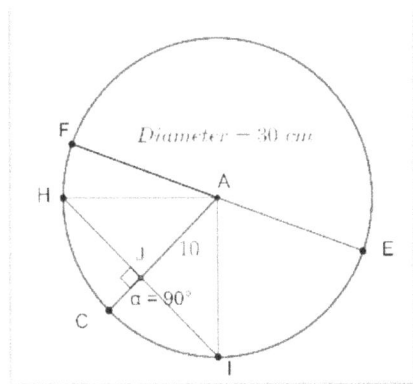

8. The length of the radius of the above circle is 15 cm

9. The length of HJ is 11.18 cm

In the right-angle triangle AJH $HJ = \sqrt{15^2 - 10^2} = \sqrt{225 - 100} = \sqrt{125} = 11.18 \; cm$

10. The length of segment JC is 15-10=5 cm

Chapter 2. C. b. Arc length and the radian, properties of angles in a circle

1. In a circle we have 2π radians

2. The <u>central angle</u> is an angle whose arms go from the centre of the circle to the circle itself

The next problems use the figure shown below

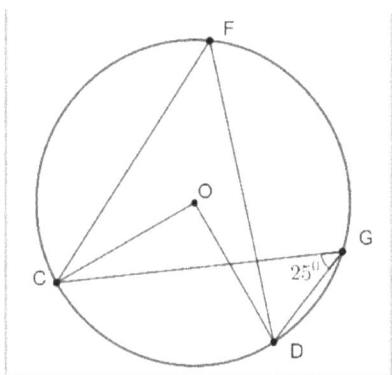

3. The measure of the angle CFD is 25^0

4. The measure of the angle COD is 50^0

5. The angle CFD is called inscribed angle.

The next problems use the figure shown below

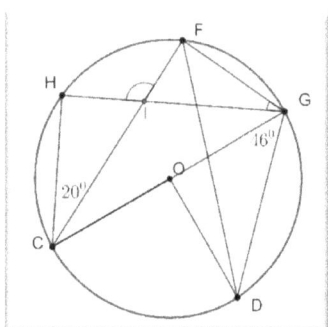

6. The measure of angle CFD is 46^0

7. The measure of angle COD is 92^0

8. The measure of angle HIF is 110^0

The triangle CHI is right-angle with angle CHI 90^0

So, $\angle HIC = 180^0 - 90^0 - 20^0 = 70^0 = \angle FIG$

$\angle HIC + \angle CIG + \angle FIG + \angle HIF = 360^0$

$70^0 + \angle CIG + 70^0 + \angle HIF = 360^0$

$\angle CIG = \angle HIF$

$140^0 + 2 * \angle HIF = 360^0$

$\angle HIF = \frac{360^0 - 140^0}{2} = \frac{220^0}{2} = 110^0$

9. The measure of the angle FGI is 20^0

The triangle IFG is right-angle with angle IFG 90^0

$\angle FIG = 70^0$

$\angle FGI = 180^0 - 90^0 - 70^0 = 20^0$

10. All inscribed angles subtended by a semicircle are right angles.

CHAPTER 3

Chapter 3. A. a.   Sine, Cosine, Tangent, Cotangent, solving right-angle triangles, special ratios

1. The shadow of a 20 m tall tree with the light rays of 23^0 with the vertical is 8.4 m long

$\tan 23^0 = \frac{shadow}{20\,m} = 0.42$

$shadow = 0.42 * 20 = 8.4\,m$

2. Eiffel Tower in Paris has the line between top and the end of the shadow 400 m long. The height of the tower is 324 m. The angle of the light rays and soil is

$\sin \propto = \frac{324}{400} = 0.81$

$\alpha = 54^0$

3. The 55.86 m tall Tower of Pisa, Italy, is lean with almost four degrees. The distance from the vertical is

$\sin 4^0 = \dfrac{horizontal\ distance}{55.86\ m} = 0.069$

$horizontal\ distance = 0.069 * 55.86 = 3.89\ m$

4. Big Ben is 96 m tall. The shade is 70 m long. The tangent of the light rays with the ground is

$\tan \alpha = \dfrac{96}{70} = 1.37$

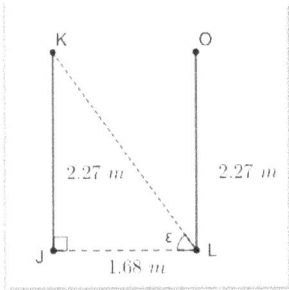

5. The height of the Caryatides is 2.27 m. they are 1.68 m apart. The tangent of the angle that the shade of one touch the next statue is

$\tan \varepsilon = \dfrac{2.27}{1.68} = 1.35$

6. $\angle \varepsilon = 53.49^0$. The distance between the top (K) and the bottom of the next (L) is

$distance = \sqrt{2.27^2 + 1.68^2} = \sqrt{7.97} = 2.82\ m$

7. The height of a pole whose shade of 4 meters makes an angle of 30^0 with the ground is

$\tan 30^0 = \dfrac{height}{4} = \dfrac{\sqrt{3}}{3}$

$height = \dfrac{\sqrt{3}}{3} * 4 = 2.3\ m$

8. A hiker climbs a 30 m long ramp that makes 60^0 with the ground. The vertical height is

$\sin 60^0 = \dfrac{height}{30\ m} = \dfrac{\sqrt{3}}{2}$

$height = \dfrac{\sqrt{3}}{2} * 30 = 15\sqrt{3}\ m$

The next problems use the figure shown below

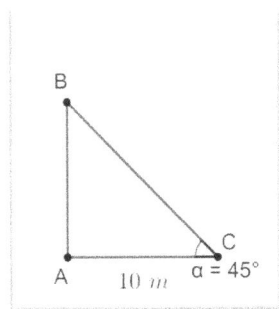

9. An electric pole breaks due to storm. The broken part bends so that the top of the pole touches the ground making an angle 45° with it. The distance between the foot of the tree to the point where the top touches the ground is 10 m. The vertical height of the pole is 10 m.

$\tan 45^0 = \dfrac{AB}{10\ m} = 1$

AB = 10 m

10. The total height of the above pole

$\sin 45^0 = \dfrac{10\ m}{BC} = \dfrac{\sqrt{2}}{2}$

$BC = \dfrac{10\ m}{\frac{\sqrt{2}}{2}} = \dfrac{10\ m}{0.705} = 14.18\ m$

Total height is 10 + 14.18 = 24.18 m

CHAPTER 4

Chapter 4.A. a. Powers – Definition, rules

1. The simplified expression of $\frac{5x^6y^9}{15x^4y^3}$ is $\frac{1}{3}x^2y^6$, $x, y \neq 0$

2. The simplified expression of $(6x^4y)(xy^2)(-3x^4y^3)$ is $-18x^9y^6$

3. The simplified expression of $(-3a^2b^3)^3$ is $-27a^6b^9$

4. The simplified expression of $(\frac{-7x^4y^3}{xy})^2(\frac{x^2yz}{7x^2z})^3$, $x, y \neq 0$ is

$$(\frac{-7x^4y^3}{xy})^2(\frac{x^2yz}{7x^2z})^3 = \frac{49x^8y^6}{x^2y^2} * \frac{x^6y^3z^3}{343x^6z^3} = \frac{x^{14}y^9z^3}{7x^8y^2z^3} = \frac{1}{7}x^6y^7 , x, y, z \neq 0$$

5. The simplified expression of $(-5x^4y^{-3}z^0)^{-3}$ is

$$(-5x^4y^{-3}z^0)^{-3} = \frac{1}{(-5x^4y^{-3}z^0)^3} = \frac{1}{-125x^{12}y^{-9}*1} = \frac{y^9}{-125x^{12}}$$

6. The result of the expression $(\frac{1}{7})^{-2}$ is 49

7. The result of the expression $(\frac{2}{5})^{-1}$ is 2.5

8. The result of the expression $15x^5 \div 5x^8$ is $\frac{3}{x^3}$

9. The simplified expression of $\frac{27x^3y^{-5}}{2x^{-5}y^{-3}} * \frac{4x^{-5}}{-9xy^{-4}} * \frac{1}{yz^{-2}}$, $x, y, z \neq 0$ is

$$\frac{27x^3y^{-5}}{2x^{-5}y^{-3}} * \frac{4x^{-5}}{-9xy^{-4}} * \frac{1}{yz^{-2}} = \frac{-3x^2y^2*2}{z^{-2}} = -6x^2y^2z^2$$

10. The simplified expression of $(\frac{8x^{-5}}{24xy^{-3}})^{-1}$, $x, y \neq 0$ is

$$(\frac{8x^{-5}}{24xy^{-3}})^{-1} = \frac{1}{\frac{8x^{-5}}{24xy^{-3}}} = \frac{24xy^{-3}}{8x^{-5}} = \frac{3x^6}{y^3}$$

Chapter 4. A. b. Radicals, rules

1. The root of $\sqrt[4]{81x^4}$ is 3x

2. The simplified expression of $\sqrt[3]{x^2y^3} * \sqrt[3]{x^4y^6}$ is

$$\sqrt[3]{x^2y^3} * \sqrt[3]{x^4y^6} = \sqrt[3]{x^{2+4}y^{3+6}} = \sqrt[3]{x^6y^9} = x^2y^3$$

3. The simplified expression of $\sqrt{\frac{9x^4y}{y^{-2}}} * \sqrt{\frac{4xy^2}{x^{-2}}}$ is

$$\sqrt{\frac{9x^4y}{y^{-2}}} * \sqrt{\frac{4xy^2}{x^{-2}}} = \sqrt{\frac{9*4x^3y^3x^3y^2}{1}} = \sqrt{36x^6y^5} = \sqrt{36} * \sqrt{x^6} * \sqrt{y^5} = 6x^3y^2\sqrt{y}$$

4. The perimeter of a rectangle with length $L = 3\sqrt{20} + 5\sqrt{5}$ and width $W = 5\sqrt{20}$ is

$$P = 2(3\sqrt{20} + 5\sqrt{5}) + 2(5\sqrt{20}) = 6\sqrt{20} + 10\sqrt{5} + 10\sqrt{20} = 16\sqrt{20} + 10\sqrt{5} = 16\sqrt{4*5} + 10\sqrt{5} = 16\sqrt{4} * \sqrt{5} + 10\sqrt{5} = 32 * \sqrt{5} + 10\sqrt{5} = 42\sqrt{5}$$

5. The area of the above rectangle is

$Area = L * W = \left(3\sqrt{20} + 5\sqrt{5}\right) * 5\sqrt{20} = 15\sqrt{20} * \sqrt{20} + 25\sqrt{5} * \sqrt{20} = 15\sqrt{20 * 20} + 25\sqrt{5 * 20} = 15\sqrt{400} + 25\sqrt{100} = 15\sqrt{4 * 100} + 25 * 10 = 15\sqrt{4} * \sqrt{100} + 250 = 15 * 2 * 10 + 250 = 300 + 250 = 550$

6. The simplified expression of $\frac{3\sqrt[3]{x^5 y}}{27\sqrt[3]{x^2 y^4}}$, $x, y \neq 0$ is

$\frac{3\sqrt[3]{x^5 y}}{27\sqrt[3]{x^2 y^4}} = \frac{1}{9}\sqrt[3]{\frac{x^5 y}{x^2 y^4}} = \frac{1}{9}\sqrt[3]{\frac{x^{5-2}}{y^{4-1}}} = \frac{1}{9}\sqrt[3]{\frac{x^{5-2}}{y^{4-1}}} = \frac{1}{9}\sqrt[3]{\frac{x^3}{y^3}} = \frac{\sqrt[3]{x^3}}{9\sqrt[3]{y^3}} = \frac{x}{9y}$

7. The simplified expression of $\sqrt[4]{\frac{625 a^6 b}{a^2 b^5}}$, $a, b \neq 0$ is

$\sqrt[4]{\frac{625 a^6 b}{a^2 b^5}} = \sqrt[4]{\frac{625 a^{6-2}}{b^{5-1}}} = \sqrt[4]{\frac{625 a^4}{b^4}} = \frac{\sqrt[4]{625 a^4}}{\sqrt[4]{b^4}} = \frac{\sqrt[4]{625} * \sqrt[4]{a^4}}{\sqrt[4]{b^4}} = \frac{5a}{b}$ $b \neq 0$

8. The numerical value of $\sqrt[3]{-81} \div \sqrt[3]{-3}$ is

$\sqrt[3]{-81} \div \sqrt[3]{-3} = \frac{\sqrt[3]{-81}}{\sqrt[3]{-3}} = \sqrt[3]{\frac{-81}{-3}} = \sqrt[3]{27} = \sqrt[3]{3^3} = 3$

9. The area of a right-angle triangle with the base $B = \sqrt{5}\ m$ and height $H = \sqrt{125}\ m$ is

$Area = \frac{B * H}{2} = \frac{\sqrt{5} * \sqrt{125}}{2} = \frac{\sqrt{5 * 5^3}}{2} = \frac{\sqrt{5^4}}{2} = \frac{25}{2} = 12.5\ m^2$

10. The simplified form of $\frac{\sqrt{\sqrt{16}}}{\sqrt{\sqrt{625}}}$ is

$\frac{\sqrt{\sqrt{16}}}{\sqrt{\sqrt{625}}} = \frac{\sqrt{4}}{\sqrt{25}} = \frac{2}{5}$

Chapter 4. A. c.   Mixed radicals, conversion of radicals

1. The mixed radical of $\sqrt{192}$ is $8\sqrt{3}$

$\sqrt{192} = \sqrt{4 * 48} = \sqrt{16 * 12} = \sqrt{16} * \sqrt{12} = 4\sqrt{12} = 4\sqrt{4 * 3} = 4 * \sqrt{4} * \sqrt{3} = 4 * 2 * \sqrt{3} = 8\sqrt{3}$

2. The mixed radical of $\sqrt{160}$ is $4\sqrt{10}$

$\sqrt{160} = \sqrt{16 * 10} = \sqrt{16} * \sqrt{10} = 4\sqrt{10}$

3. The mixed radical of $\sqrt{175}$ is $5\sqrt{7}$

$\sqrt{175} = \sqrt{25 * 7} = \sqrt{25} * \sqrt{7} = 5\sqrt{7}$

4. The mixed radical of $\sqrt{486}$ is $9\sqrt{6}$

$\sqrt{486} = \sqrt{2 * 243} = \sqrt{2 * 3 * 81} = \sqrt{81 * 6} = \sqrt{81} * \sqrt{6} = 9\sqrt{6}$

5. The mixed radical of $\sqrt{288}$ is $12\sqrt{2}$

$\sqrt{288} = \sqrt{2 * 144} = \sqrt{144} * \sqrt{2} = 12\sqrt{2}$

6. The mixed radical of $\sqrt{375}$ is $5\sqrt{15}$

$\sqrt{375} = \sqrt{5 * 75} = \sqrt{3 * 5 * 25} = \sqrt{15 * 25} = \sqrt{25} * \sqrt{15} = 5\sqrt{15}$

7. The entire radical of $8\sqrt{7}$ is $\sqrt{448}$

$8\sqrt{7} = \sqrt{64 * 7} = \sqrt{448}$

8. The entire radical of $7\sqrt{8}$ is $\sqrt{392}$

$7\sqrt{8} = \sqrt{49 * 8} = \sqrt{392}$

9. The entire radical of $6\sqrt{3}$ is $\sqrt{108}$

$6\sqrt{3} = \sqrt{36 * 3} = \sqrt{108}$

10. The entire radical of $12\sqrt{5}$ is $\sqrt{720}$

$12\sqrt{5} = \sqrt{144 * 5} = \sqrt{720}$

Chapter 4. B. a.   Associative, commutative and distributive properties

1. (6+7)+4 is equal with (6+4)+7

(6+7)+4 = 13+4 = 17

(6+4)+7 = 10+7 = 17

2. 2+(3+6)+5 is equal with (2+6)+(3+5)

2+(3+6)+5 = 2+9+5 = 16

(2+6)+(3+5) = 8+8 = 16

3. $(5 \times 7) \times 3 = 5 \times (7 \times 3)$

$(5 \times 7) \times 3 = 35 \times 3 = 105$

$5 \times (7 \times 3) = 5 \times 21 = 108$

4. $5 \times (9 + 3)$ is equal with $5 \times 9 + 5 \times 3$

$5 \times (9 + 3) = 5 \times 12 = 60$

$5 \times 9 + 5 \times 3 = 45 + 15 = 60$

5. $2 \times 3 \times (9 \times 5)$ is equal with $(2 \times 9 \times 3) \times 5$

$2 \times 3 \times (9 \times 5) = 6 \times 45 = 270$

$(2 \times 9 \times 3) \times 5 = 54 \times 5 = 270$

6. $\frac{3\times5\times(7\times9.}{2\times(4\times3)}$ is equal with $\frac{3\times(5\times7)\times9}{2\times4\times3}$

$\frac{3\times5\times(7\times9.}{2\times(4\times3)} = \frac{15\times63}{2\times12} = \frac{15\times21}{2\times4} = \frac{315}{8}$

$\frac{3\times(5\times7)\times9}{2\times4\times3} = \frac{3\times35\times9}{2\times12} = \frac{35\times9}{2\times4} = \frac{315}{8}$

7. $\frac{3\times(5+7)}{2\times(4-3)}$ is equal with $\frac{3\times5+3\times7}{2\times4-2\times3}$

$\frac{3\times(5+7)}{2\times(4-3)} = \frac{3\times12}{2\times1} = \frac{3\times6}{1\times1} = 18$

$\frac{3\times5+3\times7}{2\times4-2\times3} = \frac{15+21}{8-6} = \frac{36}{2} = 18$

8. $4 \times 5 \times (3 \times 2)$ is equal with $(4 \times 5) \times (3 \times 2)$

$4 \times 5 \times (3 \times 2) = 20 \times 6 = 120$

$(4 \times 5) \times (3 \times 2) = 20 \times 6 = 120$

9. $3 \times (7 - 4)$ is equal with $3 \times 7 - 3 \times 4$

$3 \times (7 - 4) = 3 \times 3 = 9$

$3 \times 7 - 3 \times 4 = 21 - 12 = 9$

10. $(5-7) \times 3$ is equal with $5 \times 3 - 7 \times 3$

$(5\text{-}7)\times 3 = -2 \times 3 = -6$

$5 \times 3 - 7 \times 3 = 15 - 21 = -6$

Chapter 4. B. b. Expanding brackets, cross multiplication property

1. $(3 + x)(x - 1)$ is equal with $x^2 + 2x - 3$

$(3 + x)(x - 1) = 3x - 3 + x^2 - x = x^2 + 2x - 3$

2. $(-5 + 3x)(x + 6)$ is equal with $3x^2 + 13x - 30$

$(-5 + 3x)(x + 6) = -5x - 30 + 3x^2 + 18x = 3x^2 + 13x - 30$

3. $(2x^2 - 3x)(4x + 5)$ is equal with $8x^3 - 2x^2 - 15x$

$(2x^2 - 3x)(4x + 5) = 8x^3 + 10x^2 - 12x^2 - 15x = 8x^3 - 2x^2 - 15x$

4. $(-3x^2 + 4x - 5)(2x^3 + 7)$ is equal with $-6x^5 + 8x^4 - 10x^3 - 21x^2 + 28x - 35$

$(-3x^2 + 4x - 5)(2x^3 + 7) = -6x^5 - 21x^2 + 8x^4 + 28x - 10x^3 - 35 = -6x^5 + 8x^4 - 10x^3 - 21x^2 + 28x - 35$

5. $(x^2 - 5)(2x^3 + 7x - 3)$ is equal with $2x^5 - 3x^3 - 3x^2 - 35x + 15$

$(x^2 - 5)(2x^3 + 7x - 3) = 2x^5 + 7x^3 - 3x^2 - 10x^3 - 35x + 15 = 2x^5 - 3x^3 - 3x^2 - 35x + 15$

6. The value of x from $\frac{x}{3} = \frac{7}{5}$ is 4.2

$5x = 21$

$x = \frac{21}{5} = 4.2$

7. The value of x from $\frac{x-2}{4} = \frac{2x}{7}$ is

$\frac{x-2}{4} = \frac{2x}{7}$

$7(x - 2) = 8x$

$7x - 14 = 8x$

$x = -14$

8. The ratio between the width (x+4) and length (2x-6) of a rectangle is ¾. Length is 28

$\frac{x+4}{2x-6} = \frac{3}{4} \, x \neq 3$

$4(x + 4) = 3(2x - 6)$

$4x + 16 = 6x - 18$

$2x = 34$

$x = \frac{34}{2} = 17$

$Lenght = 2 * 17 - 6 = 34 - 6 = 28$

9. The value of x from the expression $\frac{x+7}{x} = \frac{3x-2}{5}$ $(x \neq 0)$ is

$\frac{x+7}{x} = \frac{3x-2}{5}$

$5(x + 7) = x(3x - 2)$

$5x + 35 = 3x^2 - 2x$

$3x^2 - 7x - 35 = 0$

$x = \frac{-(-7)\pm\sqrt{7^2-4(3)(-35)}}{2*3} = \frac{7\pm\sqrt{49+420}}{6} = \frac{7}{6}\pm\frac{\sqrt{469}}{6} = 1.16\pm\frac{21.65}{6}$

$x_1 = 1.16 + 3.6 = 4.76$

$x_2 = 1.16 - 3.6 = -2.44$

10. Area of a triangle with the $base = x + 3$, $height = 2x + 4$ and, the ratio base/height $\frac{3}{5}$ is 30

$\frac{x+3}{2x+5} = \frac{3}{5} \quad x \neq -\frac{5}{2}$

$5(x + 3) = 3(2x + 4)$

$5x + 15 = 6x + 12$

$x = 3$

$base = 3 + 3 = 6$

$height = 2 * 3 + 4 = 10$

$Area = \frac{base*height}{2} = \frac{6*10}{2} = 30$

Chapter 4. C. a.   Logarithms – Definition, rules

1. The expression $\log_3(4 \times 6)$ will become $\log_3 4 + \log_3 6$

$\log_3(4 \times 6) = \log_3 4 + \log_3 6$

2. The expression $\frac{1}{3}\log_6 a + 5\log_6 b - 7\log_6 c$ will become

$\frac{1}{3}\log_6 a + 5\log_6 b - 7\log_6 c = \log_6 a^{\frac{1}{3}} + \log_6 b^5 - \log_6 c^7 = \log_6\left(\frac{a^{\frac{1}{3}}*b^5}{c^7}\right), \quad a,b > 0$

3. The expression $\log_3 5^7$ will become $7 \times \log_3 5$

$\log_3 5^7 = 7 \times \log_3 5$

4. The expression $\log_3 63 + \log_3 5 - \log_3 35$ equals 3

$\log_3 63 + \log_3 5 - \log_3 35 = \log_3\frac{63*5}{35} = \log_3\frac{315}{35} = \log_3 9 = \log_3 3^3 = 3\log_3 3 = 3$

5. The single logarithm of expression $3(\log_5 a + \log_5 b) - 2\log_5 b$ $\log_6 a^{\frac{1}{3}} + \log_6 b^5 -$

$\log_6 c^7 = \log_6\left(\frac{a^{\frac{1}{3}}*b^5}{c^7}\right), \quad a,b > 0$ is

$3(\log_5 a + \log_5 b) - 2\log_5 b = 3\log_5 a + 3\log_5 b - 2\log_5 b = \log_5 a^3 + \log_5 b^3 -$

$\log_5 b^2 = \log_5\frac{a^3*b^3}{b^2} = \log_5 a^3 b$

6. The single logarithm of expression $4\log_3 6 - 3$ is

$4\log_3 6 - 3 = \log_3 6^4 - 3\log_3 3 = \log_3 6^4 - \log_3 3^3 = \log_3\frac{6^4}{3^3} = \log_3\frac{(2*3)^4}{3^3} = \log_3\frac{2^4*3^4}{3^3} =$

$\log_3 2^4 * 3 = \log_3 48$

7. The single logarithm of expression $2\ln x + \ln(3x - 4), x, 3x - 4 > 0$ is

$2\ln x + \ln(3x - 4) = \ln x^2(3x - 4) = \ln(3x^3 - 4x^2)$

8. The expression $\log_3(\frac{a^2}{b^4})$ in terms of $\log_3 a$ and $\log_3 b, a, b > 0$ is

$\log_3\frac{a^2}{b^4} = \log_3 a^2 - \log_3 b^4 = 2\log_3 a - 4\log_3 b$

9. The result of $\log_4 8$ in base 2 is

$$\log_4 8 = \frac{\log_2 8}{\log_2 4} = \frac{\log_2 2^3}{\log_2 2^2} = \frac{3\log_2 2}{2\log_2 2} = \frac{3}{2}$$

10. The result of $\log_{125} 25$ in base 5 is

$$\log_{125} 25 = \frac{\log_5 25}{\log_5 125} = \frac{\log_5 5^2}{\log_5 5^3} = \frac{2\log_5 5}{3\log_5 5} = \frac{2}{3}$$

Chapter 4. C. a.   Logarithms - Exponential and logarithmic equations

1. The solution of the equation $620 = 2 * 3^{x+1}$ is $x = \log_3 310 - 1$

$620 = 2 * 3^{x+1}$

$310 = 3^{x+1}$

$\log_3 310 = \log_3 3^{x+1}$

$\log_3 310 = (x + 1) * \log_3 3$

$\log_3 310 - 1 = x$

2. The solution of the equation $3 = \log_5 x + \log_5(x - 3)$ is $x = 12.78$

$3 = \log_5 x + \log_5(x - 3)\,, x > 3$

$3 * \log_5 5 = \log_5 x(x - 3)$

$\log_5 5^3 = \log_5 x(x - 3)$

$5^3 = x(x - 3)$

$125 = x^2 - 3x$

$x^2 - 3x - 125 = 0$

$x = \frac{-(-3)\pm\sqrt{9-4(1)(-125)}}{2} = \frac{3\pm\sqrt{9+500}}{2} = \frac{3}{2} \pm \frac{\sqrt{509}}{2} = 1.5 \pm \frac{22.56}{2}$

$x_1 = 1.5 + 11.28 = 12.78$

$x_2 = 1.5 - 11.28 = -9.78$ NOT A SOLUTION

3. The solution of the equation $5^{x-4} = 25$ is $x = 6$

$5^{x-4} = 25$

$5^{x-4} = 5^2$

$x - 4 = 2$

$x = 6$

4. The solution of the equation $49^{x-2} = 343^{3x+5}$ is $x = -2.71$

$49^{x-2} = 343^{3x+5}$

$7^{2(x-2)} = 7^{3(3x+5)}$

$2(x - 2) = 3(3x + 5)$

$2x - 4 = 9x + 15$

$7x = -19$

$x = \frac{-19}{7} = -2.71$

5. The solution of the equation $\log_4 36 = \log_4(x + 3) + \log_4(x - 3)\,, x > 0$ is $x = 6.7$

$\log_4 36 = \log_4(x + 3) + \log_4(x - 3)$

$\log_4 36 = \log_4(x + 3)(x - 3)$

$36 = (x + 3)(x - 3)$

$36 = x^2 - 9$

$x^2 - 9 - 36 = 0$

$x^2 - 45 = 0$

$x^2 = 45$

$x = \sqrt{45}$

$x = \pm 6.7$

Only $x = 6.7$ is the solution (is bigger than 3)

6. The solution of the equation $\log_5(3x + 9) - \log_5(x + 3) = \log_5(x - 4)$, $x > 4$ is $x = 7$

$\log_5(3x + 9) - \log_5(x + 3) = \log_5(x - 4)$

$\log_5 \frac{3x+9}{x+3} = \log_5(x - 4)$

$\frac{3(x+3)}{x+3} = x - 4$

$3 = x - 4$

$x = 7$

7. The solution of the equation $3^{x+3} = 7^{2x+5}$ is $x = -0.953$

$3^{x+3} = 7^{2x+5}$

$\ln 3^{x+3} = \ln 7^{2x+5}$

$(x + 3)\ln 3 = (2x + 5)\ln 7$

$x\ln 3 + 3\ln 3 = 2x\ln 7 + 5\ln 7$

$3\ln 3 - 5\ln 7 = 2x\ln 7 - x\ln 3$

$x(2\ln 7 - \ln 3) = 3\ln 3 - 5\ln 7$

$x = \frac{3(1.0986)-5(1.9549.}{2(1.9549)-1.0986} = \frac{3.2958-5.977}{3.9098-1.0986} = \frac{-2.68}{2.811} = -0.953$

8. The solution of the equation $5 * 3^x = 2^{x-2}$ is $x = -7.4$

$5 * 3^x = 2^{x-2}$

$\ln(5 * 3^x) = \ln 2^{x-2}$

$\ln 5 + \ln 3^x = \ln 2^{x-2}$

$\ln 5 + x\ln 3 = (x - 2)\ln 2$

$\ln 5 + x\ln 3 = x\ln 2 - 2\ln 2$

$x\ln 3 - x\ln 2 = -2\ln 2 - \ln 5$

$x(\ln 3 - \ln 2) = -2\ln 2 - \ln 5$

$x = \frac{-2\ln 2-\ln 5}{\ln 3-\ln 2} = \frac{-2(0.693)-1.609}{1.0986-0.693} = \frac{-2.995}{0.4038} = -7.417$

9. The solution of the equation $\log x + \log(x - 3) = \log 5 + \log(x + 5)$ (x>3)

$\log x + \log(x - 3) = \log 5 + \log(x + 5)$

$\log[x(x - 3)] = \log 5(x + 5)$

$x(x - 3) = 5(x + 5)$

$$x^2 - 3x = 5x + 25$$

$$x^2 - 8x - 25 = 0$$

$$x = \frac{8 \pm \sqrt{64 - 4(1)(-25)}}{2} = \frac{8}{2} \pm \frac{\sqrt{164}}{2} = 4 \pm \frac{12.8}{2}$$

$$x_1 = 4 + 6.4 = 10.4$$

$$x_1 = 4 - 6.4 = -2.4 \text{ NOT A SOLUTION}$$

10.

The solution of the equation $\log(x^2 + x - 6) - \log(x + 3) = 1, x > 2$ is $x = 12$

$$\log(x^2 + x - 6) - \log(x + 3) = 1$$

$$\log \frac{x^2 + x - 6}{x + 3} = \log 10$$

$$\frac{x^2 + x - 6}{x + 3} = 10$$

$$\frac{(x - 2)(x + 3)}{x + 3} = 10$$

$$x - 2 = 10$$

$$x = 12$$

Chapter 4. D. a. Polynomials – Definition, operations with polynomials

1. The simplified polynomial of $(2x^2 - 3x - 5) + (-4x + 3)$ is

$$(2x^2 - 3x - 5) + (-4x + 3) = 2x^2 - 3x - 4x - 5 + 3 = 2x^2 - 7x - 2$$

2. The simplified polynomial of $(-4x^2 + 2x) - (-3x^2 - 5x + 7)$ is

$$(-4x^2 + 2x) - (-3x^2 - 5x + 7) = -4x^2 + 2x + 3x^2 + 5x - 7 = -x^2 + 7x - 7$$

3. The perimeter of a rectangle with Length $=2x - 3$ and Width $= 3x + 7$ is

$$Perimeter = 2x - 3 + 3x + 7 + 2x - 3 + 3x + 7 = 10x + 8$$

4. The value of the above perimeter for $x = 5\,m$ is

$$Perimeter = 10x + 8 = 10(5) + 8 = 58\,m$$

5. The area of a triangle with the base $=5x + 4$ and the height $= 3x + 1$ is

$$Area = \frac{base * height}{2} = \frac{(5x+4)(3x+1)}{2} = \frac{1}{2}(15x^2 + 5x + 12x + 4) = \frac{1}{2}(15x^2 + 17x + 4)$$

6. The area of the figure shown below is

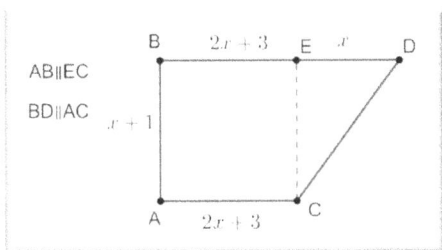

AB∥EC
BD∥AC

$$Area = Area_{ABEC} + Area_{CDE} = (x + 1)(2x + 3) +$$

$$\frac{1}{2}(x + 1)(x) = 2x^2 + 3x + 2x + 3 + 0.5x^2 + 0.5x =$$

$$2.5x^2 + 5.5x + 3$$

7. The result of $x^2 + 3x + 4$ divided by $x + 1$ is $x + 2 + \frac{2}{x+1}$

$$x \neq -1$$

$$
\begin{array}{r}
x + 2 \\
x + 1)\overline{x^2 + 3x + 4} \\
\underline{-(x^2 + x)} \\
= \quad 2x + 4 \\
\underline{-(2x + 2)} \\
===+2
\end{array}
$$

8. The result of $2x^2 - 5x + 6$ divided by $x - 2$ is $2x - 1 \; remainder \frac{4}{x-2} \; x \neq 2$

$$
\begin{array}{r}
2x - 1 \\
x - 2)\overline{2x^2 - 5x + 6} \\
\underline{-(2x^2 - 4x)} \\
= \quad -x + 6 \\
\underline{-(-x + 2)} \\
====+4
\end{array}
$$

9. The result of $6x^3 - 3x - 4$ divided by $3x^2 + x - 4$ is: $2x - \frac{2}{3} \; reminder \frac{5\frac{2}{3}x - 6\frac{2}{3}}{3x^2 + x - 4} \; , 3x^2 + x - 4 \neq 0$

$$
\begin{array}{r}
2x - \frac{2}{3} \\
3x^2 + x - 4)\overline{6x^3 \qquad - 3x - 4} \\
\underline{-(6x^3 + 2x^2 - 8x)} \\
= \quad -2x^2 + 5x - 4 \\
\underline{-(-2x^2 - \frac{2}{3}x + \frac{8}{3})} \\
==== 5\frac{2}{3}x - 6\frac{2}{3}
\end{array}
$$

10. The length of a rectangle with area $= 4x^2 + 2x$ and width $= 2x$ is:

$Area = 4x^2 + 2x = 2x(length)$

$Length = \frac{Area}{width} = \frac{4x^2 + 2x}{2x} = 2x + 1, x \neq 0$

Chapter 4. D. b.   Polynomials – Rationalizing the denominator, special binomial products

1. The rationalized expression of $\frac{2}{\sqrt{7}}$ is

$\frac{2}{\sqrt{7}} = \frac{2*\sqrt{7}}{\sqrt{7}*\sqrt{7}} = \frac{2*\sqrt{7}}{7}$

2. The rationalized expression of $\frac{2\sqrt{3}}{\sqrt{11}}$ is

$\frac{2\sqrt{3}}{\sqrt{11}} = \frac{2\sqrt{3}*\sqrt{11}}{\sqrt{11}*\sqrt{11}} = \frac{2\sqrt{33}}{11}$

3. The rationalized expression of $\frac{\sqrt{7}+3\sqrt{5}}{\sqrt{3}}$ s

$$\frac{\sqrt{7}+3\sqrt{5}}{\sqrt{3}} = \frac{(\sqrt{7}+3\sqrt{5})*\sqrt{3}}{\sqrt{3}*\sqrt{3}} = \frac{\sqrt{7}*\sqrt{3}+3\sqrt{5}*\sqrt{3}}{3} = \frac{\sqrt{21}+3\sqrt{15}}{3}$$

4. The rationalized expression of $\frac{\sqrt{5}}{\sqrt{3}+\sqrt{2}}$ is

$$\frac{\sqrt{5}}{\sqrt{3}+\sqrt{2}} = \frac{\sqrt{5}*(\sqrt{3}-\sqrt{2})}{(\sqrt{3}+\sqrt{2})(\sqrt{3}-\sqrt{2})} = \frac{\sqrt{5}*\sqrt{3}-\sqrt{5}*\sqrt{2}}{3-2} = \sqrt{15}-\sqrt{10}$$

5. The area of a rectangle is $\sqrt{7}+2\sqrt{5}$ and length $\sqrt{3}-1$. The width is:

$$Area = (\sqrt{7}+2\sqrt{5}) = (\sqrt{3}-1)*width$$

$$width = \frac{Area}{length} = \frac{\sqrt{7}+2\sqrt{5}}{\sqrt{3}-1} = \frac{(\sqrt{7}+2\sqrt{5})*(\sqrt{3}+1)}{(\sqrt{3}-1)*(\sqrt{3}+1)} = \frac{\sqrt{7}*\sqrt{3}+\sqrt{7}+2\sqrt{5}*\sqrt{3}+2\sqrt{5}}{3-1} = \frac{\sqrt{21}+\sqrt{7}+2\sqrt{15}+2\sqrt{5}}{2}$$

6. The area of the rectangle ABDE in the figure shown below is

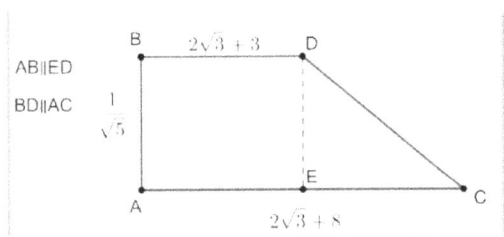

$$Area = base * height = \frac{2\sqrt{3}+3}{\sqrt{5}} = \frac{(2\sqrt{3}+3)\sqrt{5}}{\sqrt{5}*\sqrt{5}} = $$

$$\frac{2\sqrt{15}+3\sqrt{5}}{5}$$

7. The area of the right-angle triangle CDE in the figure shown above is

$$Area = \frac{base*height}{2} = \frac{5}{2\sqrt{5}} = \frac{5\sqrt{5}}{2\sqrt{5}*\sqrt{5}} = \frac{5\sqrt{5}}{2*5} = \frac{\sqrt{5}}{2}$$

8. The rationalized expression of $\frac{7}{\sqrt{3}+\sqrt{2}}$ is

$$\frac{7}{\sqrt{3}+\sqrt{2}} = \frac{7*(\sqrt{3}-\sqrt{2})}{(\sqrt{3}+\sqrt{2})(\sqrt{3}-\sqrt{2})} = \frac{7*(\sqrt{3}-\sqrt{2})}{3-2} = 7\sqrt{3}-7\sqrt{2}$$

9. The rationalized expression of $\frac{8-2\sqrt{15}}{\sqrt{5}-\sqrt{3}}$ is

$$\frac{8-2\sqrt{15}}{\sqrt{5}-\sqrt{3}} = \frac{5-2\sqrt{15}+3}{\sqrt{5}-\sqrt{3}} = \frac{(\sqrt{5})^2-2\sqrt{5}*\sqrt{3}+(\sqrt{3})^2}{\sqrt{5}-\sqrt{3}} = \frac{(\sqrt{5}-\sqrt{3})^2}{\sqrt{5}-\sqrt{3}} = \sqrt{5}-\sqrt{3}$$

10. The expression $\frac{15\sqrt{2}}{\sqrt{10}} - \frac{5\sqrt{25}}{\sqrt{125}}$ can be written as $k*\sqrt{5}$. The value of k is 2

$$\frac{15\sqrt{2}}{\sqrt{10}} - \frac{5\sqrt{25}}{\sqrt{125}} = \frac{15}{\sqrt{\frac{10}{2}}} - \frac{5}{\sqrt{\frac{125}{25}}} = \frac{15}{\sqrt{5}} - \frac{5}{\sqrt{5}} = \frac{10}{\sqrt{5}} = \frac{10\sqrt{5}}{\sqrt{5}*\sqrt{5}} = \frac{10\sqrt{5}}{5} = 2\sqrt{5}$$

Chapter 4. E. a.   Linear equations – Straight line graphs

1. The slope of the line $y = 3x - 1$ is

2. The y intercept of the line $2x + 3y = 5$ is $b = \frac{5}{3}$

$$2x + 3y = 5$$

$$3y = -2x + 5$$

$$y = -\frac{2}{3}x + \frac{5}{3}$$

3. The slope of the line $-3x + 5y = -1$ is $m = \frac{3}{5}$

$$5y = 3x - 1$$

$$y = \frac{3}{5}x - \frac{1}{5}$$

4. The y intercept of the line $2(x - 5) + 3(y + 2) = 2$ is $b = 2$

$2(x - 5) + 3(y + 2) = 2$

$2x - 10 + 3y + 6 = 2$

$2x + 3y - 4 = 2$

$3y = -2x + 6$

$y = -\frac{2}{3}x + 2$

5. The slope of the line $-3(x + 2) - 4(y - 7) = 6$ is $m = -\frac{3}{4}$

$-3(x + 2) - 4(y - 7) = 6$

$-3x - 6 - 4y + 28 = 6$

$-3x - 4y + 22 = 6$

$-4y = 3x - 16$

$4y = -3x + 16$

$y = -\frac{3}{4}x + 4$

The next problems use the figure shown below

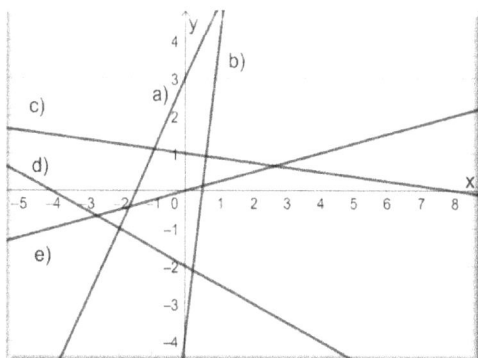

6. The line a) has the equation

Y=2x+3

7. The line b) has the equation

Y= 8x-4

8. The line c) has the equation

$Y = -\frac{1}{8}x + 1$

9. The line d) has the equation

$y = -\frac{1}{2}x - 2$

10. The line e) has the equation

$$y = \frac{1}{4}x$$

Chapter 4. F. a.   Factoring used in quadratic equations

1. The factored expression of $x^2 + 2x - 15$ is $(x - 3)(x + 5)$

$x^2 + 2x - 15$

The product of -3 and 5 is -15

The sum 5+(-3) is 2

2. The factored expression of $x^2 - xy - 6y^2$ is $(x + 2y)(x - 3y)$

$x^2 - xy - 6y^2$

The product is -6 or $2 \times (-3)$

The sum is -1 or -3+2 = -1

We multiply the second term with y, so we have

$(x + 2y)(x - 3y)$

3. The factored expression by grouping of $35x^2 + 8x - 3$ is $(5x - 1)(7x + 3)$

$35x^2 + 8x - 3 = 35x^2 - 7x + 15x - 3 = 7x(5x - 1) + 3(5x - 1) = (5x - 1)(7x + 3)$

4. The factored by decomposition of $12x^2 + 11x - 5$ is $(3x - 1)(4x + 5)$

$12x^2 + 11x - 5$

The product is $-60 = -6 \times 10 = -2 \times 3 \times 2 \times 5 = -4 \times 15$

The sum is $11 = 15\text{-}4$

So, we write $11x = 15x - 4x$

$12x^2 + 11x - 5 = 12x^2 - 4x + 15x - 5 = 4x(3x - 1) + 5(3x - 1) = (3x - 1)(4x + 5)$

5. The factored expression of $x^4 + 3x^2 + 2$ is $(x^2 + 1)(x^2 + 2)$

$x^4 + 3x^2 + 2$

Here, $x^2 = A$ so,

$A^2 + 3A + 2 = (A + 1)(A + 2)$ so,

$(x^2 + 1)(x^2 + 2)$

6. The factored expression of $16x^4 - 1$ is

$16x^4 - 1 = (4^2x^2)^2 - 1 - (4^2x^2 - 1)(4^2x^2 + 1) = (4x - 1)(4x + 1)(4^2x^2 + 1)$

7. The solution of $x^2 + 2x - 3 = 0$ by factoring is

$x^2 + 2x - 3 = 0$

$(x - 1)(x + 3) = 0$

So, is either x=1 or x=-3

8. The solution of $6x^2 - x - 12 = 0$ by factoring is

$6x^2 - x - 12 = 0$

$6 \times (-12) = -72 = -2 \times 3 \times 2 \times 2 \times 3 = -9 \times 8$

Sum is -9+8=-1

$6x^2 - x - 12 = 6x^2 - 9x + 8x - 12 = 3x(2x - 3) + 4(2x - 3) = (2x - 3)(3x + 4)$

$(2x - 3)(3x + 4) = 0$

$2x - 3 = 0$ so, $x = \frac{3}{2} = 1.5$

Or

$3x + 4 = 0$ so, $x = \frac{-4}{3} = -1.3$

9. The solution of $3x^4 - 3x^2 - 60 = 0$ is

$3x^4 - 3x^2 - 60 = 0$

$x^2 = A$

$3A^2 - 3A - 60 = 3(A^2 - A - 20) = 0$

$A^2 - A - 20 = 0$

$4 \times (-5) = 20, and\ 4 - 5 = -1$

$(A + 4)(A - 5) = 0$

Or

$(x^2 + 4)(x^2 - 5) = 0$

So, only $(x^2 - 5) = 0$ will have real solutions

$x^2 = 5$ then, $x = \mp\sqrt{5}$

10. The solution of $(x^4 - 625) = 0$ is

$x^2 = A$

So,

$(A^2 - 625) = (A - 25)(A + 25) = 0$

$(x^2 - 25)(x^2 + 25) = 0$

$(x - 5)(x + 5)(x^2 + 25) = 0$

The only real solutions are $x = \pm5$

Chapter 4. F. b.   Completing the square

1. The number to be added to $x^2 + 2x$ to make a perfect square is 1

2. The number to be added to $x^2 + 5x$ to make a perfect square is $\frac{25}{4}$

$x^2 + 5x = x^2 + 2 \times \frac{5}{2}x + (\frac{5}{2})^2$

3. The perfect square of $x^2 - 3x$ is $(x - \frac{3}{2})^2$

$x^2 - 3x = x^2 - 2 \times \frac{3}{2}x + (\frac{3}{2})^2 - (\frac{3}{2})^2 = (x - \frac{3}{2})^2 - (\frac{3}{2})^2$

4. The term outside the square of expression $x^2 - 7x + 6$ is $-\frac{25}{4}$

$x^2 - 7x + 6 = x^2 - 2 \times \frac{7}{2}x + \left(\frac{7}{2}\right)^2 - \left(\frac{7}{2}\right)^2 + 6 = (x - \frac{7}{2})^2 - \frac{49}{4} + 6 = (x - \frac{7}{2})^2 + \frac{-49+24}{4} = (x - \frac{7}{2})^2 - \frac{25}{4}$

The constant term is $\frac{25}{4} = 6.25$

5. The term outside the square of expression $5x^2 - 15x + 3$ is $-\frac{33}{4} = -8.25$

$5x^2 - 15x + 3 = 5(x^2 - 3x) + 3 = 5\left[x^2 - 2 \times \frac{3}{2}x + \left(\frac{3}{2}\right)^2 - \left(\frac{3}{2}\right)^2\right] + 3 = 5\left[x^2 - 2 \times \frac{3}{2}x + \left(\frac{3}{2}\right)^2\right] - 5 \times \left(\frac{3}{2}\right)^2 + 3 = 5(x - \frac{3}{2})^2 - 5 \times \frac{9}{4} + 3$

The constant term is $-5 \times \frac{9}{4} + 3 = -\frac{45}{4} + 3 = \frac{-45+12}{4} = -\frac{33}{4} = -8.25$

6. The term outside the square of expression $x^2 - 5x + a$ is $-9 + a$

$x^2 - 6x + a = x^2 - 2 \times \frac{6}{2}x + (\frac{6}{2})^2 - (\frac{6}{2})^2 + a = (x - 3)^2 - 9 + a$

So, the term outside the square is -9+a

7. The completed square form of $x^2 + 6x - 4$ is $(x + 3)^2$

$x^2 + 6x - 4 = x^2 + 2 \times 3x + 9 - 9 - 4 = (x + 3)^2 - 13$

8. The completed square form of $x^2 + bx - 4$ is $\left(x + \frac{b}{2}\right)^2 - (\frac{b^2+4}{4})$

$$x^2 + bx - 4 = x^2 + 2 \times \frac{b}{2}x + \left(\frac{b}{2}\right)^2 - \left(\frac{b}{2}\right)^2 - 4 = \left(x + \frac{b}{2}\right)^2 - \frac{b^2}{4} - 4 = \left(x + \frac{b}{2}\right)^2 - \left(\frac{b^2+4}{4}\right)$$

9. The completed square form of $ax^2 + 2x - 4, a \neq 0$ is $a\left(x + \frac{1}{a}\right)^2 - \left(\frac{1+4a}{a}\right)$

$$ax^2 + 2x - 4 = a\left(x^2 + 2 \times \frac{1}{a}x + \left(\frac{1}{a}\right)^2 - \left(\frac{1}{a}\right)^2\right) - 4 = a\left(x + \frac{1}{a}\right)^2 - \frac{1}{a} - 4 = a\left(x + \frac{1}{a}\right)^2 - \left(\frac{1+4a}{a}\right)$$

10. To form a perfect square k in $x^2 + kx + 3$, is $k = -6.25$

$$x^2 + 5x + k = x^2 + 2 \times \frac{5}{2}x + \left(\frac{5}{2}\right)^2 - \left(\frac{5}{2}\right)^2 + k = \left(x + \frac{5}{2}\right)^2 - \frac{25}{4} + k$$

So,

$$\frac{25}{4} + k = 0$$

CHAPTER 5

Chapter 5. A. a. Determine: if a relation is a function, the values of a function, the range

1. For each value of the domain there is only one value that belongs to the range. For domain value of 3, there are two values that belong to the range: 3, and **12**. (3,3) and (3,12).

2. For each value of the domain there is only one value that belongs to the range.

3. For each value of the domain there is only one value that belongs to the range. For domain value of -4, there are two values that belong to the range: 8, and **1**. (-4,**8**. and (-4,1)

4. For each value of the domain there is only one value that belongs to the range.

5. For x=2 ; $f(2) = 5(2) + 3 = 10 + 3 = 13$

6. For x=1 ; $f(1) = 3(1^2) - 3(1) + 2 = 3(1) - 3 + 2 = 3 - 3 + 2 = 2$

7. For s=2 ; $G(2) = \frac{2^2 - 3(2) + 7}{2+3} = \frac{4-6+7}{5} = \frac{5}{5} = 1$

8. For t=-1; $G(-1) = 3 - 2(-1) = 3 + 2 = 5$

For t=-2 ; $G(-2) = 3 - 2(-2) = 3 + 4 = 7$

For t=3 ; $G(3) = 3 - 2(3) = 3 - 6 = -3$

9. For x=-2 ; $F(-2) = (-2)^2 - 5(-2) + 1 = 4 + 10 + 1 = 15$

For x=0 ; $F(0) = (0)^2 - 5(0) + 1 = 0 - 0 + 1 = 1$

For x=3 ; $F(3) = (3)^2 - 5(3) + 1 = 9 - 15 + 1 = 7 \ not - 4$

10. For c=-3 ; $H(-3) = \frac{(-3)^2 - 2(-3)}{-3+2} = \frac{9+6}{-1} = -15$

For c=0 ; $H(0) = \frac{(0)^2 - 2(0)}{0+2} = \frac{0}{2} = 0$

For c=3 ; $H(3) = \frac{(3)^2 - 2(3)}{3+2} = \frac{9-6}{5} = \frac{3}{5}$

Chapter 5. A. b. Linear and quadratic functions and their graphs

1. The difference between each consecutive value belonging to the domain should be always the same. The difference between x=9 and x=7 is two not one as it should be. The difference between each consecutive value belonging to the range should be always the same. The difference between y=7 and y=5 is two, the difference between y=8 and y=7 is one, the difference between y=10 and y=8 is two again.

2. The difference between each consecutive value belonging to the domain and range are be always the same, not with each other.

3. The difference between each consecutive value belonging to the domain should be always the same. The difference between x=2 and x=2 is zero not one as it should be.

4. $f(x) = 2x + 3$ represents the expression of a linear function with slope of 2 and y intercept at y=3

5. The degree of the function is 2 not one. This expression represents a quadratic function.

6. $f(x) = 34x + 15$ represents the expression of a linear function with slope of 34 and y intercept at y=15

7. The graph is a straight line. It represents a linear function.

8. The graph is not a straight line. It represents a quadratic function.

9. The graph is not a straight line.

10. The expression represents a quadratic function.

11. The expression represents a quadratic function.

12. The expression doesn't represent a quadratic function. The degree of the function is 4 not 2 as it should be.

Chapter 5. A. c. Inverse functions and their graphs

1. $y = 2x + 3$ so, we interchange x with y and have:

$x = 2y + 3$; minus 3 each side

$x - 3 = 2y$; divide with 2 both sides

$\dfrac{x - 3}{2} = y$; so $f^{-1}(x) = \dfrac{x - 3}{2}$

2. $y = 3x - 5$ so, we interchange x with y and have:

$x = 3y - 5$; + 5 each side

$x + 5 = 3y$; divide with 3 both sides

$\dfrac{x + 5}{3} = y$; so $f^{-1}(x) = \dfrac{x + 5}{3}$

3. $y = x^2 - 1$, so, we interchange x with y and have

$x = y^2 - 1$; plus 1 each side

$x + 1 = y^2$; square root of both sides

$\sqrt{x + 1} = y$: $x \geq -1$; so $f^{-1}(x) = \sqrt{x + 1}$

4. $y = \frac{3}{2x+4}$; so, we interchange x with y and have

$x = \frac{3}{2y+4}$; cross multiply and have

$x(2y + 4) = 3$; use the distributivity property and have

$2xy + 4x = 3$; subtract 4x from both sides

$2xy = 3 - 4x$; divide both sides with 2x

$y = \frac{3 - 4x}{2x}$ so $f^{-1}(x) = \frac{3 - 4x}{2x}$

5. $y = \frac{5}{2x^2+4}$; so, we interchange x with y and have

$x = \frac{5}{2y^2+4}$; cross multiply and have

$x(2y^2 + 4) = 5$; use the distributivity property and have

$2xy^2 + 4x = 5$; subtract 4x from both sides

$2xy^2 = 5 - 4x$; divide both sides with 2x

$y^2 = \frac{5-4x}{2x}$; square root both sides

$y = \mp\sqrt{\frac{5-4x}{2x}}$; so $f^{-1}(x) = \mp\sqrt{\frac{5-4x}{2x}}$; $x > 0, x \leq 5/4$

6. $y = \frac{\sqrt{x-1}}{3}$; so, we interchange x with y and have

$x = \frac{\sqrt{y-1}}{3}$; multiply with 3 both sides

$3x = \sqrt{y - 1}$ square both sides

$9x^2 = y - 1$; add 1 in both sides

$9x^2 + 1 = y$ so $f^{-1}(x) = 9x^2 + 1$

7. $y = \frac{3x}{x+3}$; so, we interchange x with y and have

$x = \frac{3y}{y+3}$; cross multiply and have

$x(y + 3) = 3y$; use the distributivity property and have

$xy + 3x = 3y$; subtract xy in each side

$3x = 3y - xy$ Take y as common factor

$3x = y(3 - x)$; divide with (3-x) in both sides

$\frac{3x}{3-x} = y$ so $f^{-1}(x) = \frac{3x}{3-x}$

8. $y = \frac{1}{5x+3}$; so, we interchange x with y and have

$x = \frac{1}{5y+3}$; cross multiply and have

$x(5y + 3) = 1$; use the distributivity property and have

$5xy + 3x = 1$; subtract 3x in both sides

$5xy = 1 - 3x$; divide by 5x both sides

$y = \frac{1-3x}{5x}$; so $f^{-1}(x) = \frac{1-3x}{5x}$,

Graph the following functions and their inverse functions on the same graph.

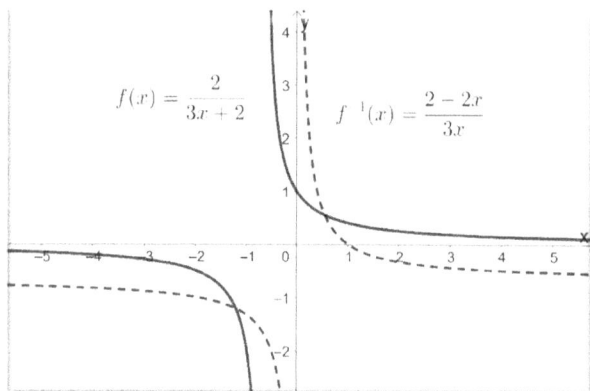

9. $f(x) = y = \frac{2}{3x+2}$ the inverse is:

$f^{-1}(x) = \frac{2-2x}{3x}$

10. $f(x) = y = \frac{2x-1}{x+2}$ the inverse is: $f^{-1}(x) = \frac{-2x-1}{x-2}$

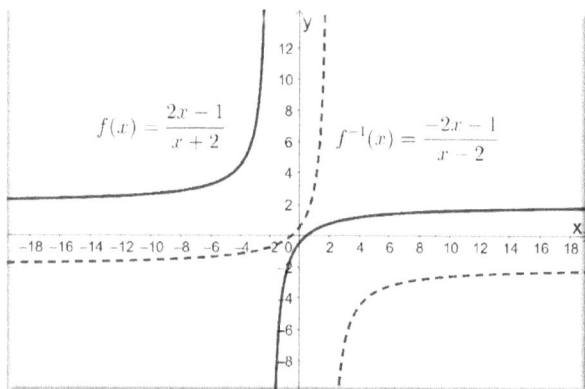

Chapter 5. B.   Piecewise functions

1. $f(x) = 2x^2 + 5x - 3$, $x \in R$ is a parabola.

2. $f(x) = \begin{cases} 2x \ for \ x < 0 \\ 3 \ for \ x \geq 0 \end{cases}$ is defined by multiple sub-functions. Each of these sub-functions is

spread over to a certain interval or part of the function's domain.

3. $f(x) = 4x + 3, x \in R$ is a straight line.

4. $f(x) = 7x - 3, x \in R$ is a straight line.

5. $f(x) = \begin{cases} 2x^4 \ for \ x < -2 \\ 3x + 6 \ for \ x \geq -2 \end{cases}$ is defined by multiple sub-functions. Each of these sub-functions is

spread over to a certain interval or part of the function's domain.

6. $f(x) = 3\sin(x - 2) + 3, x \in R$ is a periodic continuous function.

7. $f(x) = \begin{cases} x^2 \ for \ x \leq -2 \\ 4 \ for -2 < x < 2 \\ 2x \ for \ x \geq 2 \end{cases}$ is defined by multiple sub-functions. Each of these sub-functions is

spread over to a certain interval or part of the function's domain.

8. $f(x) = \sqrt{x-1}$ is a continuous function for x>1

Graph function 5 and 7.

Function 5

$f(x) = \begin{cases} 075x^4 \ for \ x < 2 \\ 3x + 6 \ for \ x \geq 2 \end{cases}$

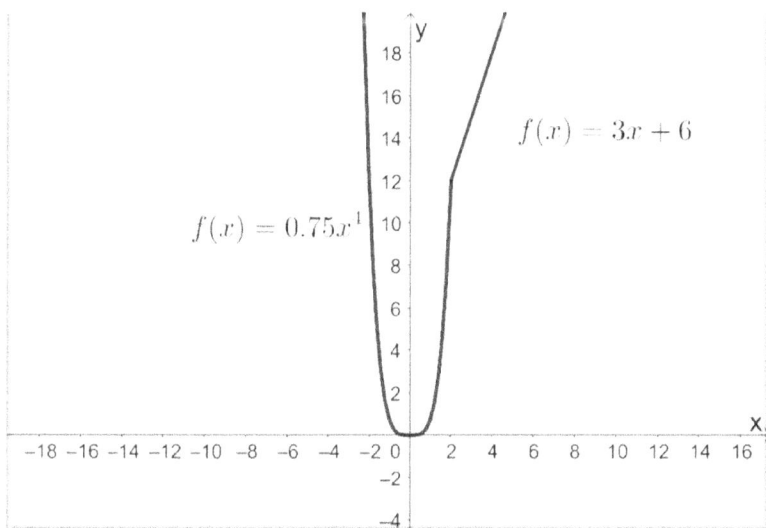

Function 7

$f(x) = \begin{cases} x^2 \ for \ x \leq -2 \\ 4 \ for -2 < x < 2 \\ 2x \ for \ x \geq 2 \end{cases}$

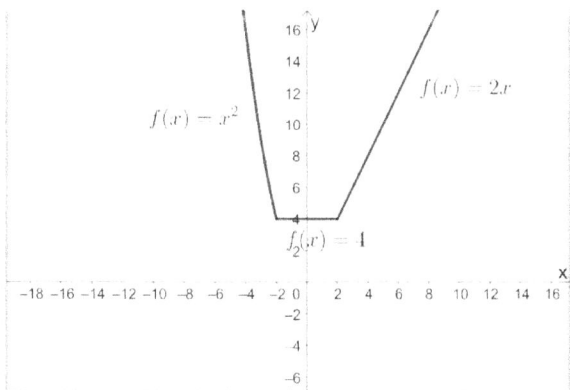

Chapter 5. C.   Trigonometric functions

1. The length of the part of a trigonometric function that repeats, measured along the x axis is called period.

2. One radian is the measure of an angle subtended at the center of a circle by an arc which is equal in length to the radius of the circle.

3. $\pi\ radians = 180^0$

4. In Quadrant 1 when α is between zero and 90 degrees or $\frac{\pi}{2}$, sin (α) is positive. See below.

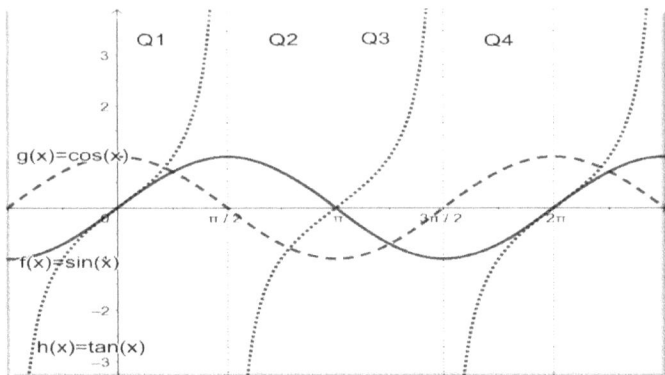

5. In Quadrant 3 where when α is between 180^0 or $\pi\ rad$ and 270^0 or $\frac{3\pi}{2}\ rad$, tan (α) is positive. See below.

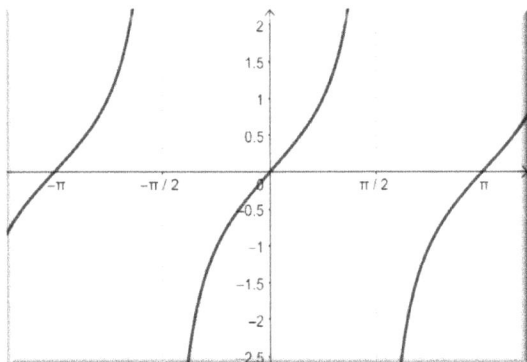

6. If $\cos(\alpha_1) = k, k \geq 0$ the other value of α that is solution of the equation is: $\alpha_2 = 2\pi - \alpha_1$ (radians). See below.

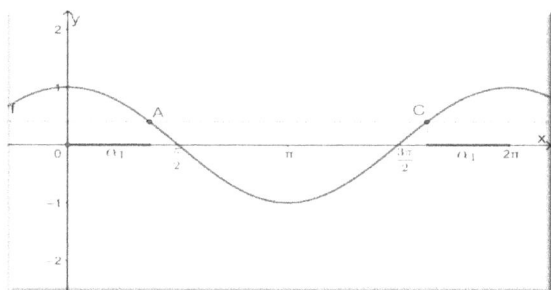

7. In a right-angle triangle $\cos(\alpha) = \frac{adjacent}{hypotenuse}$

8. In a right-angle triangle $\tan(\alpha) = \frac{opposite}{adjacent}$

9. $\sin\left(\frac{\pi}{2}\right) = 1$.

10. $\cos\left(\frac{\pi}{4}\right) = \frac{\sqrt{2}}{2} = \frac{1.41}{2} = 0.707 \; not \; 1$

The values for the special angles in a right-angle triangle are given below.

	30^0	60^0	45^0
Sin(Φ)	$\frac{1}{2}$	$\frac{\sqrt{3}}{2}$	$\frac{\sqrt{2}}{2}$
Cos(Φ)	$\frac{\sqrt{3}}{2}$	$\frac{1}{2}$	$\frac{\sqrt{2}}{2}$
Tan(Φ)	$\frac{\sqrt{3}}{3}$	$\sqrt{3}$	1

Chapter 5. D. a.   Graphing sine and cosine functions

1. For each of values of α, the values of $\sin(\alpha)$ are:

α	0	$\frac{\pi}{6} = 30^0$	$\frac{\pi}{2} = 90^0$	$\frac{\pi}{3} = 60^0$	π
$\sin(\alpha)$	0	$\frac{1}{2}$	1	$\frac{\sqrt{3}}{2}$	0

2. The graph of $\sin(\alpha)$ is:

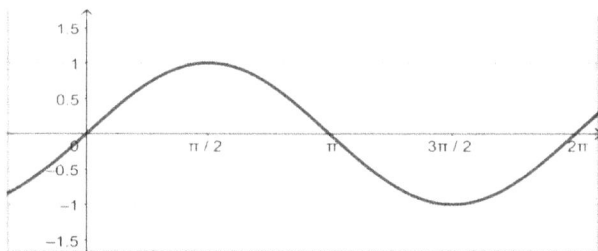

3. The minimum of $\cos(\alpha)$ is -1

4. The maximum of $\sin(\alpha)$ is +1

5. The graph of $\cos(\alpha)$ is:

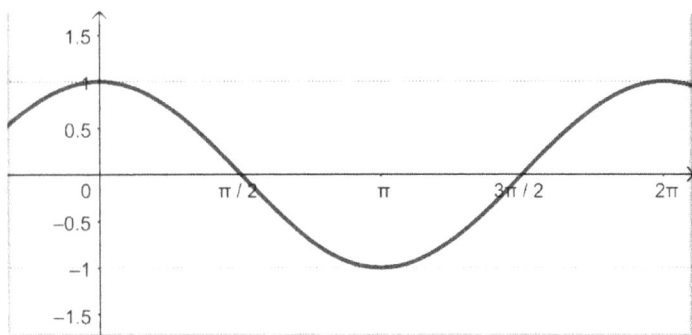

6. For each of values of α, the values of $\cos(\alpha)$ are:

α	0	$\frac{\pi}{6} = 30^0$	$\frac{\pi}{2}$	$\frac{\pi}{3} = 60^0$	π
$\cos(\alpha)$	1	$\frac{\sqrt{3}}{2}$	0	$\frac{1}{2}$	-1

Chapter 5. D. b. Graphing tangent and cotangent functions

1. The tangent function is defined for angle 180^0 ($= 0$) and *undefined for* 270^0

2. The tangent graph looks like the one below for $-\frac{\pi}{2} < \alpha < \frac{\pi}{2}$

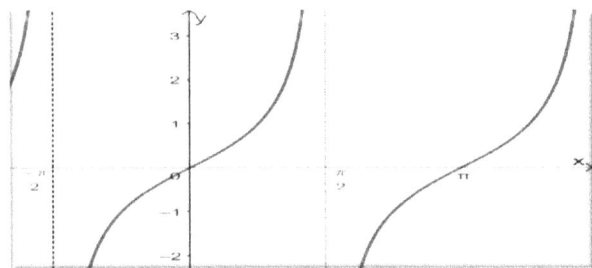

3. The cotangent function is zero for $\frac{\pi}{2} + k\pi, k\ integer$

4. The tangent function is undefined for $\frac{\pi}{2} + k\pi, k\ integer$.

5. For 60^0 the tangent is $\sqrt{3}$

6. Tangent of 45^0 is 1

7. Cotangent of 30^0 is $\sqrt{3}$

8. The tangent function does not have an amplitude.

9. Cotangent of 45^0 is 1

10. The formula of tangent in terms of sine and cosine is $\tan(\alpha) = \frac{\sin(\alpha)}{\cos(\alpha)}$

11. Tangent of 30^0 is $\frac{\sqrt{3}}{3}$

12. Cotangent of 60^0 is $\frac{\sqrt{3}}{3}$

Chapter 5. E. Inverse Trigonometric Functions

1. Inverse trig functions do the opposite of the "regular" trig functions.

2. The angle α in right angle triangle PSQ is: 53.2^0

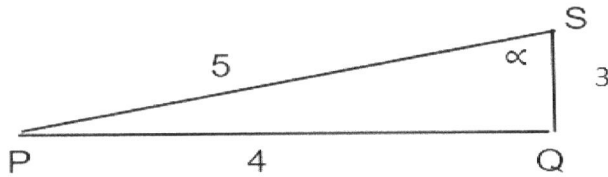

$$\sin \alpha = \frac{4}{5} = 0.8 \; ; so \; \alpha = \sin^{-1}(0.8) = 53.2^0$$

3. The angle α in right angle triangle PSQ is: 36.8^0

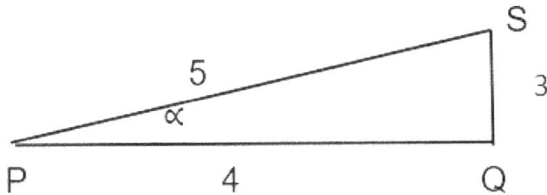

$$\cos \alpha = \frac{4}{5} = 0.8 \; ; so \; \alpha = \cos^{-1}(0.8) = 36.8^0$$

4. $Tan^{-1}(\sqrt{3})$ is 60^0

5. The domain of $f(x) = \sin^{-1}(x)$ is [-1,1]

6. The range (principal value) of $f(x) = \cos^{-1}(x)$ is all real numbers.

7. The domain of $f(x) = \tan^{-1}(x)$ is all real numbers.

8. The angle α in the figure bellow is 29.7^0

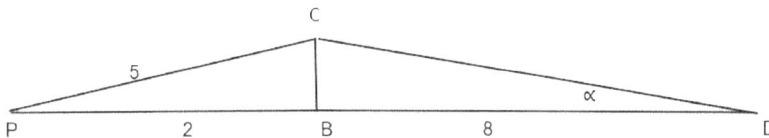

Step 1

In the right-angle triangle PBC we are using the Pythagorean theorem. So, we have:

$$CB^2 = CP^2 - PB^2 = 5^2 - 2^2 = 25 - 4 = 21 \; ; so \; CB = \sqrt{21} = 4.58$$

Step 2

In the right-angle triangle CBD $\tan \alpha = \frac{4.58}{8} = 0.57 \; ; so \; \tan^{-1}(0.57) = 29.7^0$

9. The $\cos^{-1}\left(\frac{\sqrt{2}}{2}\right)$ is 45^0

10. The $\sin^{-1}(0.3)$ is 17.45^0

Chapter 5. F.   Graphs of Inverse Trigonometric Functions

1. The domain of $f(x) = \cos^{-1}(x)$ is [-1,1]

2. Graph the function $f(x) = \cos^{-1}(x)$

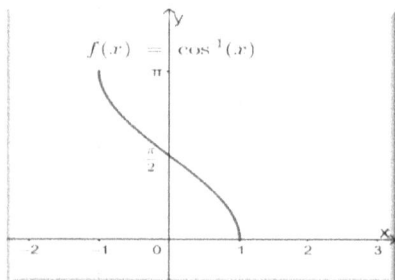

3. Graph the function $f(x) = sin^{-1}(x)$

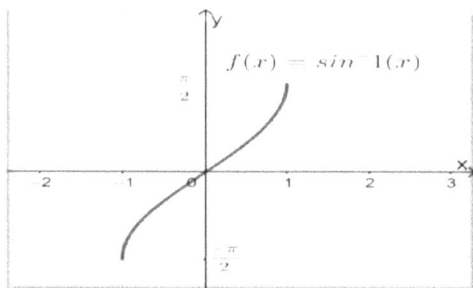

4. The value of $f(1) = cos^{-1}(1)$ is 0

5. The graph of $f(x) = cos^{-1}(x)$ looks like the one in problem 2 above.

6. The value of $f(0) = cos^{-1}(0)$ is $\frac{\pi}{2}$

CHAPTER 6

Chapter 6. A. a.   Limits – Table of values, and graphically

1. We can show limits through graphs or tables of values.

2. If x=2 doesn't belong to the domain, the function could have limit for f(2).

3. The table below represent the value of $f(x) = 0.3x^2 - 2$ around -0.8 for x around 2

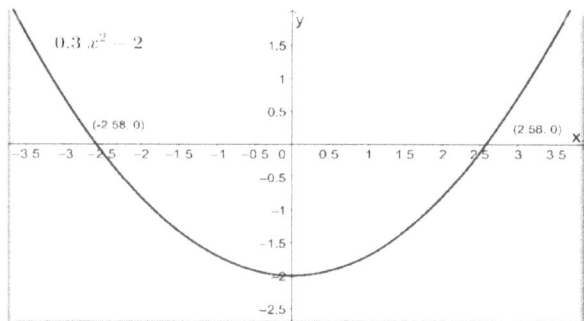

X	$f(x) = 0.3x^2 - 2$
1.8	-1.028
1.9	-0.917
1.999	-0.8012
2.001	-0.7988
2.01	-0.7879
2.1	-0.677

4. As x is approaching 5, f(x) is approaching 28

X	$f(x) = x^2 + 3$
4.8	26.04
4.99	27.9
4.999	27.99
5.001	28.01
5.01	28.1
5.1	29.1

5. The value of $f(x) = x^2 + 3$ for x around 5 is not around 7.**6.**
When x approaches 5 the value of the function goes towards 28.

6. The table below represents the value of $f(x) = x^2 + 2x$ around 8 for x around 2.

X	$f(x) = x^2 + 2x$
1.89	7.35
1.999	7.98
2.001	8.006
2.01	8.06
2.1	8.61

Chapter 6. A. b.   Limits – Algebraically

1. $\lim\limits_{x\to 0}(\frac{6x^2-3}{x+3}) = \frac{\lim\limits_{x\to 0}(6x^2-3)}{\lim\limits_{x\to 0}(x+3)} = \frac{-3}{3} = -1$

2. $\lim\limits_{x\to 3}(\frac{9x^2-1}{3x-1}) = \lim\limits_{x\to 3}\left[\frac{(3x-1)(3x+1)}{(3x-1)}\right] = \lim\limits_{x\to 3}(3x+1) = 10$

3. $\lim\limits_{x\to 2}\frac{x^2+2x-8}{x-2} = \lim\limits_{x\to 2}\left[\frac{(x-2)(x+4)}{x-2}\right] = \lim\limits_{x\to 2}(x+4) = 6$

4. $\lim\limits_{x\to 3}\left(\frac{1}{x-3} - \frac{6}{x^2-9}\right) = \lim\limits_{x\to 3}\left[\frac{1}{x-3}\left(1 - \frac{6}{x+3}\right)\right] = \lim\limits_{x\to 3}\left[\frac{1}{x-3}(\frac{x+3-6}{x+3})\right] = \lim\limits_{x\to 3}\left[\frac{1}{x-3}\left(\frac{x-3}{x+3}\right)\right] = \lim\limits_{x\to 3}\frac{1}{x+3} = \frac{1}{6}$

5. $\lim\limits_{x\to\infty}\frac{4x+3}{x^2+4x-2} = \lim\limits_{x\to\infty}\left[\frac{x^2(\frac{4}{x}+\frac{3}{x^2})}{x^2(1+\frac{4}{x}-\frac{2}{x^2})}\right] = \frac{\lim\limits_{x\to\infty}(\frac{4}{x})+\lim\limits_{x\to\infty}(\frac{3}{x^2})}{\lim\limits_{x\to\infty}1+\lim\limits_{x\to\infty}(\frac{4}{x})-\lim\limits_{x\to\infty}(\frac{2}{x^2})} = \frac{0+0}{1+0+0} = \frac{0}{1} = 0$

6. $\lim\limits_{x\to\infty}\left(\sqrt{2x+1} - \sqrt{2x}\right) = \lim\limits_{x\to\infty}\left(\sqrt{2x+1} - \sqrt{2x}\right)(\frac{\sqrt{2x+1}+\sqrt{2x}}{\sqrt{2x+1}+\sqrt{2x}}) = \lim\limits_{x\to\infty}\left(\frac{2x+1-2x}{\sqrt{2x+1}+\sqrt{x}}\right) = \lim\limits_{x\to\infty}\frac{1}{\sqrt{2x+1}+\sqrt{x}} = 0$

7. $\lim\limits_{x\to 3}(\frac{16x^2-1}{4x-1}) = \lim\limits_{x\to 3}\left[\frac{(4x-1)(4x+1)}{4x-1}\right] = \lim\limits_{x\to 3}(4x+1) = 13$

8. $\lim\limits_{x\to 1}\left(\frac{x^4-1}{x-1}\right) = \lim\limits_{x\to 1}\left[\frac{(x^2-1)(x^2+1)}{x-1}\right] = \lim\limits_{x\to 1}\left[\frac{(x-1)(x+1)(x^2+1)}{x-1}\right] = \lim\limits_{x\to 1}(x+1)(x^2+1) =$
$\lim\limits_{x\to 1}(x+1) * \lim\limits_{x\to 1}(x^2+1) = 2 * 2 = 4$

9. $\lim\limits_{x\to 1}\left(\frac{x^4-1}{x^2-1}\right) = \lim\limits_{x\to 1}\left[\frac{(x^2-1)(x^2+1)}{(x-1)(x+1)}\right] = \lim\limits_{x\to 1}\left[\frac{(x-1)(x+1)(x^2+1)}{(x-1)(x+1)}\right] = \lim\limits_{x\to 1}(x^2+1) = 2$

10. $\lim\limits_{h\to 0}\frac{f(x+h)-f(x)}{h}$ when $f(x) = 3x^2 + 4x$

$f(x+h) = 3(x+h)^2 + 4(x+h) = 3(x^2+2xh+h^2) + 4x + 4h = 3x^2 + 6xh + 3h^2 + 4x + 4h$

$f(x) = 3x^2 + 4x$

So: $f(x+h) - f(x) = 3x^2 + 6xh + 3h^2 + 4x + 4h - (3x^2 + 4x) = 3x^2 + 6xh + 3h^2 + 4x + 4h -$
$3x^2 - 4x = 6xh + 4h + 3h^2$

Then, $\frac{f(x+h)-f(x)}{h} = \frac{6xh+4h+3h^2}{h} = \frac{h(6x+4+3h)}{h} = 6x + 4 + 3h$

So $\lim\limits_{h\to 0}\frac{f(x+h)-f(x)}{h} = \lim\limits_{h\to 0}(6x + 4 + 3h) = 6x + 4$

Chapter 6. B. Limits-One side versus two sides

In the graph below $g(x) = -(0.5x - 3)^2 + 3 \; for \; x < 5$
$$g(x) = (0.4x - 3)^2 + 1 \; for \; x \geq 5$$

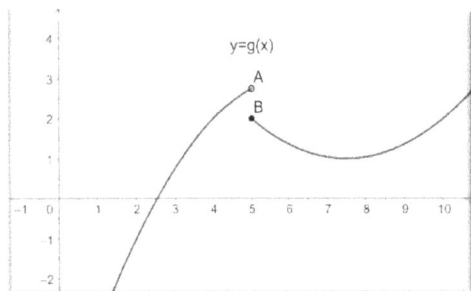

1. $\lim\limits_{x \to 5^-} g(x) = 2.75$

2. $\lim\limits_{x \to 5^+} g(x) = 2$

3. As $\lim\limits_{x \to 5^-} g(x) = 2.75$ and $\lim\limits_{x \to 5^+} g(x) = 2$

$\lim\limits_{x \to 5} g(x) \neq 5$

The limit does not exist because, there are two different values of g(x) when x gets near to 5.

In the graph below

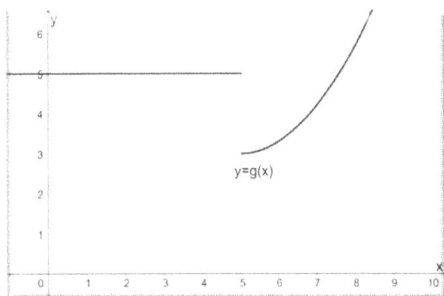

4. $\lim\limits_{x \to 5^-} g(x) = 5$

5. $\lim\limits_{x \to 5^+} g(x) = 3$

6. As $\lim\limits_{x \to 5^-} g(x) = 6$ and $\lim\limits_{x \to 5^+} g(x) = 2$ $\lim\limits_{x \to 5} g(x) \neq 5$

The limit does not exist because, there are two different values of g(x) when x gets near to 5.

7. $\lim\limits_{x \to 5^+} \dfrac{5x}{x-5} = \dfrac{25}{positive \; very \; small \; number} = +\infty$

8. $\lim\limits_{x \to 5^-} \dfrac{5x}{x-4} = \lim\limits_{x \to 5^-} \dfrac{5(5)}{1} = 25$

9. $\lim\limits_{x \to 3^-} \dfrac{5x}{x-5} = \dfrac{\lim\limits_{x \to 3^-} 5x}{\lim\limits_{x \to 3^-} (x-5)} = \dfrac{15}{-2} = -7.5$

10. $\lim\limits_{x \to 3^-} \dfrac{x}{x-5} = -1.5$

$$\lim\limits_{x \to 3^-} \dfrac{x}{x-5} = \dfrac{\lim\limits_{x \to 3^-} x}{\lim\limits_{x \to 3^-} (x-5)} = \dfrac{3}{-2} = -1.5$$

Chapter 6. C. Limits-End behavior

Determine what is the end behavior of these functions using limits.

1. $f(x) = 2x^4 - 3x^2 + 15x - 24$: when x goes to $+\infty$ it goes up in first quadrant.
Point a) in the Theory section. The polynomial function has an even degree and the coefficient of the highest degree term is positive.

2. $f(x) = -4x^4 + 2x^2 - 5x$: when x goes to $-\infty$ it goes down in the third quadrant.
Point b) in the Theory section. The polynomial function has an even degree and the coefficient of the highest degree term is negative.

3. $f(x) = -4x^3 + 6x^2 - x$: when x goes to $-\infty$ it goes up in the second quadrant.

Point c) in the Theory section. The polynomial function has an odd degree and the coefficient of the highest degree term is negative.

4. $f(x) = x^3 - 36x^2 - x + 73$: when x goes to -∞ it goes down in the third quadrant

Point c) in the Theory section. The polynomial function has an odd degree and the coefficient of the highest degree term is positive.

5. $f(x) = \frac{2x^3+3x^2-4x+5}{6x^3-5x+4}$: when x goes to +∞ $f(x)$ goes towards $\frac{1}{3}$

$$\lim_{x\to\infty} \frac{2x^3+3x^2-4x+5}{6x^3-5x+4} = \lim_{x\to\infty} \frac{x^3(2+\frac{3}{x}-\frac{4}{x^2}+\frac{5}{x^3})}{x^3(6-\frac{5}{x^2}+\frac{4}{x^3})} = \lim_{x\to\infty} \frac{(2+\frac{3}{x}-\frac{4}{x^2}+\frac{5}{x^3})}{(6-\frac{5}{x^2}+\frac{4}{x^3})} = \frac{\lim_{x\to\infty}(2+\frac{3}{x}-\frac{4}{x^2}+\frac{5}{x^3})}{\lim_{x\to\infty}(6-\frac{5}{x^2}+\frac{4}{x^3})} = \frac{2+0-0+0}{6-0+0} = \frac{2}{6} = \frac{1}{3}$$

6. $f(x) = \frac{x^3+2x^2-3x+4}{6x^4-5x^3+4x^2-3x}$: when x goes to -∞ $f(x)$ goes towards 0

$$\lim_{x\to-\infty} \frac{x^3+2x^2-3x+4}{6x^4-5x^3+4x^2-3x} = \lim_{x\to-\infty} \frac{x^3(1+\frac{2}{x}-\frac{3}{x^2}+\frac{4}{x^3})}{x^4(6-\frac{5}{x}+\frac{4}{x^2}-\frac{3}{x^3})} = \lim_{x\to-\infty} \frac{(1+\frac{2}{x}-\frac{3}{x^2}+\frac{4}{x^3})}{x(6-\frac{5}{x}+\frac{4}{x^2}-\frac{3}{x^3})} = \frac{\lim_{x\to-\infty}(1+\frac{2}{x}-\frac{3}{x^2}+\frac{4}{x^3})}{\lim_{x\to-\infty}x(6-\frac{5}{x}+\frac{4}{x^2}-\frac{3}{x^3})} = \frac{1+0-0+0}{\infty*(6-0+0-0)} = 0$$

7. $f(x) = \frac{x^5+2x^2-3x+4}{6x^4-5x^3+4x^2-3x}$: when x goes to +∞ $f(x)$ goes up in the first quadrant.

When the degree of the polynomial at the numerator is bigger than the degree of the polynomial at the denominator, for x going towards plus infinite, the graph goes upward in the first quadrant. The graph is represented below.

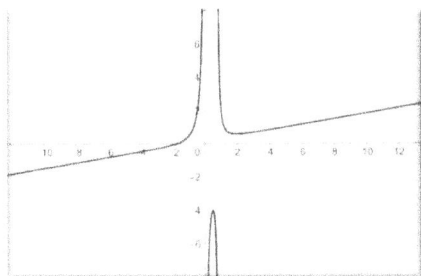

8. $f(x) = \frac{2x^3+3x^2-4x+5}{6x^2-5x+4}$: when x goes to -∞ it goes down in the third quadrant.

When the degree of the polynomial at the numerator is bigger than the degree of the polynomial at the denominator, for x going towards minus infinite, the graph goes downward in the third quadrant.

The graph is represented below.

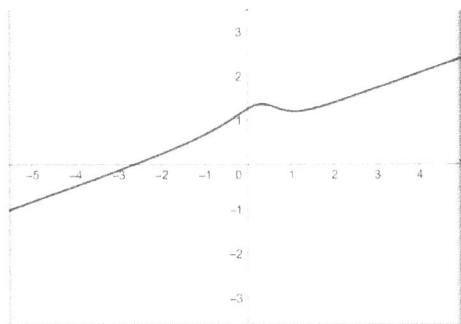

Graph the following functions and determine the end behavior for x going towards +∞

9. F(x) is going towards 0. $f(x) = \frac{x^3-2x^2+3x}{x^4+1}$

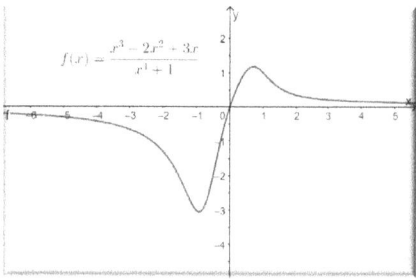

10. F(x) is going towards 0. $f(x) = \frac{2x^2+3x-2}{x^3+1}$

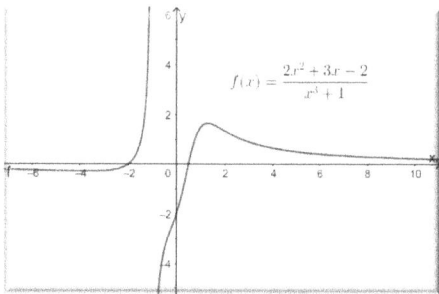

Chapter 6. D.   Limits-Intermediate limits theorem

1. $f(c) = 6$ for at least one value of c that is between -1 and 5

2. $f(c) = 2$ for at least one value of c that is between -1 and 5

3. $f(c) = -2$ for at least one value of c that is between -1 and 5

-2 does not belong between 2 and 7

4. $f(c) = 3$ for at least one value of c that is between 6 and 10

6 and 10 are outside the closed interval [-1,5]

Using the table below of a continuous function determine if the following questions are correct.

x	-2	-1	0	3	5	7
$f(x)$	-1	0	3	4	5	9

5. The function $f(x) = 3$ for at least one x that belong to $-1 \leq x \leq 3$. In this case x=0

6. The function $f(x) = 5$ for at least one x that belong to $-1 \leq x \leq 7$. In this case x=5

7. The function $f(x) = 0$ for at least one x that belong to $0 \leq x \leq 3$

$f(x) = 0$ for x=**-1**. X=-1 is outside the interval [0,3]

If $f(x) = \frac{2x+1}{-3x+7}$ determine if the questions below are correct.

8. There will not be a value c such that $f(c) = 1$ for $3 \leq c \leq 9$

So $f(3) = \frac{2(3)+1}{-3(3)+7} = \frac{7}{-2} = -3.5; f(9) = \frac{2(9)+1}{-3(9)+7} = \frac{19}{-20} = -0.95$

F(x) is always between -3.5 and -0.95

There is no value of c between 3 and 9 for which $f(c) = 1$

9. There will be a value c such that $f(c) = -2$ for $3 \leq c \leq 7$

So $f(3) = \frac{2(3)+1}{-3(3)+7} = \frac{7}{-2} = -3.5$; $f(7) = \frac{2(7)+1}{-3(7)+7} = \frac{15}{-14} = -1.07$

F(x) is always between -3.5 and -1.07

Yes, there will be a value c such that $f(c) = -2$ for $3 \leq c \leq 7$

10. There will be a value c such that $f(c) = 2$ for $-2 \leq c \leq 2$

So $f(-2) = \frac{2(-2)+1}{-3(-2)+7} = \frac{-3}{13} = -0.23$; $f(2) = \frac{2(2)+1}{-3(2)+7} = \frac{5}{1} = 5$

Yes, there will be a value c such that $f(c) = 2$ for $-2 \leq c \leq 2$

Chapter 6. E.   Limits-Left and right limits

Using the graph below, determine which answer is correct. In the table at the bottom of the page cross all the letters of the correct answer.

Where: $f(x) = \begin{cases} 0.5x + 1, x \leq 2 \\ 2x - 3, x > 2 \end{cases}$

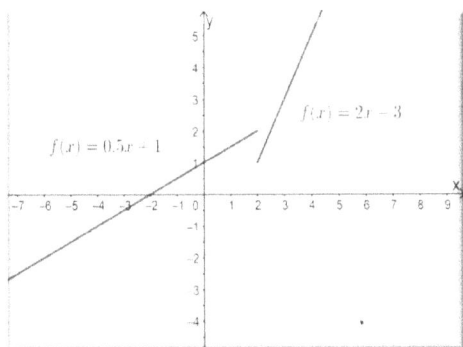

1. $\lim\limits_{x \to 0^-} f(x) = 0.5(0) + 1 = 1$

2. $\lim\limits_{x \to 0^+} f(x) = 1$

3. $\lim\limits_{x \to 2^-} f(x) = 0.5(2) + 1 = 1 + 1 = 2$

4. $\lim\limits_{x \to 2^+} f(x) = 2(2) - 3 = 4 - 3 = 1$

Using the graph below, where: $f(x) = \begin{cases} 0.3x + 2, x < 1 \\ 0.5x^2 + 0.5, x \geq 1 \end{cases}$

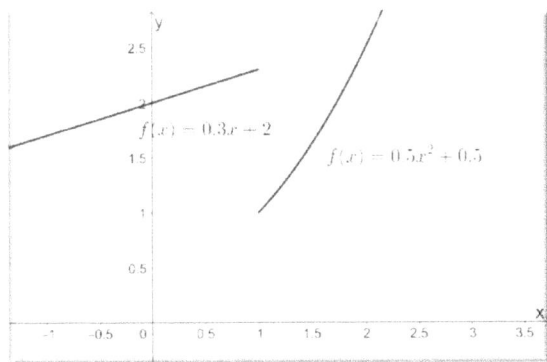

5. $\lim\limits_{x \to 1^-} f(x) = 0.3(1) + 2 = 0.3 + 2 = 2.3$

6. $\lim\limits_{x \to 1^+} f(x) = 0.5(1)^2 + 0.5 = 0.5 + 0.5 = 1$

7. As $\lim\limits_{x \to 1^-} f(x) = 2.3$ and $\lim\limits_{x \to 1^+} f(x) = 1$ The limit does not exist because, there are two different values of f(x) when x gets near to 1.

8. $\lim\limits_{x \to 3^-} f(x) = 0.5(3)^2 + 0.5 = 0.5(9. +0.5 = 4.5 + 0.5 = 5$

9. $\lim\limits_{x \to 3^+} f(x) = 5$

10. $\lim\limits_{x \to 7^-} f(x) = 0.5(7)^2 + 0.5 = 0.5(49. +0.5 = 24.5 + 0.5 = 25$

Chapter 6. F.   Limits-Limits to infinity

1. $\lim\limits_{x\to\infty} \dfrac{x^2+2x-45}{3x^3-5x^2+7x-3} = \lim\limits_{x\to\infty} \dfrac{x^2(1+\frac{2}{x}-\frac{45}{x^2})}{x^3(3-\frac{5}{x}+\frac{7}{x^2}-\frac{3}{x^3})} = \lim\limits_{x\to\infty} \dfrac{(1+\frac{2}{x}-\frac{45}{x^2})}{x(3-\frac{5}{x}+\frac{7}{x^2}-\frac{3}{x^3})} =$

$\dfrac{\lim\limits_{x\to\infty}(1)+\lim\limits_{x\to\infty}(\frac{2}{x})-\lim\limits_{x\to\infty}(\frac{45}{x^2})}{\lim\limits_{x\to\infty}(x)[\lim\limits_{x\to\infty}(3)-\lim\limits_{x\to\infty}(\frac{5}{x})+\lim\limits_{x\to\infty}(\frac{7}{x^2})-\lim\limits_{x\to\infty}(\frac{3}{x^3})]} = \dfrac{1+0+0}{\lim\limits_{x\to\infty}(x)(3-0+0-0)} = 0$

2. $\lim\limits_{x\to-\infty} \dfrac{x^5+5x-5}{2x^3-56x^2+8x-3} = \lim\limits_{x\to-\infty} \dfrac{x^5(1+\frac{5}{x^4}-\frac{5}{x^5})}{x^3(2-\frac{56}{x}+\frac{8}{x^2}-\frac{3}{x^3})} = \lim\limits_{x\to-\infty} \dfrac{x^2(1+\frac{5}{x^4}-\frac{5}{x^5})}{(2-\frac{56}{x}+\frac{8}{x^2}-\frac{3}{x^3})} =$

$\dfrac{\lim\limits_{x\to-\infty}(x^2)[\lim\limits_{x\to-\infty}(1)+\lim\limits_{x\to-\infty}(\frac{5}{x^4})-\lim\limits_{x\to-\infty}(\frac{5}{x^5})]}{\lim\limits_{x\to-\infty}(2)-\lim\limits_{x\to-\infty}(\frac{56}{x})+\lim\limits_{x\to-\infty}(\frac{8}{x^2})-\lim\limits_{x\to-\infty}(\frac{3}{x^3})} = \dfrac{\lim\limits_{x\to-\infty}(x^2)(1+0-0)}{2-0+0-0} = \infty$

3. $\lim\limits_{x\to\infty} \dfrac{3x^2+2x-45}{3x^3-5x^2+7x-3} = \lim\limits_{x\to\infty} \dfrac{x^2(3+\frac{2}{x}-\frac{45}{x^2})}{x^3(3-\frac{5}{x}+\frac{7}{x^2}-\frac{3}{x^3})} =$

$\lim\limits_{x\to\infty} \dfrac{(3+\frac{2}{x}-\frac{45}{x^2})}{x(3-\frac{5}{x}+\frac{7}{x^2}-\frac{3}{x^3})} \lim\limits_{x\to\infty} \dfrac{\lim\limits_{x\to\infty}(3)+\lim\limits_{x\to\infty}(\frac{2}{x})-\lim\limits_{x\to\infty}(\frac{45}{x^2})}{\lim\limits_{x\to\infty}(x)[\lim\limits_{x\to\infty}(3)-\lim\limits_{x\to\infty}(\frac{5}{x})+\lim\limits_{x\to\infty}(\frac{7}{x^2})-\lim\limits_{x\to\infty}(\frac{3}{x^3})]} = \dfrac{3+0-0}{\lim\limits_{x\to\infty}(x)(3-0+0-0)} = \dfrac{3}{\lim\limits_{x\to\infty}(x)(3)} = 0$

4. $\lim\limits_{x\to-\infty} \dfrac{x-4}{2x^2-8x-2} = \lim\limits_{x\to-\infty} \dfrac{x(1-\frac{4}{x})}{x^2(2-\frac{8}{x}-\frac{2}{x^2})} = \lim\limits_{x\to-\infty} \dfrac{(1-\frac{4}{x})}{x(2-\frac{8}{x}-\frac{2}{x^2})} = \dfrac{\lim\limits_{x\to-\infty}(1)-\lim\limits_{x\to-\infty}(\frac{4}{x})}{\lim\limits_{x\to-\infty}(x)[\lim\limits_{x\to-\infty}(2)-\lim\limits_{x\to-\infty}(\frac{8}{x})-\lim\limits_{x\to-\infty}(\frac{2}{x^2})]} =$

$\dfrac{1-0}{\lim\limits_{x\to-\infty}(x)(2-0-0)} = \dfrac{1}{\lim\limits_{x\to-\infty}(x)(2)} = 0$

5. $\lim\limits_{x\to\infty} \dfrac{x^2+2x-45}{3x^2+37x-35} = \lim\limits_{x\to\infty} \dfrac{x^2(1+\frac{2}{x}-\frac{45}{x^2})}{x^2(3+\frac{37}{x}-\frac{35}{x^2})} = \lim\limits_{x\to\infty} \dfrac{(1+\frac{2}{x}-\frac{45}{x^2})}{(3+\frac{37}{x}-\frac{35}{x^2})} = \dfrac{\lim\limits_{x\to\infty}(1)+\lim\limits_{x\to\infty}(\frac{2}{x})-\lim\limits_{x\to\infty}(\frac{45}{x^2})}{\lim\limits_{x\to\infty}(3)+\lim\limits_{x\to\infty}(\frac{37}{x})-\lim\limits_{x\to\infty}(\frac{35}{x^2})} = \dfrac{1+0-0}{3+0-0} = \dfrac{1}{3}$

6. $\lim\limits_{x\to-\infty} \dfrac{3x^3+27x^2-4x+5}{2x^3-54x^2+27x-31} = \lim\limits_{x\to-\infty} \dfrac{x^3(3+\frac{27}{x}-\frac{4}{x^2}+\frac{5}{x^3})}{x^3(2-\frac{54}{x}+\frac{27}{x^2}-\frac{31}{x^3})} =$

$\lim\limits_{x\to-\infty} \dfrac{(3+\frac{27}{x}-\frac{4}{x^2}+\frac{5}{x^3})}{(2-\frac{54}{x}+\frac{27}{x^2}-\frac{31}{x^3})} = \dfrac{\lim\limits_{x\to-\infty}(3)+\lim\limits_{x\to-\infty}(\frac{27}{x})-\lim\limits_{x\to-\infty}(\frac{4}{x^2})+\lim\limits_{x\to-\infty}(\frac{5}{x^3})}{\lim\limits_{x\to-\infty}(2)-\lim\limits_{x\to-\infty}(\frac{54}{x})+\lim\limits_{x\to-\infty}(\frac{27}{x^2})-\lim\limits_{x\to-\infty}(\frac{31}{x^3})} = \dfrac{3+0-0+0}{2-0+0-0} = \dfrac{3}{2}$

7. $\lim\limits_{x\to\infty} \dfrac{9x^2-4}{5x^2+7x-3} = \lim\limits_{x\to\infty} \dfrac{x^2(9-\frac{4}{x^2})}{x^2(5+\frac{7}{x}-\frac{4}{x^2})} = \lim\limits_{x\to\infty} \dfrac{(9-\frac{4}{x^2})}{(5+\frac{7}{x}-\frac{4}{x^2})} = \dfrac{\lim\limits_{x\to\infty}(9)-\lim\limits_{x\to\infty}(\frac{4}{x^2})}{\lim\limits_{x\to\infty}(5)+\lim\limits_{x\to\infty}(\frac{7}{x})-\lim\limits_{x\to\infty}(\frac{4}{x^2})} = \dfrac{9-0}{5+0-0} = \dfrac{9}{5}$

8. $\lim\limits_{x\to\infty} \dfrac{3x+7}{\sqrt{x^2-3x+4}} = \lim\limits_{x\to\infty} \dfrac{x(3+\frac{7}{x})}{\sqrt{x^2(1-\frac{3}{x}+\frac{4}{x^2})}} = \lim\limits_{x\to\infty} \dfrac{x(3+\frac{7}{x})}{x\sqrt{(1-\frac{3}{x}+\frac{4}{x^2})}} = \lim\limits_{x\to\infty} \dfrac{(3+\frac{7}{x})}{\sqrt{(1-\frac{3}{x}+\frac{4}{x^2})}} = \dfrac{\lim\limits_{x\to\infty}(3)+\lim\limits_{x\to\infty}(\frac{7}{x})}{\sqrt{\lim\limits_{x\to\infty}(1)-\lim\limits_{x\to\infty}\frac{3}{x}+\lim\limits_{x\to\infty}(\frac{4}{x^2})}} =$

$\dfrac{3+0}{\sqrt{1-0+0}} = \dfrac{3}{1} = 3$

9. $\lim\limits_{x\to-\infty} \dfrac{5x-6}{\sqrt{x^2-5}} = \lim\limits_{x\to-\infty} \dfrac{x(5-\frac{6}{x})}{\sqrt{x^2(1-\frac{5}{x^2})}} = \lim\limits_{x\to-\infty} \dfrac{x(5-\frac{6}{x})}{x\sqrt{(1-\frac{5}{x^2})}} = \lim\limits_{x\to-\infty} \dfrac{(5-\frac{6}{x})}{\sqrt{(1-\frac{5}{x^2})}} = \dfrac{\lim\limits_{x\to-\infty}(5)-\lim\limits_{x\to-\infty}(\frac{6}{x})}{\sqrt{\lim\limits_{x\to-\infty}(1)-\lim\limits_{x\to-\infty}(\frac{5}{x^2})}} = \dfrac{5-0}{\sqrt{1-0}} = 5$

10. $\lim\limits_{x\to-\infty} \dfrac{3x^2-2x+7}{3x^3-6x^2+8x-1} = \lim\limits_{x\to-\infty} \dfrac{x^2(3-\frac{2}{x}+\frac{7}{x^2})}{x^3(3-\frac{6}{x}+\frac{8}{x^2}-\frac{1}{x^3})} =$

$\lim\limits_{x\to-\infty} \dfrac{(3-\frac{2}{x}+\frac{7}{x^2})}{x(3-\frac{6}{x}+\frac{8}{x^2}-\frac{1}{x^3})} = \dfrac{\lim\limits_{x\to-\infty}(3)-\lim\limits_{x\to-\infty}(\frac{2}{x})+\lim\limits_{x\to-\infty}\frac{7}{x^2}}{\lim\limits_{x\to-\infty}(x)[\lim\limits_{x\to-\infty}(3)-\lim\limits_{x\to-\infty}(\frac{6}{x})+\lim\limits_{x\to-\infty}(\frac{8}{x^2})-\lim\limits_{x\to-\infty}(\frac{1}{x^3})]} = \dfrac{3-0+0}{\lim\limits_{x\to-\infty}(x)(3-0+0-0)} = \dfrac{3}{\lim\limits_{x\to-\infty}(x)(3)} = 0$

11. $\lim\limits_{x\to\infty} \dfrac{2x+3}{\sqrt{x^3-27}} = \lim\limits_{x\to\infty} \dfrac{x(2+\frac{3}{x})}{\sqrt{x^2(x-\frac{27}{x^2})}} = \lim\limits_{x\to\infty} \dfrac{x(2+\frac{3}{x})}{x\sqrt{(x-\frac{27}{x^2})}} = \lim\limits_{x\to\infty} \dfrac{(2+\frac{3}{x})}{\sqrt{(x-\frac{27}{x^2})}} = \dfrac{\lim\limits_{x\to\infty}(2)+\lim\limits_{x\to\infty}(\frac{3}{x})}{\sqrt{\lim\limits_{x\to\infty}(x)-\lim\limits_{x\to\infty}(\frac{27}{x^2})}} = \dfrac{2+0}{\sqrt{\lim\limits_{x\to\infty}(x)-0}} = 0$

12. $\lim\limits_{x\to\infty} \dfrac{x^2+2x-45}{4x^2-3x^2+7x-3} = \lim\limits_{x\to\infty} \dfrac{x^2(1+\frac{2}{x}-\frac{45}{x^2})}{x^2(4-3+\frac{7}{x}-\frac{3}{x^2})} = \lim\limits_{x\to\infty} \dfrac{(1+\frac{2}{x}-\frac{45}{x^2})}{(4+\frac{7}{x}-\frac{3}{x^2})} = \dfrac{\lim\limits_{x\to\infty}(1)+\lim\limits_{x\to\infty}(\frac{2}{x})-\lim\limits_{x\to\infty}(\frac{45}{x^2})}{\lim\limits_{x\to\infty}(4)+\lim\limits_{x\to\infty}(\frac{7}{x})-\lim\limits_{x\to\infty}(\frac{3}{x^2})} = \dfrac{1+0-0}{4+0-0} = \dfrac{1}{4}$

Chapter 6. G.   Limits-Continuity

We know that for a function f(x) to be continuous in x=a:

- f(a) has to be defined

- $\lim\limits_{x\to a} f(x)$ exists

- $\lim\limits_{x\to a} f(x) = f(a)$

The following function is represented below: $f(x) = \begin{cases} 2x-1, x \geq 2 \\ 3x-3, 1.5 < x < 2 \\ 0.5x, x \leq 1.5 \end{cases}$

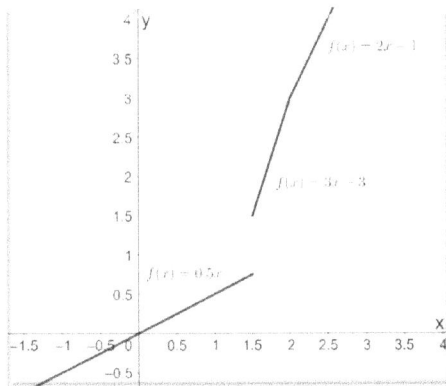

1. $\lim\limits_{x\to 4} f(x) = 2(4) - 1 = 7$

$\lim\limits_{x\to 4^-}[f(x)] = \lim\limits_{x\to 4^+}[f(x)] = 7$

The function f(x) is continuous in x=4

2. $\lim\limits_{x\to 2^-}(3x-3) = 3(2) - 3 = 3$

$\lim\limits_{x\to 2^+}(2x-1) = 2(2) - 1 = 3$

So $\lim\limits_{x\to 2^-}[f(x)] = \lim\limits_{x\to 2^+}[f(x)] = f(2)$

The function f(x) is continuous in x=2

3. $\lim\limits_{x\to 2} f(x) = 2(2) - 1 = 3$

$\lim\limits_{x\to 2^-}[f(x)] = 3(2) - 3 = 3$

$\lim\limits_{x\to 2^+}[f(x)] = 2(2) - 1 = 3$

$\lim\limits_{x\to 2^-}[f(x)] = \lim\limits_{x\to 2^+}[f(x)] = 3$

The function f(x) is continuous in x=1

4. The function f(x) is continuous in x=-7

5. $\lim\limits_{x\to 1.5^-} f(x) = \lim\limits_{x\to 1.5^-}[0.5(1.5)] = 0.75$

$\lim\limits_{x\to 1.5^+} f(x) = \lim\limits_{x\to 1.5^-}[3(1.5) - 3] = 1.3$

The limit for x=1.5 does not exist.

The function f(x) is not continuous in x=1.5

The function g(x) is defined as follows: $g(x) = \begin{cases} \frac{9x^2-1}{3x-1} \ for \ x \neq 1/3 \\ 0 \ for \ x = 1/3 \end{cases}$

6. The function g(x) exists for x=1/3. For x=1/3 g(x)=0

7. $\lim\limits_{x\to 1/3^-} g(x) = \lim\limits_{x\to 1/3^+} g(x) = \dfrac{9(\frac{1}{3})^2-1}{3(\frac{1}{3})-2} = \dfrac{1-1}{1-2} = 0 = g(\tfrac{1}{3})$

The limit $\lim\limits_{x\to 1/3} g(x)$ does exist for x=1/3

8. The function g(x) is not continuous for x=0.33

- g(0.66) is not defined

- $\lim_{x \to 0.66^-} g(x) \frac{9(0.66)^2 - 1}{3(0.66) - 2} = \frac{2.96}{-0.02} = -\infty \neq \lim_{x \to 0.66^+} g(x) = +\infty$

- There is a vertical asymptote at x=0.66

The graph of g(x) is represented below.

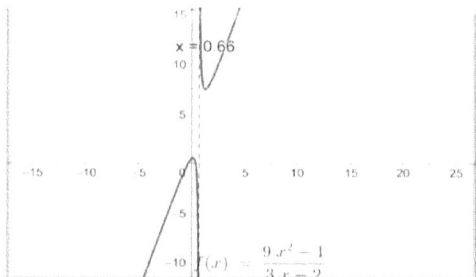

9. The function $f(x) = \begin{cases} 3x \ if \ -1 \leq x \leq 2 \\ x + 2 \ if \ x > 2 \end{cases}$ is

continuous over interval [-1,2]

A linear function f(x)=3x is continuous for all real values of x. It is continuous on interval [-1,2]. The left-hand limit at 2 is 6 ; it exist and equals the value of the function at x=2.

10. The function $f(x) = \begin{cases} 5x \ if \ -1 \leq x \leq 3 \\ 2x + 2 \ if \ x > 3 \end{cases}$ is not continuous over [-5,5]

$\lim_{x \to 3^-} f(x) = 5(3) = 15; \ \lim_{x \to 3^+} f(x) = 2(3) + 2 = 8$

$\lim_{x \to 3} f(x)$ does not exist

So, f(x) is not continuous over [-5,5].

CHAPTER 7

Chapter 7. B.   Differentiation-Definition of derivatives

It is known that the formula of the derivative is: $f'(x) = \lim_{h \to 0} \frac{f(x+h) - f(x)}{h}$

Using this formula calculate the following derivatives.

1. $f(x) = 2x + 45 \ so \ f'(x) = 2$

$f'(x) = \lim_{h \to 0} \frac{f(x+h) - f(x)}{h} = \lim_{h \to 0} \frac{2(x+h) + 45 - 2x - 45}{h} = \lim_{h \to 0} \frac{2x + 2h + 45 - 2x - 45}{h} = \lim_{h \to 0} \frac{2h}{h} = \lim_{h \to 0} \frac{2}{1} = 2$

2. $f(x) = 3x^2 + 5x - 4 \ so \ f'(x) = 6x - 5$

$f'(x) = \lim_{h \to 0} \frac{f(x+h) - f(x)}{h} = \lim_{h \to 0} \frac{3(x+h)^2 + 5(x+h) - 4 - (3x^2 + 5x - 4)}{h} = \lim_{h \to 0} \frac{3(x^2 + 2hx + h^2) + 5x + 5h - 4 - 3x^2 - 5x + 4}{h} =$

$\lim_{h \to 0} \frac{3x^2 + 6hx + 3h^2 + 5x + 5h - 4 - 3x^2 - 5x + 4}{h} = \lim_{h \to 0} \frac{6hx + 3h^2 + 5h}{h} = \lim_{h \to 0} \frac{h(6x + 3h + 5)}{h} = \lim_{h \to 0} (6x + 3h + +5) = 6x + 5$

3. $f(x) = Ax^2 + Bx - 37 \ so \ f'(x) = 2Ax + B$

$$f'(x) = \lim_{h \to 0} \frac{f(x+h)-f(x)}{h} = \lim_{h \to 0} \frac{A(x+h)^2+B(x+h)-37-[Ax^2+Bx-37]}{h} =$$

$$\lim_{h \to 0} \frac{A(x^2+2xh+h^2)+Bx+Bh-37-Ax^2-Bx+37}{h} = \lim_{h \to 0} \frac{Ax^2+2Axh+Ah^2+Bx+Bh-37-Ax^2-Bx+37}{h} =$$

$$\lim_{h \to 0} \frac{2Axh+Ah^2+Bh}{h} = \lim_{h \to 0} \frac{h(2Ax+Ah+B)}{h} = \lim_{h \to 0} (2Ax + Ah + B) = 2Ax + B$$

Using Δx notation instead of h find the derivative of the following functions:

4. $f(x) = 5x^2 - 4x + 3$ so $f'(x) = 10x - 4$

$$f'(x) = \lim_{\Delta x \to 0} \frac{f(x+\Delta x)-f(x)}{\Delta x} = \lim_{\Delta x \to 0} \frac{5(x+\Delta x)^2-4(x+\Delta x)+3-(5x^2-4x+3)}{\Delta x} =$$

$$\lim_{\Delta x \to 0} \frac{5[x^2+2x\Delta x+(\Delta x)^2]-4(x+\Delta x)+3-(5x^2-4x+3)}{\Delta x} == \lim_{\Delta x \to 0} \frac{5x^2+10x\Delta x+5(\Delta x)^2-4x-4\Delta x+3-5x^2+4x-3)}{\Delta x} =$$

$$\lim_{\Delta x \to 0} \frac{10x\Delta x+5(\Delta x)^2-4\Delta x}{\Delta x} = \lim_{\Delta x \to 0} \frac{\Delta x(10x+5\Delta x-4)}{\Delta x} = \lim_{\Delta x \to 0} (10x + 5\Delta x - 4) = 10x - 4$$

5. $f(x) = 7x - 27$ so $f'(x) = 7$

$$f'(x) = \lim_{\Delta x \to 0} \frac{f(x+\Delta x)-f(x)}{\Delta x} = \lim_{\Delta x \to 0} \frac{7(x+\Delta x)-27-(7x-27)}{\Delta x} = \lim_{\Delta x \to 0} \frac{7x+7\Delta x-27-7x+27}{\Delta x} = \lim_{\Delta x \to 0} \frac{7\Delta x}{\Delta x} = \lim_{\Delta x \to 0} 7 = 7$$

6. $f(x) = x^{-1}$ so $f'(x) = -x^{-2}$

$$f'(x) = \lim_{\Delta x \to 0} \frac{f(x+\Delta x)-f(x)}{\Delta x} == \lim_{\Delta x \to 0} \frac{(x+\Delta x)^{-1}-x^{-1}}{\Delta x} == \lim_{\Delta x \to 0} \frac{\frac{1}{x+\Delta x}-x^{-1}}{\Delta x} = \lim_{\Delta x \to 0} \frac{\frac{1}{x+\Delta x}-\frac{x^{-1}(x+\Delta x)}{x+\Delta x}}{\Delta x} =$$

$$\lim_{\Delta x \to 0} \frac{\frac{1-x^{-1}(x+\Delta x)}{x+\Delta x}}{\Delta x} = \lim_{\Delta x \to 0} \frac{\frac{1-1-x^{-1}\Delta x}{x+\Delta x}}{\Delta x} = \lim_{\Delta x \to 0} \frac{\frac{-x^{-1}\Delta x}{x+\Delta x}}{\Delta x} = \lim_{\Delta x \to 0} \frac{-x^{-1}\Delta x}{\Delta x(x+\Delta x)} = \lim_{\Delta x \to 0} \frac{-x^{-1}}{(x+\Delta x)} = \lim_{\Delta x \to 0} \frac{-x^{-1}}{(x+\Delta x)} =$$

$$-\frac{x^{-1}}{x} = -x^{-2}$$

7. $f(x) = 3x^{-1} + x$ so $f'(x) = 1 - 3x^{-2}$

$$f'(x) = \lim_{\Delta x \to 0} \frac{f(x + \Delta x) - f(x)}{\Delta x} = \lim_{\Delta x \to 0} \frac{3(x + \Delta x)^{-1} + (x + \Delta x) - 3x^{-1} - x}{\Delta x} =$$

$$= \lim_{\Delta x \to 0} \frac{3(x+\Delta x)^{-1}+x+\Delta x-3x^{-1}-x}{\Delta x} = \lim_{\Delta x \to 0} \frac{3(x+\Delta x)^{-1}+\Delta x-3x^{-1}}{\Delta x} = \lim_{\Delta x \to 0} \frac{\frac{3}{x+\Delta x}+\Delta x-3x^{-1}}{\Delta x} =$$

$$\lim_{\Delta x \to 0} \frac{\frac{3}{x+\Delta x}+\frac{(\Delta x-3x^{-1})(x+\Delta x)}{x+\Delta x}}{\Delta x} = \lim_{\Delta x \to 0} \frac{\frac{3}{x+\Delta x}+\frac{x\Delta x+\Delta x^2-3-3x^{-1}\Delta x}{x+\Delta x}}{\Delta x} = \lim_{\Delta x \to 0} \frac{\frac{x\Delta x+\Delta x^2-3x^{-1}\Delta x}{x+\Delta x}}{\Delta x} = \lim_{\Delta x \to 0} \frac{x\Delta x+\Delta x^2-3x^{-1}\Delta x}{\Delta x(x+\Delta x)} =$$

$$\lim_{\Delta x \to 0} \frac{\Delta x(x+\Delta x-3x^{-1})}{\Delta x(x+\Delta x)} = \lim_{\Delta x \to 0} \frac{x+\Delta x-3x^{-1}}{(x+\Delta x)} = \frac{x-3x^{-1}}{x} = (x - 3x^{-1})x^{-1} = 1 - 3x^{-2}$$

Using the formula $f'(x) = \lim_{h \to 0} \frac{f(x+h)-f(x)}{h}$ find the following derivatives.

8. $f(x) = x^{-1} + x^2$ so $f'(x) = -x^{-2} + 2x = 2x - x^{-2}$

$$f'(x) = \lim_{h \to 0} \frac{f(x+h)-f(x)}{h} = \lim_{h \to 0} \frac{(x+h)^{-1}+(x+h)^2-(x^{-1}+x^2)}{h} = \lim_{h \to 0} \frac{\frac{1}{x+h}+(x+h)^2-(x^{-1}+x^2)}{h} =$$

$$= \lim_{h \to 0} \frac{\frac{1}{x+h}+\frac{(x+h)^3}{x-h}-\frac{(x^{-1}+x^2)(x+h)}{x+h}}{h} = \lim_{h \to 0} \frac{1+(x+h)^3-(1+x^2h+x^3+x^{-1}h)}{h(x+h)} =$$

$$= \lim_{h \to 0} \frac{1+(x+h)^2(x+h)-(1+x^2h+x^3+x^{-1}h)}{h(x+h)} = \lim_{h \to 0} \frac{1+(x^2+2xh+h^2)(x+h)-(1+x^2h+x^3+x^{-1}h)}{h(x+h)} =$$

$$= \lim_{h \to 0} \frac{1+x^3+x^2h+2x^2h+2xh^2+xh^2+h^3-1-x^2h-x^3-x^{-1}h}{h(x+h)} = \lim_{h \to 0} \frac{2x^2h+2xh^2+xh^2+h^3-x^{-1}h}{h(x+h)} =$$

$$= \lim_{h \to 0} \frac{h(2x^2+3xh+h^2-x^{-1})}{h(x+h)} = \lim_{h \to 0} \frac{2x^2+3xh+h^2-x^{-1}}{x+h} = \frac{2x^2-x^{-1}}{x} = 2x - x^{-2}$$

9. $f(x) = x^2 + 3x$ so for $x = 2$, $f'(2) = 7$

$$f'(x) = \lim_{h \to 0} \frac{f(x+h)-f(x)}{h} = \lim_{h \to 0} \frac{(x+h)^2+3(x+h)-(x^2+3x)}{h} = \lim_{h \to 0} \frac{x^2+2xh+h^2+3x+3h-x^2-3x}{h} ==$$

$$\lim_{h \to 0} \frac{2xh+h^2+3h}{h} = \lim_{h \to 0} \frac{h(2x+h+3)}{h} = \lim_{h \to 0} (2x + h + 3) = 2x + 3$$

For x=2; $f'(2) = 2(2) + 3 = 7$

10. $f(x) = x^2 + 33$ so, for $x = 3$, $f'(3) = 6$

$$f'(x) = \lim_{h \to 0} \frac{f(x+h)-f(x)}{h} = \lim_{h \to 0} \frac{(x+h)^2+33-(x^2+33)}{h} = \lim_{h \to 0} \frac{x^2+2xh+h^2+33-x^2-33}{h} = \lim_{h \to 0} \frac{2xh+h^2}{h} =$$

$$\lim_{h \to 0} \frac{h(2x+h)}{h} = \lim_{h \to 0} (2x + h) = 2x$$

For x=3; $f'(3) = 2(3) = 6$

Chapter 7. C.   Differentiation-Notation

1. Gottfried Leibniz's notation is: $\dfrac{dy}{dx}$

2. Another notation is: $f(x) = y$

3. Joseph Louis Lagrange's notation is $f'(x)$

4. This is formula for calculating distance d=V*t

5. Leonhard Euler's notation for the second derivative is: D_x^2

6. There is no such formula F=k/e

7. One of Isaac Newton's notation is: $\dot{y}$ for first derivative.

8. Henry's formula is: SO=CCER. There is no such formula.

9. Another formula used by Isaac Newton was: $\dot{x}$

10. Isaac Newton also used this notation for first derivative: $\Box \dot{y}$

Chapter 3 D, a.   Rate of change- Average versus Instantaneous

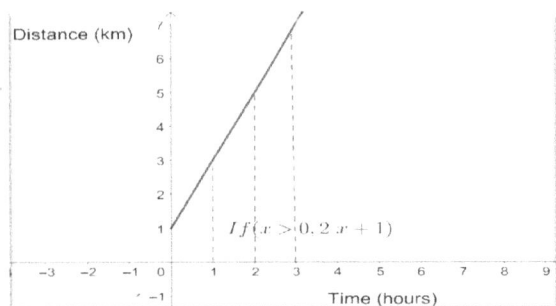

1. The average rate of change (speed) between 1 and 3 hours is: 2km/h

$$speed = \frac{7km-3km}{3hours-1hour} = \frac{4km}{2hours} = 2km/h$$

2. The average rate of change (speed) between 0 and 3 hours is: 3km/h

$$speed = \frac{7km-1km}{3hours-0hour} = \frac{6km}{3hours} = 2km/h$$

A tire rolls by the relation between distance and time as follows: $s(t) = 2t^2 + 4t + 3$

3. The instantaneous velocity (first derivate) at t=2 seconds is: 12m/s

$$s'(t) = 4t + 4, for\ t = 2, s'(2) = 4(2) + 4 = 12m/s$$

4. The instantaneous velocity (first derivate) at t=5 seconds is: 24m/s

$s'(t) = 4t + 4, for\ t = 2, s'(5) = 4(5) + 4 = 24m/s$

If $R(t) = \frac{3000t^2+500t}{5} + 20,000$ represent the revenue that a company earns in time.

5. After 10 days, the revenue is increasing with a speed of $32,100 per day.

$R'(t) = \frac{1}{5}(6000t + 500) + 20,000\ so, R'(10) = \frac{1}{5}[6000(10) + 500] + 20,000 = \frac{1}{5}[60,000 + 500] +$

$20,000 = \frac{1}{5}[60,500] + 20,000 = 12,100 + 20,000 = \$32,100$ per day.

6. After 30 days, the revenue is increasing with a speed of $56,100

$R'(30) = \frac{1}{5}[6000(30) + 500] + 20,000 = \frac{1}{5}[180,000 + 500] + 20,000 = \frac{1}{5}[180,500] + 20,000 =$

$36,100 + 20,000 = \$56,100$ per day.

The following questions follow the graph below. The graph represents the function:

$f(x) = -0.2x^2 + 2x + 1$

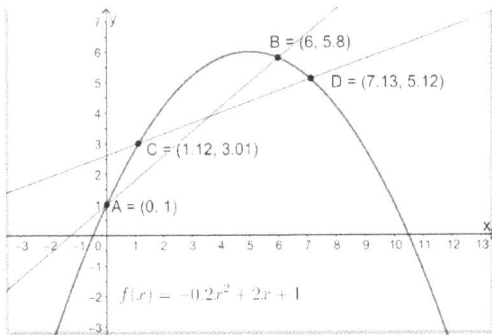

7. The average rate of change between point A and B is $\frac{5}{6}$

$Av. rate\ of\ change = \frac{5.8-1}{6-0} = \frac{4.8}{6}$

8. The average rate of change between point C and D is

$Av. rate\ of\ change = \frac{5.12-3.01}{7.13-1.12} = \frac{2.11}{6.01}$

9. The instantaneous rate of change (first derivate) at point C equals 1.552

$f'(x) = -0.2(2)x + 2 = -0.4x + 2\ ;\ f'(1) = -0.4(1.12) + 2 = -0.448 + 2 = 1.552$

10. The instantaneous rate of change (first derivate) at point D equals -0.852

$f'(x) = -0.2(2)x + 2 = -0.4x + 2\ ;\ f'(7) = -0.4(7.13) + 2 = -2.852 + 2 = -0.852$

Chapter 7. D. b.   Rate of change- Slope of secant and tangent lines

The next questions follow the graph below. The function represented below is:

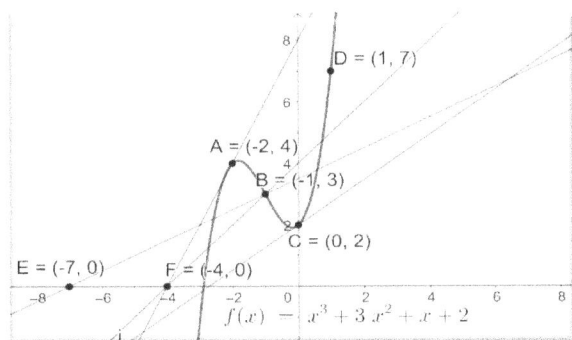

$f(x) = x^3 + 3x^2 + x + 2$

1. The slope of the secant that passes through the points E and B is: $\frac{1}{2}$

$m = \frac{y_2-y_1}{x_2-x_1} = \frac{3-0}{-1-(-7)} = \frac{3}{6} = \frac{1}{2}$

2. The slope of the secant that passes through the points F and B is 1

$m = \frac{y_2-y_1}{x_2-x_1} = \frac{3-0}{-1-(-4)} = \frac{3}{3} = 1$

3. The slope of the tangent to the graph in point A equals 1

$f(x) = x^3 + 3x^2 + x + 2$; so $f'(x) = 3x^2 + 6x + 1$

$f'(-2) = 3(-2)^2 + 6(-2) + 1 = 12 - 12 + 1 = 1$

4. The slope of the tangent to the graph in point C equals 1

$f(x) = x^3 + 3x^2 + x + 2$; so $f'(x) = 3x^2 + 6x + 1$

$f'(0) = 3(0)^2 + 6(0) + 1 = 1$

5. The slope of the tangent to the graph in point D equals 10

$f(x) = x^3 + 3x^2 + x + 2$; so $f'(x) = 3x^2 + 6x + 1$

$f'(1) = 3(1)^2 + 6(1) + 1 = 3 + 6 + 1 = 10$

6. The slope of the tangent to the graph in point B equals -2

$f(x) = x^3 + 3x^2 + x + 2$; so $f'(x) = 3x^2 + 6x + 1$

$f'(1) = 3(-1)^2 + 6(-1) + 1 = 3 - 6 + 1 = -2$

7. The slope of the secant that passes through the points A and B is: -1

$m = \frac{y_2 - y_1}{x_2 - x_1} = \frac{3-4}{-1-(-2)} = \frac{-1}{1} = -1$

8. The slope of the secant that passes through the points B and C is: -1

$m = \frac{y_2 - y_1}{x_2 - x_1} = \frac{2-3}{0-(-1)} = \frac{-1}{1} = -1$

9. The slope of the secant that passes through the points B and D is: 2

$m = \frac{y_2 - y_1}{x_2 - x_1} = \frac{7-3}{1-(-1)} = \frac{4}{2} = 2$

10. The slope of the secant that passes through the points C and D is: 5

$m = \frac{y_2 - y_1}{x_2 - x_1} = \frac{7-2}{1-0} = \frac{5}{1} = 5$

Chapter 7. E. Transcendental Functions – Logarithmic, Exponential, Trigonometric

1. If $f(x) = 2e^x - 7\ln(x)$; then $f'(x) = 2e^x - \frac{7}{x}$

$f'(x) = (2e^x)' - [7\ln(x)]' = 2e^x - \frac{7}{x}$

2. If $f(x) = 3\ln(x) + 5e^x - ex$; then $f'(x) = \frac{3}{x} + 5e^x - e$

$f'(x) = [3\ln(x)]' + [5e^x]' - (ex)' = \frac{3}{x} + 5e^x - e$ (e in the term ex is a constant)

3. If $f(x) = 37 - e^x$; then $f'(x) = -e^x$

$f'(x) = (37 - e^x)' = -e^x$

4. If $f(x) = 73 + \ln(124) - e^2$; then $f'(x) = 0$

$f'(x) = [73 + \ln(124) - e^2]' = 0$ (all the terms are constants)

5. If $f(x) = -4e^x + \ln(27)$; then $f'(x) = -4e^x$

$f'(x) = [-4e^x + \ln(27)]' = -(4e^x)' + [\ln(27)]' = -4e^x = -4e^x = -4e^x$

6. If $f(x) = \cos(x) + 3x$; then $f'(x) = -\sin(x) + 3$

$f'(x) = -\sin(x) + 3$

7. If $(x) = 2\ln(x) + 3e^x + \tan(x)$; then $f'(x) = \frac{2}{x} + 3e^x + \sec^2(x)$

$f'(x) = [2\ln(x) + 3e^x + \tan(x)]' = [2\ln(x)]' + [3e^x]' + [\tan(x)]' = \frac{2}{x} + 3e^x + \sec^2(x)$

8. If $(x) = \frac{3}{7}\ln(x) + \frac{4}{9}e^x + \frac{5}{9}\tan(x)$; *then* $f'(x) = \frac{3}{7x} + \frac{4}{9}e^x + \frac{5}{9}\sec^2(x)$

$f'(x) = [\frac{3}{7}\ln(x) + \frac{4}{9}e^x + \frac{5}{9}\tan(x)]' = [\frac{3}{7}\ln(x)]' + [\frac{4}{9}e^x]' + [\frac{5}{9}\tan(x)]' = \frac{3}{7x} + \frac{4}{9}e^x + \frac{5}{9}\sec^2(x)$

9. If $(x) = \frac{5}{9}\ln(2) + \frac{1}{2}e^x + \cos(x)$; *then* $f'(x) = \frac{1}{2}e^x - \sin(x)$

$f'(x) = [\frac{5}{9}\ln(2) + \frac{1}{2}e^x + \cos(x)]' = [\frac{5}{9}\ln(2)]' + [\frac{1}{2}e^x]' + [\cos(x)]' = 0 + \frac{1}{2}e^x - \sin(x)$

10. If $(x) = 4e^x + \frac{5}{6}\ln(x) - 6\cot(x)$; *then* $f'(x) = 4e^x + \frac{5}{6x} + 6\csc^2(x)$

$f'(x) = [4e^x + \frac{5}{6}\ln(x) - 6\cot(x)]' = [4e^x]' + [\frac{5}{6}\ln(x)]' - 6\cot(x)]' = 4e^x + \frac{5}{6x} - 6[-\csc^2(x)] =$

$4e^x + \frac{5}{6x} + 6\csc^2(x)$

Chapter 7. F. a.   Differentiation rules - Power

1. If $f(x) = 3x^2 - 7x + 89$ *then* $f'(x) = 6x - 7$

$f'(x) = 2(3)x^{2-1} - 7x^{1-1} = 6x - 7x^0 = 6x - 7(1) = 6x - 7$

2. If $f(x) = \frac{2}{5}x^2 + 27x - 63$

$f'(x) = 2\left(\frac{2}{5}\right)x^{2-1} + 27x^{1-1} - 0 = \left(\frac{4}{5}\right)x^2 + 27x^0 = \frac{4}{5}x + 27$

3. If $f(x) = \frac{3}{7}x^4 - 2x^3 - 6x^2 + 81$

$f'(x) = 4(\frac{3}{7})x^{4-1} - 3(2)x^{3-1} - 2(6)x^{2-1} + 0 = \frac{12}{7}x^3 - 6x^2 - 12x$

4. If $f(x) = x^4 - 2x^3 - 6x^2 + 81$ *then* $f'(x) = 4x^3 - 6x^2 - 12x$

$f'(x) = 4x^{4-1} - 2(3)x^{3-1} - 2(6)x^{2-1} + 0 = 4x^3 - 6x^2 - 12x$

5. If $f(x) = \frac{1}{x}$

Using the exponents rules we have: $f(x) = \frac{1}{x} = x^{-1}$

So $f'(x) = -1x^{-1-1} = -x^{-2} = -\frac{1}{x^2}$

6. If $f(x) = \frac{1}{x^4}$ *then* $f'(x) = -4x^{-5}$

Using the exponents rules we have: $f(x) = \frac{1}{x^4} = x^{-4}$

So $f'(x) = -4x^{-4-1} = -4x^{-5} = \frac{-4}{x^5}$

7. If $f(x) = \sqrt{x^3}$

Using the exponents rules we have: $f(x) = \sqrt{x^3} = x^{\frac{3}{2}}$

So $f'(x) = \frac{3}{2}x^{\frac{3}{2}-1} = \frac{3}{2}x^{\frac{1}{2}} = \frac{3}{2}\sqrt{x}$

8. If $f(x) = \frac{\sqrt{x}}{x^4}$

Using the exponents rules we have: $f(x) = \frac{\sqrt{x}}{x^4} = x^{\frac{1}{2}} * x^{-4} = x^{\frac{1}{2}-4} = x^{\frac{1}{2}-4} = x^{-\frac{7}{2}}$

So $f'(x) = -\frac{7}{2}x^{-\frac{7}{2}-1} = -\frac{7}{2}x^{-\frac{9}{2}} = -\frac{7}{2} * \frac{1}{x^{\frac{9}{2}}} = -\frac{7}{2\sqrt{x^9}} = -\frac{7}{2x^4\sqrt{x}}$

9. If $f(x) = \frac{\sqrt{x}}{x^2} + 3x$ then $f'(x) = -\frac{7}{2x^2\sqrt{x}} + 3$

Using the exponents rules we have: $f(x) = \frac{\sqrt{x}}{x^2} + 3x = x^{\frac{1}{2}} * x^{-2} + 3x = x^{\frac{1}{2}-2} = x^{-\frac{3}{2}} + 3x$

So $f'(x) = -\frac{3}{2}x^{-\frac{3}{2}-1} + 3 = -\frac{3}{2}x^{-\frac{5}{2}} + 3 = -\frac{3}{2} * \frac{1}{x^{\frac{5}{2}}} + 3 = -\frac{3}{2\sqrt{x^5}} + 3 = -\frac{3}{2x^2\sqrt{x}} + 3$

10. If $f(x) = \frac{x^3}{x^{-2}}$ then $f'(x) = 5x^4$

Using the exponents rules we have: $f(x) = \frac{x^3}{x^{-2}} = x^3 * x^2 = x^5$

So $f'(x) = 5x^{5-1} = 5x^4$

Chapter 7. F. b.   Differentiation rules – Product

1. If $f(x) = 2x^2 \cos(x)$; then $f'(x) = 4x\cos(x) - 2x^2\sin(x)$

$f'(x) = (2x^2)' \cos(x) + 2x^2[\cos(x)]' = 4x\cos(x) + 2x^2[-\sin(x)] = 4x\cos(x) - 2x^2\sin(x)$

2. If $f(x) = 3x^4\sqrt{x}$; then $f'(x) = 12\frac{3}{2}x^3\sqrt{x}$

$f(x) = 3x^4\sqrt{x}$; then $f'(x) = (3x^4)'\sqrt{x} + 3x^4(\sqrt{x})' = 12x^3\sqrt{x} + 3x^4\frac{1}{2}x^{-\frac{1}{2}} = 12x^3\sqrt{x} + \frac{3}{2}\frac{x^4}{\sqrt{x}} =$

$12x^3\sqrt{x} + \frac{3}{2}\frac{x^4\sqrt{x}}{\sqrt{x}\sqrt{x}} = 12x^3\sqrt{x} + \frac{3}{2}x^3\sqrt{x} = 12\frac{3}{2}x^3\sqrt{x}$

Where; $(\sqrt{x})' = [(x)^{\frac{1}{2}}]' = \frac{1}{2}x^{-\frac{1}{2}}$

3. If $f(x) = 4x^2 \tan(x)$; then $f'(x) = (8x)\tan(x) + 4x^2\sec^2(x)$

$f'(x) = (4x^2)' \tan(x) + 4x^2[\tan(x)]' = (8x)\tan(x) + 4x^2\sec^2(x)$

4. If $f(x) = \frac{2}{5}x^2e^x$; then $f'(x) = \frac{2}{5}xe^x(2+x)$

$f'(x) = (\frac{2}{5}x^2)'e^x + \frac{2}{5}x^2[e^x]' = \frac{4}{5}xe^x + \frac{2}{5}x^2e^x = \frac{2}{5}xe^x(2+x)$

5. If $f(x) = 2^x \sin(x)$; then $f'(x) = 2^x \ln 2 + 2^x\cos(x)$

$f'(x) = (2^x)' \sin(x) + 2^x[\sin(x)]' = 2^x \ln 2 \sin(x) + 2^x\cos(x)$

6. If $f(x) = 3\ln|x|x^2$; then $f'(x) = 3x + 6x\ln|x|$

$f'(x) = (3\ln|x|)'x^2 + 3\ln|x|(x^2)' = \frac{3}{x} * x^2 + 3\ln|x| * 2x = 3x + 6x\ln|x|$

7. If $f(x) = \ln|x|\sqrt{x}$; then $f'(x) = x^{-\frac{1}{2}}(1 + \frac{1}{2}\ln|x|)$

$f'(x) = (\ln|x|)'\sqrt{x} + \ln|x|(\sqrt{x})' = \frac{\sqrt{x}}{x} + \ln|x|\frac{1}{2}x^{-\frac{1}{2}} = x^{\frac{1}{2}-1} + \ln|x|\frac{1}{2}x^{-\frac{1}{2}} = x^{-\frac{1}{2}}(1 + \frac{1}{2}\ln|x|)$

Where; $(\sqrt{x})' = [(x)^{\frac{1}{2}}]' = \frac{1}{2}x^{-\frac{1}{2}}$

8. If $f(x) = \cos(x) * e^x$; then $f'(x) = e^x[\cos(x) - \sin(x)]$

$f'(x) = [\cos(x)]'e^x + \cos(x)[e^x]' = -\sin(x)e^x + \cos(x)e^x = e^x[\cos(x) - \sin(x)]$

9. If $f(x) = e^x3^x$; then $f'(x) = e^x3^x[1 + \ln(3)]$

$f'(x) = [e^x]'3^x + e^x[3^x]' = e^x3^x + e^x3^x\ln(3) = e^x3^x[1 + \ln(3)]$

10. If $f(x) = e^x x^5$; then $f'(x) = e^x x^4(x+5)$

$f'(x) = [e^x]' x^5 + e^x [x^5]' = e^x x^5 + e^x * 5x^4 = e^x x^4(x+5)$

Chapter 7. F. c.   Differentiation rules – Quotient

1. If $f(x) = \tan(x)$; then $f'(x) = [\sec(x)]^2$

$f(x) = \dfrac{\sin(x)}{\cos(x)}$

$f'(x) = \dfrac{[\sin(x)]' \cos(x) - \sin(x)[\cos(x)]'}{[\cos(x)]^2} = \dfrac{\cos(x)\cos(x) - \sin(x)[-\sin(x)]}{[\cos(x)]^2} = \dfrac{[\cos(x)]^2 + [\sin(x)]^2}{[\cos(x)]^2} = \dfrac{1}{[\cos(x)]^2} =$

$[\sec(x)]^2$

2. If $f(x) = \dfrac{\sin(x)}{x^2}$; then $f'(x) = \dfrac{[x\cos(x) - 2\sin(x)]}{x^3}$

$f'(x) = \dfrac{[\sin(x)]' x^2 - \sin(x)[x^2]'}{[x^2]^2} = \dfrac{x^2 \cos(x) - 2x\sin(x)}{x^4} = \dfrac{x[x\cos(x) - \sin(x)2]}{x^4} = \dfrac{[x\cos(x) - 2\sin(x)]}{x^3}$

3. If $f(x) = \dfrac{x}{e^x}$; then $f'(x) = \dfrac{1-x}{e^x}$

$f'(x) = \dfrac{x' e^x - x[e^x]'}{[e^x]^2} = \dfrac{e^x - xe^x}{[e^x]^2} = \dfrac{e^x(1-x)}{[e^x]^2} = \dfrac{1-x}{e^x}$

4. If $f(x) = \dfrac{\sin(x)}{x^2+1}$; then $f'(x) = \dfrac{\cos(x)(x^2+1) - \sin(x)*2x}{[x^2+1]^2}$

$f'(x) = \dfrac{[\sin(x)]'(x^2+1) - \sin(x)[x^2+1]'}{[x^2+1]^2} = \dfrac{\cos(x)(x^2+1) - \sin(x)*2x}{[x^2+1]^2}$

5. If $f(x) = \dfrac{x^2+2x}{\sqrt{x}}$; then $f'(x) = \dfrac{\sqrt{x}(1.5x-3)}{x}$

$f'(x) = \dfrac{[x^2+2x]'\sqrt{x} - (x^2+2x)(\sqrt{x})'}{(\sqrt{x})^2} = \dfrac{(2x+2)\sqrt{x} - (x^2+2x)\frac{1}{2\sqrt{x}}}{x}$

$(\sqrt{x})' = (x^{\frac{1}{2}})' = \dfrac{x^{\frac{1}{2}-1}}{2} = \dfrac{x^{-\frac{1}{2}}}{2} = \dfrac{1}{2\sqrt{x}}$

$f'(x) = \dfrac{(2x+2)\sqrt{x} - (x^2+2x)\frac{x^{-\frac{1}{2}}}{2}}{x} = \dfrac{2x\sqrt{x} - 2\sqrt{x} - \frac{x^{\frac{3}{2}}}{2} - \frac{2x^{\frac{1}{2}}}{2}}{x} = \dfrac{2x\sqrt{x} - 2\sqrt{x} - \frac{x\sqrt{x}}{2} - \sqrt{x}}{x} = \dfrac{\sqrt{x}(2x-2-0.5x-1)}{x} = \dfrac{\sqrt{x}(1.5x-3)}{x}$

6. If $f(x) = 2x\sec(x) = \dfrac{2x}{\cos(x)}$; then $f'(x) = \dfrac{2[1 - x\tan(x)]}{\cos(x)}$

$f'(x) = \dfrac{(2x)'\cos(x) - 2x[\cos(x)]'}{[\cos(x)]^2} = \dfrac{2\cos(x) - 2x[-\sin(x)]}{[\cos(x)]^2} = \dfrac{2\cos(x) + 2x\sin(x)}{[\cos(x)]^2} = \dfrac{2[\cos(x) - x\sin(x)]}{[\cos(x)]^2} =$

$\dfrac{2\cos(x)[1 - \frac{x\sin(x)}{\cos(x)}]}{[\cos(x)]^2} = \dfrac{2[1 - x\tan(x)]}{\cos(x)}$

7. If $f(x) = \dfrac{3x^2-4x+2}{\sin(x)}$; then $f'(x) = \dfrac{(6x-4)\sin(x) - (3x^2-4x+2)\cos(x)}{[\sin(x)]^2}$

$f'(x) = \dfrac{(3x^2-4x+2)'\sin(x) - (3x^2-4x+2)[\sin(x)]'}{[\sin(x)]^2} = \dfrac{(6x-4)\sin(x) - (3x^2-4x+2)\cos(x)}{[\sin(x)]^2}$

8. $f(x) = \dfrac{x^2+2x-3}{5x+4}$; then $f'(x) = \dfrac{5x^2+8x+23}{(5x+4)^2}$

$f'(x) = \dfrac{(x^2+2x-3)'(5x+4) - (x^2+2x-3)(5x+4)'}{(5x+4)^2} = \dfrac{(2x+2)(5x+4) - (x^2+2x-3)*5}{(5x+4)^2} = \dfrac{10x^2+8x+10x+8 - 5x^2-10x+15}{(5x+4)^2} =$

$\dfrac{5x^2+8x+23}{(5x+4)^2}$

9. If $f(x) = \frac{x^2}{lnx}$; then $f'(x) = \frac{x(2lnx-1)}{[\ln(x)]^2}$

$$f'(x) = \frac{(x^2)'\ln(x)-x^2[\ln(x)]'}{[\ln(x)]^2} = \frac{(2x)lnx-\frac{x^2}{x}}{[\ln(x)]^2} = \frac{(2x)lnx-x}{[\ln(x)]^2} = \frac{x(2lnx-1)}{[\ln(x)]^2}$$

10. $f(x) = \frac{2x^2-4x+6}{2x}$; then $f'(x) = 1 - \frac{3}{x^2}$

$$f'(x) = \frac{(2x^2-4x+6)'2x-(2x^2-4x+6)*2}{4x^2} = \frac{(4x-4)*2x-4x^2+8x-12}{4x^2} = \frac{8x^2-8x-4x^2+8x-12}{4x^2} = \frac{4x^2-12}{4x^2} = \frac{4(x^2-3)}{4x^2} =$$

$$\frac{x^2-3}{x^2} = 1 - \frac{3}{x^2}$$

Chapter 7. F. d.   Differentiation rules – Chain

1. If $f(x) = (x^2 - 4x + 5)^2$; then $f'(x) = 2(x^2 - 4x + 5)(2x - 4)$

$f'(x) = [x^2 - 4x + 5)^2]'(x^2 - 4x + 5)' = 2(x^2 - 4x + 5)(2x - 4)$

2. If $f(x) = (3x^2 + 7x)^{\frac{3}{2}}$; then $f'(x) = \frac{3}{2}\sqrt{3x^2 + 7x}\,(6x + 7)$

$f'(x) = \frac{3}{2}(3x^2 + 7x)^{\frac{3}{2}-1}(3x^2 + 7x)' = \frac{3}{2}(3x^2 + 7x)^{\frac{1}{2}}(6x + 7) = \frac{3}{2}\sqrt{3x^2 + 7x}\,(6x + 7)$

3. If $f(x) = \sqrt{x^3 - 5x}$; then $f'(x) = \frac{(3x^2-5)}{2\sqrt{x^3-5x}}$

$f(x) = \sqrt{x^3 - 5x} = (x^3 - 5x)^{\frac{1}{2}}$ so:

$f'(x) = \frac{1}{2}(x^3 - 5x)^{\frac{1}{2}-1}(x^3 - 5x)' = \frac{1}{2}(x^3 - 5x)^{\frac{1}{2}-1}(3x^2 - 5) = \frac{1}{2}(x^3 - 5x)^{-\frac{1}{2}}(3x^2 - 5) = \frac{(3x^2-5)}{2\sqrt{x^3-5x}}$

4. If $f(x) = \sqrt{e^x - 2x}$; then $f'(x) = \frac{(e^x-2)}{2\sqrt{(e^x-2x}}$

$f(x) = \sqrt{e^x - 2x} = (e^x - 2x)^{\frac{1}{2}}$ so:

$f'(x) = [(e^x - 2x)^{\frac{1}{2}}]'(e^x - 2x)' = \frac{1}{2}(e^x - 2x)^{\frac{1}{2}-1}(e^x - 2) = \frac{1}{2}(e^x - 2x)^{-\frac{1}{2}}(e^x - 2) = \frac{(e^x-2)}{2\sqrt{(e^x-2x}}$

5. If $f(x) = \sqrt{\sin(x)}$; then $f'(x) = \frac{1}{2}\frac{\cos(x)}{2\sqrt{\sin(x)}}$

$f(x) = \sqrt{\sin(x)} = [\sin(x)]^{\frac{1}{2}}$

$f'(x) = \frac{1}{2}[\sin(x)]^{\frac{1}{2}-1}[\sin(x)]' = \frac{1}{2}[\sin(x)]^{-\frac{1}{2}}\cos(x) = \frac{\cos(x)}{2\sqrt{\sin(x)}}$

6. If $f(x) = 2[\ln(4x - 3)]$; then $f'(x) = \frac{8}{4x-3}$

$f'(x) = 2[\ln(4x - 3)]'(4x - 3)' = \frac{2(4)}{4x-3} = \frac{8}{4x-3}$

7. If $f(x) = [\ln(e^x - e^{-x})]$; then $f'(x) = \frac{e^x+e^{-x}}{(e^x-e^{-x})}$

$f'(x) = [\ln(e^x - e^{-x})]'(e^x - e^{-x})' = \frac{e^x+e^{-x}}{(e^x-e^{-x})}$

8. If $f(x) = \sin(x - 1)$; then $f'(x) = \cos(x - 1)$

$f'(x) = \cos(x - 1)(x - 1)' = \cos(x - 1)$

9. If $f(x) = e^x + 2\ln(x + 7)$; then $f'(x) = e^x + \frac{2}{x+7}$

$$f'(x) = (e^x)' + [2\ln(x+7)]'(x+7)' = e^x + \frac{2}{x+7}$$

10. If $f(x) = \ln(x^2 - 4x)$; *then* $f'(x) = \frac{2(x-2)}{x(x-4)}$

$$f'(x) = [\ln(x^2-4x)]'(x^2-4x)' = \frac{2x-4}{x^2-4x} = \frac{2(x-2)}{x(x-4)}$$

Chapter 7. G. Higher order differentiation

1. If $f(x) = x^3 - 2x^2 + 3x - 4$; *then* $f''(x) = 6x - 4$

$$f'(x) = 3x^2 - 4x + 3$$

$$f''(x) = 6x - 4$$

2. If $f(x) = 7x^3 + 6x^2 + 5x - 4$; *then* $f'''(x) = 42$

$$f'(x) = 21x^2 + 12x + 5$$

$$f''(x) = 42x + 12$$

$$f'''(x) = 42$$

3. If $f(x) = 3x^5 + 4\sqrt{x} - 5x - 6$; *then* $f'''(x) = 180x^2 + \frac{3}{x^2\sqrt{x}}$

$$f'(x) = 15x^4 + 4x^{\frac{1}{2}-1} - 5 = 15x^4 + 4x^{-\frac{1}{2}} - 5$$

$$f''(x) = 60x^3 + 4\left(-\frac{1}{2}\right)x^{-\frac{1}{2}-1} = 60x^3 - 2x^{-\frac{3}{2}}$$

$$f'''(x) = 180x^2 - 2\left(-\frac{3}{2}\right)x^{-\frac{3}{2}-1} = 180x^2 + (3)x^{-\frac{3}{2}-1} = 180x^2 + (3)x^{-\frac{5}{2}} = 180x^2 + \frac{3}{x^2\sqrt{x}}$$

4. If $f(x) = 3\sqrt{5x-1} - 4x + 5$; *then* $f''(x) = \frac{15}{2\sqrt{5x-1}} - \frac{75x^2}{4(5x-1)\sqrt{5x-1}}$

We apply the chain rule for differentiating $\sqrt{5x-1}$, where $u(x) = 5x - 1$ *and* $f(u) = \sqrt{u}$

$$f'(x) = [3(5x-1)^{\frac{1}{2}}]' - (4x)' = 3[\frac{1}{2}(5x-1)^{\frac{1}{2}-1}](5x-1)' - 4 = \frac{3}{2}(5x-1)^{-\frac{1}{2}}(5x) - 4 =$$

$$\frac{15x}{2}(5x-1)^{-\frac{1}{2}} - 4$$

$$f''(x) = [\frac{15x}{2}(5x-1)^{-\frac{1}{2}}]' - (4)' = \left(\frac{15x}{2}\right)'(5x-1)^{-\frac{1}{2}} + \frac{15x}{2}[(5x-1)^{-\frac{1}{2}}]' = \frac{15}{2}(5x-1)^{-\frac{1}{2}} +$$

$$\frac{15x}{2}[-\frac{1}{2}(5x-1)^{-\frac{1}{2}-1}(5x-1)'] = \frac{15}{2}(5x-1)^{-\frac{1}{2}} + \frac{15x}{2}[-\frac{3}{2}(5x-1)^{-\frac{3}{2}}(5x)] = \frac{15}{2}(5x-1)^{-\frac{1}{2}} -$$

$$\frac{75x^2}{4}(5x-1)^{-\frac{3}{2}} = \frac{15}{2\sqrt{5x-1}} - \frac{75x^2}{4(5x-1)\sqrt{5x-1}}$$

Where:

For $[\frac{15x}{2}(5x-1)^{-\frac{1}{2}}]'$ we applied the product rule

For $[(5x-1)^{-\frac{1}{2}}]'$ we applied the chain rule

5. If $f(x) = \sqrt{7x^2 + 6x}$; *then* $f''(x) = \frac{-36}{(7x^2+6x)\sqrt{7x^2+6x}}$

$$f(x) = \sqrt{7x^2 + 6x} = (7x^2 + 6x)^{\frac{1}{2}}$$

$$f'(x) = [(7x^2+6x)^{\frac{1}{2}}]' = \frac{1}{2}(7x^2+6x)^{\frac{1}{2}-1}(7x^2+6x)' = \frac{1}{2}(7x^2+6x)^{-\frac{1}{2}}(14x+6)$$

Where:

For $[(7x^2 + 6x)^{\frac{1}{2}}]'$ we applied chain rule.

$f''(x) = [\frac{1}{2}(7x^2 + 6x)^{-\frac{1}{2}}]'(14x + 6) + \frac{1}{2}(7x^2 + 6x)^{-\frac{1}{2}}(14x + 6)' = -\frac{1}{4}(7x^2 + 6x)^{-\frac{1}{2}-1}(7x^2 +$

$6x)'(14x + 6) + \frac{1}{2}(7x^2 + 6x)^{-\frac{1}{2}}(14) = -\frac{1}{4}(7x^2 + 6x)^{-\frac{3}{2}}(14x + 6)(14x + 6) + 7(7x^2 + 6x)^{-\frac{1}{2}} =$

$-\dfrac{(14x+6)^2}{4(7x^2+6x)\sqrt{7x^2+6x}} + \dfrac{7}{\sqrt{7x^2+6x}} = \dfrac{28(7x^2+6x)-(14x+6)^2}{(7x^2+6x)\sqrt{7x^2+6x}} = \dfrac{-36}{(7x^2+6x)\sqrt{7x^2+6x}}$

Where:

For $[\frac{1}{2}(7x^2 + 6x)^{-\frac{1}{2}}]'$ we applied the chain rule.

6. If $f(x) = \dfrac{2}{2x-3}$; then $f'''(x) = \dfrac{-48}{(2x-3)^4}$

$f(x) = \dfrac{2}{2x-3} = 2(2x - 3)^{-1}$

$f'(x) = -2(2x - 3)^{-2}(2x - 3)' = -2(2x - 3)^{-2}(2) = -4(2x - 3)^{-2}$

We applied the chain rule for $(2x - 3)^{-1}$

$f''(x) = (-4)[(2x - 3)^{-2}]' = -4(-2)(2x - 3)^{-3}(2x - 3)' = 8(2x - 3)^{-3}(2) = 16(2x - 3)^{-3}$

$f'''(x) = [16(2x - 3)^{-3}]' = 16(-3)(2x - 3)^{-4}(2x - 3)' = -24(2x - 3)^{-4}(2) = \dfrac{-48}{(2x-3)^4}$

We applied again the chain rule.

7. If $(x) = e^x \ln (x)$; then $f'''(x) = e^x \ln(x) + 3e^x x^{-1} - 3e^x x^{-2} + 2e^x x^{-3}$

We are applying the product rule.

$f'(x) = (e^x)' \ln(x) + e^x[\ln (x)]' = e^x \ln(x) + e^x x^{-1}$

$f''(x) = [e^x]' \ln(x) + e^x x^{-1} + [e^x]' x^{-1} + e^x[x^{-1}]' = e^x \ln(x) + e^x x^{-1} + e^x x^{-1} - e^x x^{-2} =$

$e^x \ln(x) + 2e^x x^{-1} - e^x x^{-2}$

$f'''(x) = [e^x]' \ln(x) + e^x[\ln (x)]' + 2[e^x]' x^{-1} + 2e^x[x^{-1}]' - [e^x]' x^{-2} - e^x[x^{-2}]' = e^x \ln(x) +$

$e^x x^{-1} + 2e^x x^{-1} - 2e^x x^{-2} - e^x x^{-2} - e^x(-2)x^{-3} = e^x \ln(x) + 3e^x x^{-1} - 3e^x x^{-2} + 2e^x x^{-3}$

8. If $f(x) = \dfrac{1}{2x} + \sqrt{x}$; then $f'''(x) = \dfrac{-3}{x^4} + \dfrac{3}{8x^2\sqrt{x}}$

$f(x) = \dfrac{1}{2x} + \sqrt{x} = \dfrac{1}{2}x^{-1} + x^{\frac{1}{2}}$

$f'(x) = -\dfrac{1}{2}x^{-2} + \dfrac{1}{2}x^{\frac{1}{2}-1} = -\dfrac{1}{2}x^{-2} + \dfrac{1}{2}x^{-\frac{1}{2}}$

$f''(x) = x^{-3} - \dfrac{1}{4}x^{-\frac{1}{2}-1} = x^{-3} - \dfrac{1}{4}x^{-\frac{3}{2}}$

$f'''(x) = -3x^{-4} + \dfrac{3}{8}x^{-\frac{3}{2}-1} = -3x^{-4} + \dfrac{3}{8}x^{-\frac{5}{2}} = \dfrac{-3}{x^4} + \dfrac{3}{8x^2\sqrt{x}}$

9. If $f(x) = \sin(x) + x^3$; then $f''(x) = -\sin(x) + 6x$

$f'(x) = [\sin(x)]' + (x^3)' = \cos(x) + 3x^2$

$f''(x) = [\cos(x)]' + (3x^2)' = -\sin(x) + 6x$

10. If $f(x) = \cos(x) + e^x$; then $f''(x) = -\cos(x) + e^x$

$f'(x) = [\cos(x)]' + (e^x)' = -\sin(x) + e^x$

$f''(x) = [-\sin(x)]' + (e^x)' = -\cos(x) + e^x$

Chapter 7. H.   Implicit differentiation

1. If $xy + x = 2$; $y' = \frac{-1-y}{x}$

Using implicit differentiation, we have:

$(xy)' + x' = 0$

$(x)'y + xy' + 1 = 0$

$y + xy' = -1$

$xy' = -1 - y, so\ y' = \frac{-1-y}{x}$

2. If $x^2 + y^2 = 10$; $y' = -\frac{x}{y}$

Using implicit differentiation, we have:

$(x^2)' + (y^2)' = 0$

$2x + 2y(y)' = 0$

$2y(y)' = -2x, so\ y' = \frac{-2x}{2y} = -\frac{x}{y}$

3. If $2x^2 + y^3 = 3$; $y'' = \frac{-12y^3-32x^2}{9y^5}$

Using implicit differentiation, we have:

$(2x^2)' + (y^3)' = 0$

$4x + 3y^2(y)' = 0$

$3y^2(y)' = -4x, so\ y' = \frac{-4x}{3y^2}$

Then, using the quotient rule, we have:

$y'' = \frac{(-4x)'(3y^2)-(-4x)(3y^2)'}{(3y^2)^2} = \frac{-4(3y^2)+4x(6yy')}{9y^4} = \frac{-12y^2+24xy(\frac{-4x}{3y^2})}{9y^4} = \frac{-12y^2-\frac{32x^2}{y}}{9y^4} = \frac{-12y^3-32x^2}{9y^5}$

4. If $4x^2 + 2y^2 = 9$; $y'' = \frac{-2y^2-4x^2}{y^3}$

Using implicit differentiation, we have:

$(4x^2)' + (2y^2)' = 0$

$8x + 4y(y)' = 0$

$4y(y)' = -8x, so, y' = \frac{-8x}{4y} = -\frac{2x}{y}$

Then, using the quotient rule, we have:

$y'' = \frac{(-2x)'y-(-2x)\,y'}{y^2} = \frac{-2y+2x(\frac{-2x}{y})}{y^2} = \frac{-2y^2-4x^2}{y^3}$

5. At point (2,3) the tangent slope to the curve $2x^2 + xy = 2\ is -5.5$

Using implicit differentiation, we have:

$(2x^2)' + (xy)' = 0$

$4x + (x)'y + x(y)' = 0$

$4x + y + x(y)' = 0$

$x(y)' = -4x - y\ so, y' = \frac{-4x-y}{x}$

We substitute x=2 and y=3, so we have:

$y' = \frac{-4(2)-3}{2} = \frac{-8-3}{2} = \frac{-11}{2} = -5.5$

At the point (2,3), the tangent slope to the curve $2x^2 + xy = 2$ *is* -5.5

6. At point (3,4) of the curve $2x^2 + 3xy - y^2 = 38$ the slope of the tangent line is: $= -1.41$

Using implicit differentiation, we have:

$(2x^2)' + (3xy)' + (y^2)' = 0$

$4x + 3[(x)'y + x(y)'] + 2y(y)' = 0$

$4x + 3(y + xy') + 2y(y)' = 0$

$4x + 3y + 3xy' + 2y(y)' = 0$

$3xy' + 2y(y)' = -4x - 3y$

$y'(3x + 2y) = -4x - 3y$

$y' = \frac{-4x-3y}{3x+2y}$

The slope of the tangent at point (3,4) is: $y' = \frac{-4(3)-3(4)}{3(3)+2(4)} = \frac{-12-12}{9+8} = \frac{-24}{17} = -1.41$

7. The slope of the tangent line to the graph of $y = \frac{4}{\pi}x - \sin(xy)$ *at* $\left(\frac{\pi}{2}, 1\right)$ *is* $\frac{4}{\pi}$

Using implicit differentiation, we have:

$y' = (\frac{4}{\pi}x)' - [\sin(xy)]' = \frac{4}{\pi} - \cos(xy)(xy)' = \frac{4}{\pi} - \cos(xy)(y + xy')$

We are replacing $x = \frac{\pi}{2}$ and y=1 in

$y' = \frac{4}{\pi} - \cos(xy)(y + xy')$

So,

$y' = \frac{4}{\pi} - \cos\left(\frac{\pi}{2} * 1\right)\left(1 + \frac{\pi}{2}y'\right); but \; \cos\left(\frac{\pi}{2}\right) = 0 \; then \; y' = \frac{4}{\pi}$

8. If $3y^2 + \ln(x) = 2y - \cos(x)$, $y' = \frac{\sin(x)-\frac{1}{x}}{6y-2}$

Using implicit differentiation, we have:

$(3y^2)' + [\ln(x)]' = (2y)' - [\cos(x)]'$

$6y(y)' + \frac{1}{x} = 2y' + \sin(x)$

$6y(y)' - 2y' = +\sin(x) - \frac{1}{x}$

$y'(6y - 2) = \sin(x) - \frac{1}{x}$

$y' = \frac{\sin(x)-\frac{1}{x}}{6y-2}$

9. If $\cot(y) + 2x = 5y - y^2$, $y' = \frac{-2}{-csc^2(y)-5+2y}$

$[\cot(y)]' + (2x)' = (5y)' - (y^2)'$

$-csc^2(y)(y)' + 2 = 5y' - 2y\,y'$

$-csc^2(y)(y)' - 5y' + 2y\,y' = -2$

$y'[-csc^2(y) - 5 + 2y] = -2$

$y' = \frac{-2}{-csc^2(y)-5+2y}$

10. If $(y-1)^2 = 6y + x^3 + 2x$, $y' = \frac{3x^2+2}{2(y-1)-6}$

Using implicit differentiation, we have:

$[(y-1)^2]' = (6y)' + (x^3)' + (2x)'$

$2(y-1)(y-1)' = 6y' + 3x^2 + 2$

$2(y-1)y' = 6y' + 3x^2 + 2$

$2(y-1)y' - 6y' = 3x^2 + 2$

$y'[2(y-1) - 6] = 3x^2 + 2$

$y' = \frac{3x^2+2}{2(y-1)-6}$

Chapter 7. I. a.   Relating graph of f(x) to f '(x) and f "(x)

1. If the first derivative is positive, the function is increasing.

2. If the second derivative is positive, the function is concave up.

3. The graph of the function $f(x) = \frac{2}{3}x^3 - 2x^2 - 6x + 7$ is decreasing between -1 and 3

$f'(x) = \left(\frac{2}{3}x^3\right)' - (2x^2)' - (6x)' + (7)' = \frac{6}{3}x^2 - 4x - 6 = 2x^2 - 4x - 6 = 2(x^2 - 2x - 3) = 2(x+1) * (x-3)$

$f'(x) = 2(x+1)(x-3)$ is zero for x=-1 and x=3

We check the sign of the first derivative by considering x values less than -1, between -1 and 3, and bigger than 3.

$f'(-2) = 2[(-2) + 1][(-2) - 3] = 2(-1)(-5) = +10$

The first derivative is positive for values of x less than -1

$f'(0) = 2[(0) + 1][(0) - 3] = 2(1)(-3) = -6$

The first derivative is negative between -1 and 3

$f'(4) = 2[(4) + 1][(4) - 3] = 2(5)(1) = +10$

The first derivative is positive for values of x bigger than 3

The sign diagram for $f'(x)$ is shown below:

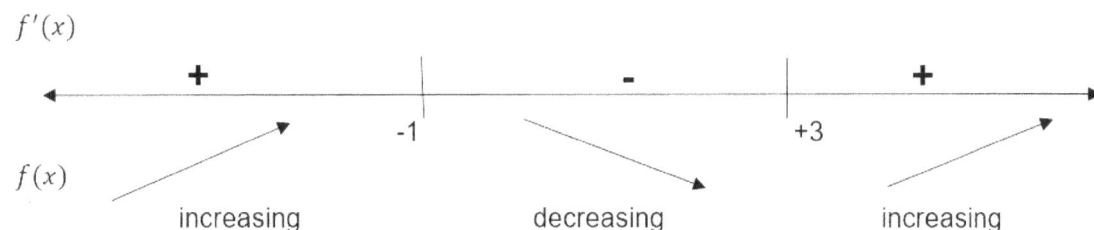

$f'(x)$

| + | | - | | + |

-1   +3

$f(x)$

increasing   decreasing   increasing

The graph is shown below.

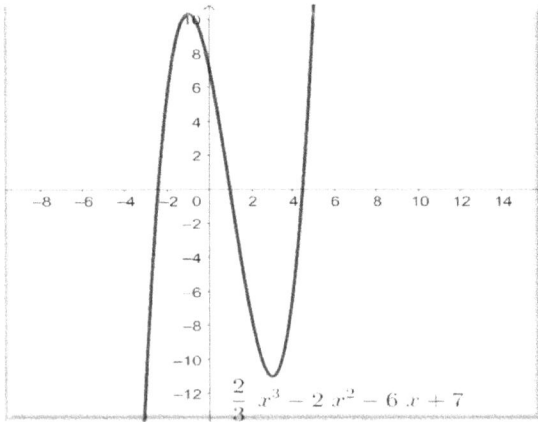

4. The graph of the function $f(x) = 2x^3 - 7x^2 + 7$ is decreasing between 0 and 2.33

$f'(x) = 6x^2 - 14x = 2x(3x - 7)$

$f'(x) = 2x(3x - 7) = 0$ for x=0 and x=7/3=2.33

$f'(-1) = 2(-1)[3(-1) - 7] = -2(-10) = +20$

The first derivative is positive for values of x less than zero.

$f'(1) = 2(1)[3(1) - 7] = 2(-4) = -8$

The first derivative is negative for values of x between 0 and 2.33

$f'(3) = 2(3)[3(3) - 7] = 6(2) = +12$

The first derivative is positive for values of x bigger than 2.33

The sign of the first derivative and the behavior of the graph is shown below

$f'(x)$

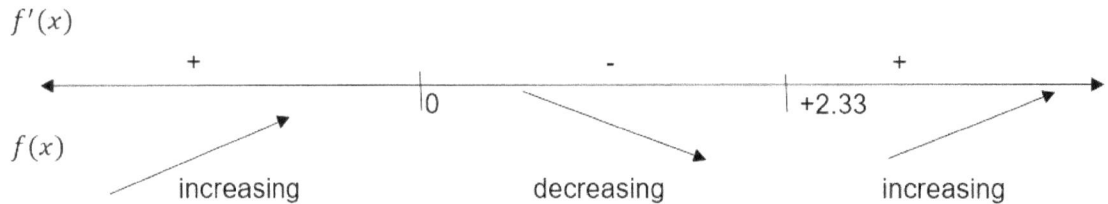

The graph of the function is shown below.

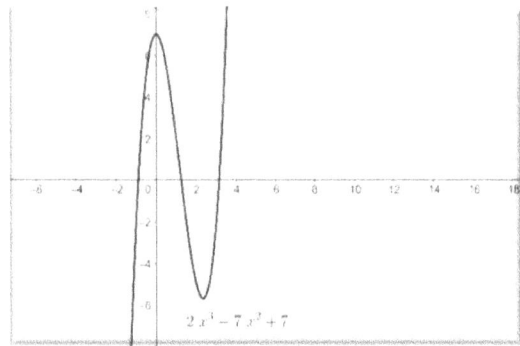

5. The graph of the function $f(x) = \frac{2x-5}{x-1}$ is concave up for values of x less than 1

First, the graph of the function will intersect x axis at x=2.5

Second, there is a vertical asymptote at x=1

$\lim_{x \to 1^-} \frac{2x-5}{x-1} = +\infty$

$\lim_{x \to 1^+} \frac{2x-5}{x-1} = -\infty$

Third, $\lim_{x \to \infty} \frac{2x-5}{x-1} = 2 = \lim_{x \to -\infty} \frac{2x-5}{x-1}$. The graph will have a horizontal asymptote at y=2

Using the quotient rule, we have:

$f'(x) = \frac{(2x-5)'(x-1)-(2x-5)(x-1)'}{(x-1)^2} = \frac{2(x-1)-(2x-5)}{(x-1)^2} = \frac{2x-2-2x+5}{(x-1)^2} = \frac{3}{(x-1)^2}$ is always positive

To verify the concavity, we have to calculate the second derivate.

$f''(x) = [3(x-1)^{-2}]' = 3(-2)(x-1)^{-3}(x-1)' = -6(x-1)^{-3} = \frac{-6}{(x-1)^3}$

$f''(0) = 6, and\ f''(2) = -6$

The sign and concavity of the graph of the function $f(x) = \frac{2x-5}{x-1}$ is shown below.

$f'(x)$

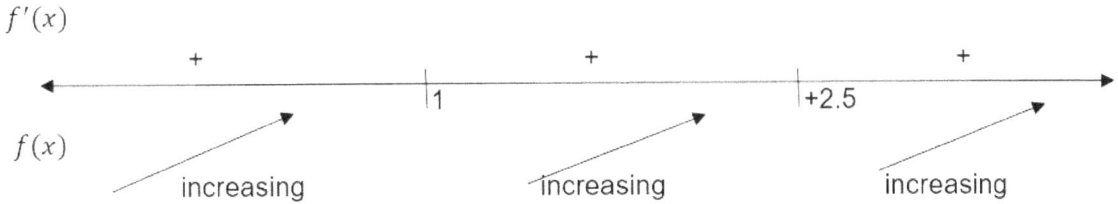

$f(x)$

increasing   increasing   increasing

The concavity is shown below

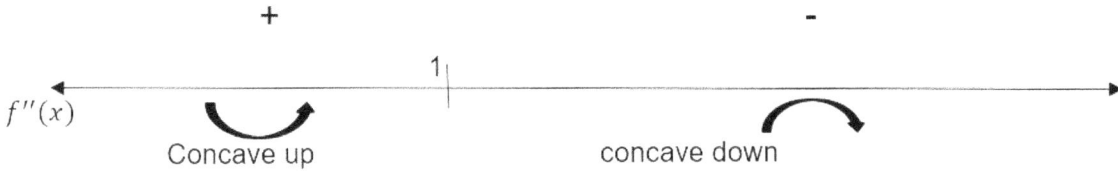

$f''(x)$

Concave up   concave down

The graph is shown below.

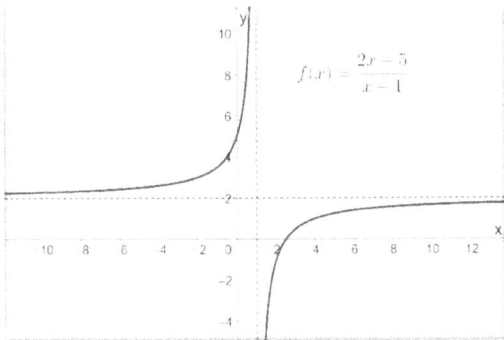

$f(x) = \frac{2x-5}{x-1}$

6. The function $f(x) = \frac{x^2}{x^2-4}$ is concave down

between -2 and 2

$$f(x) = \frac{x^2}{x^2-4} = \frac{x^2}{(x-2)(x+2)} \; 0$$

First, the graph of the function will intersect x axis

at x=0, because $x^2 = 0$ for x=0

Second, there are two vertical asymptotes at x=-2

and x=2

The sign of the function is shown below

	-2	0	2	
x^2	+	+	+	+
$x^2 - 4$	+	-	-	+
Sign of function	+	-	-	+

$$\lim_{x\to-2^-} \frac{x^2}{x^2-4} = +\infty$$

$$\lim_{x\to-2^+} \frac{x^2}{x^2-4} = -\infty$$

$$\lim_{x\to2^-} \frac{x^2}{x^2-4} = -\infty$$

$$\lim_{x\to2^+} \frac{x^2}{x^2-4} = +\infty$$

Third, $\lim\limits_{x\to\infty} \frac{x^2}{x^2-4} = 1 = \lim\limits_{x\to-\infty} \frac{x^2}{x^2-4}$. The graph will have a horizontal asymptote at y=1

Using the quotient rule, we have:

$$f'(x) = \frac{(x^2)'(x^2-4)-(x^2)(x^2-4)'}{(x^2-4)^2} = \frac{(2x)(x^2-4)-(x^2)(2x)}{(x^2-4)^2} = \frac{2x^3-8x-2x^3}{(x^2-4)^2} = \frac{-8x}{(x^2-4)^2}$$

$f'(x)$ is zero at x=0.

The sign of the first derivative is shown below.

		-2		0		2	
$-8x$	+		+		-		-
$f'(x)$	+		+		-		-
Graph	↗		↗		↘		↘

To verify the concavity, we have to calculate the second derivate. We are using the product rule.

$$f''(x) = [-8x(x^2 - 4)^{-2}]' = (-8x)'(x^2 - 4)^{-2} + (-8x)[(x^2 - 4)^{-2}]' = -8(x^2 - 4)^{-2} -$$

$$8x(-2)(x^2 - 4)^{-3}(2x) = -8(x^2 - 4)^{-2} + 32x^2(x^2 - 4)^{-3} = \frac{32x^2}{(x^2-4)^3} - \frac{8}{(x^2-4)^2} = \frac{32x^2}{(x^2-4)^3} - \frac{8(x^2-4)}{(x^2-4)^3} =$$

$$\frac{32x^2-8(x^2-4)}{(x^2-4)^3} = \frac{32x^2-8x^2+32}{(x^2-4)^3} = \frac{24x^2+32}{(x^2-4)^3}$$

$24x^2 + 32 = 0$ so $x = \sqrt{\frac{-32}{24}}$ not a real number.

$24x^2 + 32$ is positive for any x values.

For x=-3

$[(-3)^2 - 4]^3 = [9 - 4]^3 = 5 * 5 * 5 = 125$ positive

For x=0

$[(0)^2 - 4]^3 = [0 - 4]^3 = (-4) * (-4) * (-4) = -64$ negative

For x=3

$[(3)^2 - 4]^3 = [9 - 4]^3 = 5 * 5 * 5 = 125$ positive

The concavity is shown below

		-2			2	
$(x^2 - 4)^3$	+		-		+	
$f''(x)$	‿		⌒		‿	

The graph is represented below

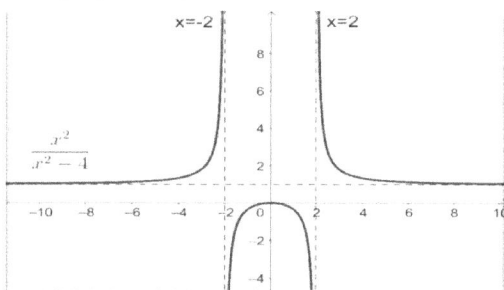

7. The function $f(x) = \frac{2x}{x+3}$ is concave down for x bigger than **-3.**

First, the graph of the function will intersect x axis at x=0.

Second, there is one vertical asymptote at x=-3.

		-3		0	
$2x$		-		-	+
$x + 3$		-		+	+
Sign of function		+		-	+

$$\lim_{x \to -3^-} \frac{2x}{x+3} = +\infty$$

$$\lim_{x \to -3^+} \frac{2x}{x+3} = -\infty$$

Third, $\lim_{x \to \infty} \frac{2x}{x+3} = 2 = \lim_{x \to -\infty} \frac{2x}{x+3}$. The graph will have a horizontal asymptote, at y=2

Using the quotient rule, we have:

$$f'(x) = \frac{(2x)'(x+3)-(2x)(x+3)'}{(x+3)^2} = \frac{2(x+3)-2x}{(x+3)^2} = \frac{2x+6-2x}{(x+3)^2} = \frac{6}{(x+3)^2} \; ; \; \frac{6}{(x+3)^2}$$ is always positive.

The sign of the first derivative is shown below.

$f'(x)$	+
Graph	

To verify the concavity, we have to calculate the second derivate.

$$f''(x) = \frac{6}{(x+3)^2} = 6(x+3)^{-2} = 6(-2)(x+3)^{-3} = \frac{-12}{(x+3)^3}$$

The concavity results from the graph below.

		-3	
-12		-	-
$(x + 3)^3$		-	+
$f''(x)$		+	-
Concavity			

The graph is shown below.

8. The function $f(x) = \frac{1}{4x^2-9}$ is concave down between x=-1.5 and x=1.5

First, the graph of the function will never intersect x axis.

$$f(x) = \frac{1}{4x^2-9} = \frac{1}{(2x-3)(2x+3)}$$

Second, there are two vertical asymptotes at x=-1.5and x=1.5

	-1.5		1.5	
$2x-3$	-		-	+
$2x+3$	-		+	+
Sign of function	+		-	+

$$\lim_{x \to -1.5^-} \frac{1}{4x^2-9} = +\infty$$

$$\lim_{x \to -1.5^+} \frac{1}{4x^2-9} = -\infty$$

$$\lim_{x \to 1.5^-} \frac{1}{4x^2-9} = -\infty$$

$$\lim_{x \to -1.5^+} \frac{1}{4x^2-9} = +\infty$$

Third, $\lim_{x \to \infty} \frac{1}{4x^2-9} = 0 = \lim_{x \to -\infty} \frac{1}{4x^2-9}$. The graph will have a horizontal asymptote, at y=0

We have:

$$f'(x) = [(4x^2-9.^{-1}]' = -(4x^2-9)^{-2}(4x^2-9)' = -(4x^2-9)^{-2}(8x) = \frac{-8x}{(4x^2-9)^2}$$ is zero at x=0

The sign of the first derivative is shown below.

	0	
$-8x$	+	-
$(4x^2-9)^2$	+	+
$f'(x)$	-	+
Graph	↘	↗

To verify the concavity, we have to calculate the second derivate. We use the product rule.

$$f''(x) = [\frac{-8x}{(4x^2-9)^2}]' = [-8x(4x^2-9)^{-2}]' = (-8x)'(4x^2-9)^{-2} + (-8x)[(4x^2-9)^{-2}]' =$$

$$-8(4x^2-9)^{-2} - 8x(-2)(4x^2-9)^{-3}(4x^2-9)' = -8(4x^2-9)^{-2} + 8x(2)(4x^2-9)^{-3}(8x) =$$

$$\frac{-8}{(4x^2-9)^2} + \frac{128x^2}{(4x^2-9)^3} = \frac{-8(4x^2-9)+128x^2}{(4x^2-9)^3} = \frac{-32x^2+72+128x^2}{(4x^2-9)^3} = \frac{96x^2+72}{(4x^2-9)^3} = \frac{24(4x^2+3)}{(4x^2-9)^3}$$

The concavity results from the graph below.

	-1.5		1.5	
$24(4x^2+3)$	+	+	+
$(4x^2-9)^3$	+	-	+
$f''(x)$	+	-	+
Concavity	⌣	⌢	⌣

The graph is shown below.

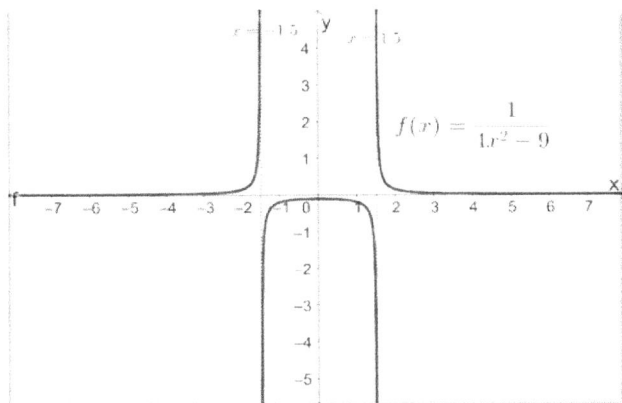

9. The function $f(x) = \frac{x}{x^2-4}$ is concave up between x=-2 and x=0 and x>2

First, the graph of the function will intersect x axis at =0.

$$f(x) = \frac{x}{x^2-4} = \frac{x}{(x-2)(x+2)}$$

Second, there are two vertical asymptotes at x=-2 and x=2

		-2		0		2	
x		-		-		+	+
$x-2$		-		-		-	+
$x+2$		-		+		+	+
Sign of function		-		+		-	+

$$\lim_{x \to -2^-} \frac{x}{x^2-4} = -\infty$$

$$\lim_{x \to -2^+} \frac{x}{x^2-4} = +\infty$$

$$\lim_{x \to 2^-} \frac{x}{x^2-4} = -\infty$$

$$\lim_{x \to 2^+} \frac{x}{x^2-4} = +\infty$$

Third, $\lim_{x \to \infty} \frac{x}{x^2-4} = 0 = \lim_{x \to -\infty} \frac{x}{x^2-4}$. The graph will have a horizontal asymptote, at y=0

We have:

$$f'(x) = [x(x^2-4)^{-1}]' = x'(x^2-4)^{-1} + x[((x^2-4)^{-1}]' = (x^2-4)^{-1} + x(-1)(x^2-4)^{-2}(x^2-$$

$$4)' = (x^2-4)^{-1} - x(x^2-4)^{-2}(2x) = \frac{1}{x^2-4} - \frac{2x^2}{(x^2-4)^2} = \frac{x^2-4}{(x^2-4)^2} - \frac{2x^2}{(x^2-4)^2} = \frac{x^2-4-2x^2}{(x^2-4)^2} = \frac{-x^2-4}{(x^2-4)^2}$$ is

never zero.

The sign of the first derivative is shown below.

$-x^2-4$	-
$(x^2-4)^2$	+
$f'(x)$	-
Graph	

To verify the concavity, we have to calculate the second derivate.

$$f''(x) = [\frac{-x^2-4}{(x^2-4)^2}]' = [(-x^2-4)(x^2-4)^{-2}]' = (-x^2-4)'(x^2-4)^{-2}] + (-x^2-4)[(x^2-4)^{-2}]' =$$

$$-2x(x^2-4)^{-2} + (-x^2-4)(-2)(x^2-4)^{-3}(2x) = \frac{-2x}{(x^2-4)^2} + \frac{4x(x^2+4)}{(x^2-4)^3} = \frac{-2x(x^2-4)}{(x^2-4)^3} + \frac{4x(x^2+4)}{(x^2-4)^3} =$$

$$\frac{-2x^3+8x+4x^3+16x}{(x^2-4)^3} = \frac{2x^3+24x}{(x^2-4)^3} = \frac{2x(x^2+12)}{(x^2-4)^3}$$

The concavity is shown below

	-2		0		2	
x	-		-		+	+
x^2+12	+		+		+	+
$(x^2-4)^3$	+		-		-	+
$f''(x)$	-		+		-	+
Concavity	⌢		⌣		⌢	⌣

The graph is shown below.

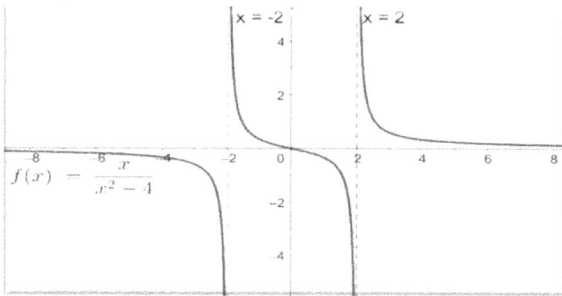

10. The function $f(x) = x^5 - 2x^3$ is concave up for $-0.77 \le x \le 0$ and $0.77 \le x \le \sqrt{2}$

First, the graph of the function will intersect x axis at x=0 ;, x=$-\sqrt{2}$ and, x=$\sqrt{2}$

$$f(x) = x^3(x^2-2)$$

Second, there are no asymptotes.

	$-\sqrt{2}$		0		$\sqrt{2}$	
x^3	-		-		+	+
$x-\sqrt{2}$	-		-		-	+
$x+\sqrt{2}$	-		+		+	+
Sign of fuction	-		+		-	+

Third,

$$\lim_{x\to\infty}(x^5 - 2x^3) = +\infty.$$

$$\lim_{x\to-\infty}(x^5 - 2x^3) = -\infty$$

We have:

$$f'(x) = (x^5 - 2x^3)' = 5x^4 - 6x^2 = x^2(5x^2 - 6)$$ is zero at x=0, x=-1.09 and, x=1.09

The sign of the first derivative is shown below.

	-1.09	0		1.09	
x^2	+	+	+	+
$5x^2 - 6$	+	-	-	+
$f'(x)$	+	-	-	+
Graph	↗	↘	↘	↗

To verify the concavity, we have to calculate the second derivate.

$f''(x) = [x^2(5x^2 - 6)]' = (x^2)'(5x^2 - 6) + x^2(5x^2 - 6)' = 2x(5x^2 - 6) + x^2(10x) = 10x^3 - 12x + 10x^3 = 4x(5x^2 - 3)$ is zero at x=0, x=-0.77 and x=0.77

	-√2	-0.77	0	0.77	√2	
$4x$	-	-	-	+	+	+
$5x^2 - 3$	+	+	-	-	+	+
Sign of $f''(x)$	-	-	+	-	+	+
Concavity	⌢	⌢	⌣	⌢	⌣	⌣

The graph of the function $f(x) = x^5 - 2x^3$ is shown below.

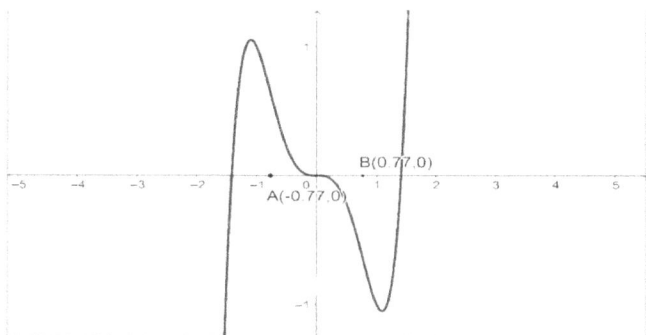

Chapter 7. I. b.   Differentiability, mean value theorem

1. A differentiable function is a function whose derivative exists all over the domain.

2. In a planar arc between two points, there is at least one point at which the tangent to the arc is parallel to the secant through the arc's two points.

3. If $f(x) = 3x^2 - 4x + 1$ and the points (-3,40) and, (1,0) the value of x where the tangent at the graph is parallel with the line that goes through (-3,40) and, (1,0) is: x=-1

$f(-3) = 3(-3)^2 - 4(-3) + 1 = 27 + 12 + 1 = 40$

$f(1) = 3(1)^2 - 4(1) + 1 = 3 - 4 + 1 = 0$

We are calculating the slope of the line that passes through (-3,40) and, (1,0)

$slope = \frac{f(1)-f(-3)}{1-(-3)} = \frac{0-40}{1+3} = \frac{-40}{4} = -10$

We calculate the first derivative.

$f'(x) = (3x^2 - 4x + 1)' = 6x - 4$

We equalize $f'(x)$ with the value of the slope, in this case -10

$f'(x) = -10$

$f'(x) = 6x - 4 = -10$, so $x = \frac{-10+4}{6} = -1$

4. If $f(x) = x^2 + 5x - 3$ and the points (-1,-7) and, (2,11) the value of x where the tangent at the graph is parallel with the line that goes through (-1,-7) and, (2,11) is: $x = \frac{1}{2}$

$f(-1) = (-1)^2 + 5(-1) - 3 = 1 - 5 - 3 = -7$

$f(2) = (2)^2 + 5(2) - 3 = 4 + 10 - 3 = 11$

We are calculating the slope of the line that passes through (-3,40) and, (1,0)

$slope = \frac{f(2)-f(-1)}{2-(-1)} = \frac{11-(-7)}{2+1} = \frac{18}{3} = 6$

We calculate the first derivative.

$f'(x) = (x^2 + 5x - 3)' = 2x + 5$

We equalize $f'(x)$ with the value of the slope, in this case 6

$f'(x) = 6$

$f'(x) = 2x + 5 = 6$, so $x = \frac{6-5}{2} = \frac{1}{2}$

5. If $f(x) = x^5 - 2x^3$ and the points (-1.6,-2.**28**. and, (1.6,2.**28**. the values of x where the tangents at the graph are parallel with the line that goes through (-1.6,-2.**28**. and, (1.6,2.**28**. are: x=∓1.18

$f(-1.6) = (-1.6)^5 - 2(-1.6)^3 = -10.48 - 2(-4.096) = -2.28$

$f(1.6) = (1.6)^5 - 2(1.6)^3 = 10.48 - 2(4.096) = 2.28$

$slope = \frac{f(1.6)-f(-1.6)}{1.6-(-1.6)} = \frac{2.28-(-2.28.}{3.2} = \frac{4.57}{3.2} = 1.43$

$f'(x) = (x^5 - 2x^3)' = 5x^4 - 6x^2 = x^2(5x^2 - 6) = 1.43$

If we substitute x^2 with A, we have:

$f'(x) = A(5A - 6) = 1.43$

$5A^2 - 6A - 1.43 = 0$

$A_{1,2} = \frac{6 \mp \sqrt{36-4(5)(-1.43)}}{2(5)} = \frac{6 \mp \sqrt{64.6}}{10} = \frac{6 \mp 8.03}{10}$

$A_1 = \frac{6-8.03}{10} = -0.203$ It will give non real x values.

$A_2 = \frac{6+8.03}{10} = 1.403$

$x^2 = 1.403$, so $x = \mp 1.18$

6. If $f(x) = 4x^3 + 3x^2$ the values of x where the tangents of 1 at the graph are: x=-0.13 and x=0.63

$f'(x) = (4x^3 - 3x^2)' = 12x^2 - 6x = 1$

$12x^2 - 6x - 1 = 0$

$$x_{1,2} = \frac{6 \mp \sqrt{36-4(12)(-1)}}{2*12} = \frac{6 \mp \sqrt{84}}{24} = \frac{6 \mp 9.16}{24}$$

$$x_1 = \frac{6+9.19}{24} = \frac{15.19}{24} = 0.63$$

$$x_2 = \frac{6-9.19}{24} = \frac{-3.19}{24} = -0.13$$

7. If $f(x) = 6x^4 + 7$ the values of x where the tangent of -3 at the graph is: $x = -\frac{1}{2}$

$$f'(x) = (6x^4 + 7)' = 24x^3$$

$$24x^3 = -3, so\; x = \sqrt[3]{\frac{-3}{24}} = \sqrt[3]{\frac{-1}{8}} = -\frac{1}{2}$$

8. If $f(x) = \frac{1}{x^2} - \frac{2}{5}$ the value of x where the tangent of 2 at the graph is: x=-1

$$f'(x) = (\frac{1}{x^2} - \frac{2}{5})' = (x^{-2} - 0.4)' = -2x^{-3} = -\frac{2}{x^3} = 2$$

$$f'(x) = -\frac{1}{x^3} = 1, so\; -1 = x^3$$

$$x^3 = -1$$

$$x = -1$$

9. If $f(x) = \ln(x) + 10$ the value of x where the tangent of 5 at the graph is: x= $\frac{1}{5}$

$$f'(x) = [\ln(x) - \frac{2}{5}]' = \frac{1}{x}$$

$$\frac{1}{x} = 5, so\; x = \frac{1}{5}$$

10. If $f(x) = x^3 + 2$ and the points (-0.5,1.875) and, (0.5,2.125) the values of x where the tangents at the graph are parallel with the line that goes through (-0.5,1.875) and, (0.5,2.125) are: x=$\mp$0.28

$$f(-0.5) = (-0.5)^3 + 2 = -0.125 + 2 = 1.875$$

$$f(0.5) = (0.5)^3 + 2 = 0.125 + 2 = 2.125$$

$$\frac{f(0.5)-f(-0.5)}{0.5-(-0.5)} = \frac{2.125-1.875}{1} = 0.25$$

$$f'(x) = 3x^2 = 0.25$$

$$x^2 = \frac{0.25}{3} \; so, x = \mp\sqrt{\frac{0.25}{3}} = \mp 0.28$$

Chapter 7. I. c. Newton's method

1. The square root of 135, using Newton's method is: 11.62

$$x^2 = 135$$

$$f(x) = x^2 - 135$$

$$f'(x) = 2x$$

We decide the initial value as 5.

$$f(5) = 5^2 - 135 = 25 - 135 = -110$$

$$x_1 = x_0 - \frac{f(x_0)}{f'(x_0)}$$

$x_1 = 5 - \frac{f(x_0)}{f'(x_0)} = 5 - \frac{-110}{2(5)} = 5 + \frac{110}{10} = 5 + 11 = 16$

$x_2 = 16 - \frac{f(16)}{f'(16)} = 16 - \frac{16^2 - 135}{2(16)} = 16 - \frac{256 - 135}{32} = 16 - 3.78 = 12.21$

$x_3 = 12.21 - \frac{f(12.21)}{f'(12.21)} = 12.21 - \frac{12.21^2 - 135}{2(12.21)} = 12.21 - \frac{149.08 - 135}{24.42} = 12.21 - 0.57 = 11.63$

$x_4 = 11.63 - \frac{f(11.63)}{f'(11.63)} = 11.63 - \frac{11.63^2 - 135}{2(11.63)} = 11.63 - \frac{135.29 - 135}{23.26} = 11.63 - 0.01 = 11.62$

2. The square root of 432, using Newton's method is: 25.43

$x^2 = 432$

$f(x) = x^2 - 432$

$f'(x) = 2x$

We decide the initial value as 15.

$f(15) = 15^2 - 432 = 225 - 432 = -207$

$x_1 = x_0 - \frac{f(x_0)}{f'(x_0)}$

$x_1 = 15 - \frac{f(x_0)}{f'(x_0)} = 15 - \frac{-207}{2(15)} = 15 + \frac{207}{30} = 15 + 69 = 81$

$x_2 = 81 - \frac{f(81)}{f'(81)} = 81 - \frac{81^2 - 432}{2(81)} = 81 - \frac{6561 - 432}{162} = 81 - 37.83 = 43.16$

$x_3 = 43.16 - \frac{f(43.16)}{f'(43.16)} = 43.16 - \frac{43.16^2 - 432}{2(43.16)} = 43.16 - \frac{1863 - 432}{86.32} = 43.16 - 16.57 = 26.58$

$x_4 = 26.58 - \frac{f(26.58)}{f'(26.58)} = 26.58 - \frac{26.58^2 - 432}{2(26.58)} = 26.58 - \frac{706.49 - 432}{53.16} = 26.58 - 5.16 = 21.41$

$x_5 = 21.41 - \frac{f(21.41)}{f'(21.41)} = 21.41 - \frac{21.41^2 - 432}{2(21.41)} = 21.41 - \frac{458.38 - 432}{42.82} = 21.41 - 0.61 = 20.79$

$x_5 = 20.79 - \frac{f(20.79)}{f'(20.79)} = 20.79 - \frac{20.79^2 - 432}{2(20.79)} = 20.79 - \frac{432.22 - 432}{41.58} = 20.79 - 0.005 = 20.78$

3. The solution of $x^3 - 2x^2 = 4$ is: 5

$f(x) = x^3 - 2x^2 - 4$

$f'(x) = 3x^2 - 4x$

$x_1 = x_0 - \frac{f(x_0)}{f'(x_0)}$

We decide the initial value as 2.

$x_1 = 2 - \frac{2^3 - 2(2)^2 - 4}{3(2)^2 - 4(2)} = 2 - \frac{8 - 8 - 4}{12 - 8} = 2 + 1 = 3$

$x_2 = 3 - \frac{3^3 - 2(3)^2 - 4}{3(3)^2 - 4(3)} = 3 - \frac{5}{15} = 3 - 0.33 = 2.66$

$x_3 = 2.66 - \frac{2.66^3 - 2(2.66)^2 - 4}{3(2.66)^2 - 4(2.66)} = 2.66 - \frac{0.74}{10.66} = 2.66 - 0.069 = 2.59$

4. The solution of $3x^4 - 4x^3 + 2x = 7$ is: x=1.624

$f(x) = 3x^4 - 4x^3 + 2x - 7$

$f'(x) = 12x^3 - 12x^2 + 2$

$x_1 = x_0 - \frac{f(x_0)}{f'(x_0)}$

We decide the initial value as 2.

$$x_1 = 2 - \frac{3(2)^4 - 4(2)^3 + 2(2) - 7}{12(2)^3 - 12(2)^2 + 2} = 2 - \frac{3(16) - 4(8) + 2(2) - 7}{12(8) - 12(4) + 2} = 2 - \frac{13}{50} = 2 - 0.26 = 1.74$$

$$x_2 = 1.74 - \frac{3(1.74)^4 - 4(1.74)^3 + 2(1.74) - 7}{12(1.74)^3 - 12(1.74)^2 + 2} = 1.74 - \frac{2.9}{28.88} = 1.74 - 0.10064 = 1.63$$

$$x_3 = 1.63 - \frac{3(1.63)^4 - 4(1.63)^3 + 2(1.63) - 7}{12(1.63)^3 - 12(1.63)^2 + 2} = 1.63 - \frac{0.32}{22.61} = 1.63 - 0.014 = 1.625$$

$$x_4 = 1.625 - \frac{3(1.625)^4 - 4(1.625)^3 + 2(1.625) - 7}{12(1.625)^3 - 12(1.625)^2 + 2} = 1.625 - \frac{0.0058}{21.80} = 1.625 - 0.000267 = 1.624$$

5. The solution of $2x^3 + 3x - 4 = \sin(x)$ is: x=0.980

$$f(x) = 2x^3 + 3x - 4 - \sin(x)$$

$$f'(x) = 6x^2 + 3 - \cos(x)$$

$$x_1 = x_0 - \frac{f(x_0)}{f'(x_0)}$$

We decide the initial value as -1.

$$x_1 = -1 - \frac{2(-1)^3 + 3(-1) - \sin(-1)}{6(-1)^2 + 3 - \cos(-1)} = -1 + \frac{8.15}{5.45} = -1 + 1.49 = 0.49$$

$$x_2 = 0.49 - \frac{2(0.49)^3 + 3(0.49) - \sin(0.49)}{6(0.49)^2 + 3 - \cos(0.49)} = 0.49 - \frac{-2.74}{0.58} = 0.49 + 4.69 = 5.18$$

$$x_3 = 5.18 - \frac{2(5.18)^3 + 3(5.18) - \sin(5.18)}{6(5.18)^2 + 3 - \cos(5.18)} = 5.18 - \frac{291.8}{161.06} = 5.18 - 1.81 = 3.37$$

$$x_4 = 3.37 - \frac{2(3.37)^3 + 3(3.37) - \sin(3.37)}{6(3.37)^2 + 3 - \cos(3.37)} = 3.37 - \frac{83.36}{69.38} = 3.37 - 1.2 = 2.17$$

$$x_5 = 2.17 - \frac{2(2.17)^3 + 3(2.17) - \sin(2.17)}{6(2.17)^2 + 3 - \cos(2.17)} = 3.37 - \frac{22.28}{28.95} = 3.37 - 0.76 = 1.4$$

$$x_6 = 1.4 - \frac{2(1.4)^3 + 3(1.4) - \sin(1.4)}{6(1.4)^2 + 3 - \cos(1.4)} = 1.4 - \frac{4.78}{11.68} = 1.4 - 0.4 = 0.99$$

$$x_7 = 0.99 - \frac{2(0.99)^3 + 3(0.99) - \sin(0.99)}{6(0.99)^2 + 3 - \cos(0.99)} = 0.99 - \frac{1.12}{5.41} = 0.99 - 0.023 = 0.97$$

$$x_8 = 0.97 - \frac{2(0.97)^3 + 3(0.97) - \sin(0.97)}{6(0.97)^2 + 3 - \cos(0.97)} = 0.97 + \frac{0.06}{5.114} = 0.97 + 0.013 = 0.985$$

$$x_9 = 0.985 - \frac{2(0.985)^3 + 3(0.985) - \sin(0.985)}{6(0.985)^2 + 3 - \cos(0.985)} = 0.985 - \frac{0.08}{5.27} = 0.985 - 0.007 = 0.978$$

$$x_{10} = 0.978 - \frac{2(0.978)^3 + 3(0.978) - \sin(0.978)}{6(0.978)^2 + 3 - \cos(0.978)} = 0.978 + \frac{0.02}{5.18} = 0.978 + 0.004 = 0.982$$

$$x_{11} = 0.982 - \frac{2(0.982)^3 + 3(0.982) - \sin(0.982)}{6(0.982)^2 + 3 - \cos(0.982)} = 0.982 - \frac{0.013}{5.23} = 0.982 - 0.002 = 0.980$$

6. The solution of $x^5 + 4x^3 - 5x = 3$ is: x=1.18

$$f(x) = x^5 + 4x^3 - 5x - 3$$

$$f'(x) = 5x^4 + 12x^2 - 5$$

$$x_1 = x_0 - \frac{f(x_0)}{f'(x_0)}$$

We decide the initial value as 2.

$$x_1 = 2 - \frac{2^5 + 4(2)^3 - 5(2) - 3}{5(2)^4 + 12(2)^2 - 5} = 2 - \frac{51}{123} = 2 - 0.41 = 1.585$$

$$x_2 = 1.585 - \frac{(1.585)^5 + 4(1.585)^3 - 5(1.585) - 3}{5(1.585)^4 + 12(1.585)^2 - 5} = 1.585 - \frac{15.02}{56.74} = 1.585 - 0.26 = 1.32$$

$$x_3 = 1.32 - \frac{(1.32)^5 + 4(1.32)^3 - 5(1.32) - 3}{5(1.32)^4 + 12(1.32)^2 - 5} = 1.32 - \frac{3.62}{31.13} = 1.32 - 0.11 = 1.2$$

$x_4 = 1.2 - \dfrac{(1.2)^5 + 4(1.2)^3 - 5(1.2) - 3}{5(1.2)^4 + 12(1.2)^2 - 5} = 1.2 - \dfrac{0.49}{22.91} = 1.2 - 0.021 = 1.182$

$x_4 = 1.182 - \dfrac{(1.182)^5 + 4(1.182)^3 - 5(1.182) - 3}{5(1.182)^4 + 12(1.182)^2 - 5} = 1.182 - \dfrac{0.014}{21.55} = 1.182 - 0.0006 = 1.1817$

7. The solution of $(x - 3)^3 = \sin(x)$ is: x=3.138

$f(x) = (x - 3)^3 - \sin(x)$

$f'(x) = 3(x - 3)^2 - \cos(x)$

$x_1 = x_0 - \dfrac{f(x_0)}{f'(x_0)}$

We decide the initial value as 2.

$x_1 = 2 - \dfrac{(2-3)^3 - \sin(2)}{3(2-3)^2 - \cos(2)} = 2 + \dfrac{1.9}{3.41} = 2 + 0.558 = 2.558$

$x_2 = 2.558 - \dfrac{(2.558-3)^3 - \sin(2.558)}{3(2.558-3)^2 - \cos(2.558)} = 2.558 + \dfrac{0.63}{1.41} = 2.558 + 0.448 = 3.007$

$x_3 = 3.007 - \dfrac{(3.007-3)^3 - \sin(3.007)}{3(3.007-3)^2 - \cos(3.007)} = 3.007 + \dfrac{0.133}{0.99} = 3.007 + 0.135 = 3.148$

$x_4 = 3.148 - \dfrac{(3.148-3)^3 - \sin(3.148)}{3(3.148-3)^2 - \cos(3.148)} = 3.148 - \dfrac{0.0036}{1.06} = 3.148 - 0.0034 = 3.138$

$x_5 = 3.138 - \dfrac{(3.138-3)^3 - \sin(3.138)}{3(3.138-3)^2 - \cos(3.138)} = 3.138 - \dfrac{0.000056}{1.057} = 3.138 - 0.00004 = 3.138$

8. The solution of $(x - 6)^3 = \ln(x)$ is: x=7.256

$f(x) = (x - 6)^3 - \ln(x)$

$f'(x) = 3(x - 6)^2 - \dfrac{1}{x}$

$x_1 = x_0 - \dfrac{f(x_0)}{f'(x_0)}$

We decide the initial value as 10.

$x_1 = 10 - \dfrac{(10-6)^3 - \ln(10)}{3(10-6)^2 - \frac{1}{10}} = 10 - \dfrac{61.69}{47.9} = 10 - 1.28 = 8.711$

$x_2 = 8.711 - \dfrac{(8.711-6)^3 - \ln(8.711)}{3(8.711-6)^2 - \frac{1}{8.711}} = 8.711 - \dfrac{17.78}{21.94} = 8.711 - 0.81 = 7.9$

$x_3 = 7.9 - \dfrac{(7.9-6)^3 - \ln(7.9)}{3(7.9-6)^2 - \frac{1}{7.9}} = 7.9 - \dfrac{4.81}{10.72} = 7.9 - 0.448 = 7.45$

$x_4 = 7.45 - \dfrac{(7.45-6)^3 - \ln(7.45)}{3(7.45-6)^2 - \frac{1}{7.45}} = 7.45 - \dfrac{1.06}{6.2} = 7.45 - 0.17 = 7.28$

$x_5 = 7.28 - \dfrac{(7.28-6)^3 - \ln(7.28)}{3(7.28-6)^2 - \frac{1}{7.28}} = 7.28 - \dfrac{0.12}{4.79} = 7.28 - 0.025 = 7.256$

9. The solution of $2(x - 9.^3 = \ln(x) + 2x$ is: x=11.33

$f(x) = 2(x - 9.^3 - \ln(x) - 2x$

$f'(x) = 6(x - 6)^2 - \dfrac{1}{x} - 2$

$x_1 = x_0 - \dfrac{f(x_0)}{f'(x_0)}$

We decide the initial value as 12.

$x_1 = 12 - \dfrac{2(12-9.^3 - \ln(12) - 2(12)}{6(12-6)^2 - \frac{1}{12} - 2} = 12 - \dfrac{12}{213.91} = 12 - 0.12 = 11.871$

$$x_2 = 11.871 - \frac{2(11.871 - 9.^3 - \ln(11.871) - 2(11.871)}{6(11.871 - 6)^2 - \frac{1}{11.871} - 2} = 11.871 - \frac{21.13}{204.754} = 11.871 - 0.10 = 11.768$$

$$x_3 = 11.768 - \frac{2(11.768 - 9.^3 - \ln(11.768) - 2(11.768.}{6(11.768 - 6)^2 - \frac{1}{11.768} - 2} = 11.768 - \frac{16.42}{197.54} = 11.768 - 0.083 = 11.685$$

$$x_4 = 11.685 - \frac{2(11.685 - 9.^3 - \ln(11.685) - 2(11.685)}{6(11.685 - 6)^2 - \frac{1}{11.685} - 2} = 11.685 - \frac{12.887}{191.83} = 11.685 - 0.067 = 11.617$$

$$x_5 = 11.617 - \frac{2(11.617 - 9.^3 - \ln(11.617) - 2(11.617)}{6(11.617 - 6)^2 - \frac{1}{11.617} - 2} = 11.617 - \frac{10.19}{187.27} = 11.617 - 0.054 = 11.56$$

$$x_6 = 11.56 - \frac{2(11.56 - 9.^3 - \ln(11.56) - 2(11.56)}{6(11.56 - 6)^2 - \frac{1}{11.56} - 2} = 11.56 - \frac{8.11}{183.62} = 11.56 - 0.044 = 11.519$$

$$x_7 = 11.519 - \frac{2(11.519 - 9.^3 - \ln(11.519) - 2(11.519.}{6(11.519 - 6)^2 - \frac{1}{11.519} - 2} = 11.519 - \frac{6.49}{180.68} = 11.519 - 0.035 = 11.48$$

$$x_8 = 11.48 - \frac{2(11.48 - 9.^3 - \ln(11.48) - 2(11.48.}{6(11.48 - 6)^2 - \frac{1}{11.48} - 2} = 11.48 - \frac{5.22}{178.31} = 11.48 - 0.029 = 11.45$$

$$x_9 = 11.45 - \frac{2(11.45 - 9.^3 - \ln(11.45) - 2(11.45)}{6(11.45 - 6)^2 - \frac{1}{11.45} - 2} = 11.45 - \frac{4.21}{176.39} = 11.45 - 0.023 = 11.43$$

$$x_{10} = 11.43 - \frac{2(11.43 - 9.^3 - \ln(11.43) - 2(11.43)}{6(11.43 - 6)^2 - \frac{1}{11.43} - 2} = 11.43 - \frac{3.4}{174.83} = 11.43 - 0.019 = 11.41$$

$$x_{11} = 11.41 - \frac{2(11.41 - 9.^3 - \ln(11.41) - 2(11.41)}{6(11.41 - 6)^2 - \frac{1}{11.41} - 2} = 11.41 - \frac{2.76}{173.56} = 11.41 - 0.015 = 11.39$$

....

....

$$x_{22} = 11.332 - \frac{2(11.332 - 9.^3 - \ln(11.332) - 2(11.332)}{6(11.332 - 6)^2 - \frac{1}{11.33} - 2} = 11.332 - \frac{0.29}{168.54} = 11.332 - 0.0017 = 11.331$$

Chapter 7. I. d.   Problems in contextual situations, including related rates and optimization problems

For first 4 questions we assume that all variables depend on z.

1. If $x^3 - 2x + y = 1$ the first derivative with respect to z is: $3x^2 x' - 2x' + y' = 0$

If we differentiate $x^3 - 2x + y = 1$ with respect to z, we have: $3x^2 x' - 2x' + y' = 0$

2. If $2x^5 + 3x^4 - 4x + y^2 = 10$ the first derivative with respect to z is: $10x^4 x' + 12x^3 x' - 4x' + 2yy' = 0$

If we differentiate $2x^5 + 3x^4 - 4x + y^2 = 10$ with respect to z, we have:

$10x^4 x' + 12x^3 x' - 4x' + 2yy' = 0$

3. If $x^3 + 3y^2 - x^2 = \sin(x)$ the first derivative with respect to z is:

$3x^2 x' + 6yy' - 2xx' = x' \cos(x)$

If we differentiate $x^3 + 3y^2 - x^2 = \sin(x)$ with respect to z, we have:

$3x^2 x' + 6yy' - 2xx' = x' \cos(x)$

4. If A=πR^2 the first derivative with respect to z is: $A' = 2\pi R R'$

If we differentiate A=πR^2 with respect to z, we have:

$A' = 2\pi RR'$

5. We have a rectangular yard with perimeter of 300 m. It has to be fenced. The dimensions of the yard that will give us the greatest area are: x=y=75m.

Let's suppose the length equals x and width equals y.

Then we have:

$Area(A) = xy, and\ Perimeter(P) = 2x + 2y = 300$

$2x = 300 - 2y$

$x = \frac{300-2y}{2} = 150 - y$

We substitute x in formula for area.

$A = xy = (150 - y)y = 150y - y^2$

To find the maximum value of the Area we need to find the y values for which the derivative of function A(y) is zero.

$A(y) = 150y - y^2$

$A' = (150y)' - (y^2)' = 150 - 2y = 0$

$150 = 2y$

$So, y = \frac{150}{2} = 75m$

$x = 150 - y = 150 - 75 = 75m$

6. We have to build a box with the length of the base four times the width. The height is the 3 times length minus 10 cm. The dimensions that will minimize the volume are: Length= 7.2 cm , Width= 1.8 cm , height= 11.6 cm

Length=x, Width=y, height=h

$Area(A) = xy, where\ x = 4y$

$Area(A) = xy = (4y)y = 4y^2$

$h = 3x - 10 = 12y - 10$

$Volume(V) = A * h = (4y^2)h = 4y^2(3x - 10) = 2y^2(12y - 10) = 24y^3 - 20y^2$

$V' = (24y^3)' - (20y^2)' = 72y^2 - 40y = 0$

$72y^2 - 40y = 8y(9y - 5) = 0$

$y = 0$ not real

$y = Width = \frac{9}{5}\ cm = 1.8\ cm$ then:

$x = Length = 4(1.8) = 7.2\ cm\ and, h = Height = 3(7.2) - 10 = 11.6\ cm$

7. We have to build a box with a base that have length two times the width, and we have 25 m square of material. The dimensions for the maximum volume are:

Length= 2.88 m , Width= 1.44 m   and Height= 1.93 m

We have:

Length = x ; Width = y ; Height = h

$x = 2y$

$Surface\ area(SA) = 2xy + 2xh + 2yh = 25\ cm^2$

We isolate h

$2xy + 2xh + 2yh = 25$

$2xh + 2yh = 25 - 2xy = 25 - 2(2y)y = 25 - 4y^2$

$2xh + 2yh = 25 - 4y^2$

$2(2y)h + 2yh = 25 - 4y^2$

$4yh + 2yh = 25 - 4y^2$

$6yh = 25 - 4y^2$

$h = \frac{25 - 4y^2}{6y}$

So, the volume can be written as:

$V = xyh = (2y)y\left(\frac{25 - 4y^2}{6y}\right) = y(\frac{25 - 4y^2}{3})$

$V(y) = y(\frac{25 - 4y^2}{3})$

$V'(y) = y'\left(\frac{25 - 4y^2}{3}\right) + y(\frac{25 - 4y^2}{3})' = \left(\frac{25 - 4y^2}{3}\right) + \frac{y(-8y)}{3} = \frac{25 - 4y^2 - 8y^2}{3} = \frac{25 - 12y^2}{3}$

So:

$V'(y) = \frac{25}{3} - 4y^2 = 0$

$\frac{25}{3} = 4y^2$

$y^2 = \frac{25}{12}; then\ y = \mp\sqrt{\frac{25}{12}} = \mp 1.44$

y can't be negative so, y=1.44m

$x = 2y = 2 * 1.44 = 2.88\ m$

$h = \frac{25 - 4y^2}{6y} = \frac{25 - 4(1.44)^2}{6(1.44)} = \frac{25 - 8.29}{8.64} = \frac{16.71}{8.64} = 1.93\ m$

8. We have to build a tunnel that has a cylinder shape that has 50 Liters in volume. The dimensions of the tunnel in order to have the smallest surface area are: R= 29.24 cm And h = 18.62 cm.

$V = \pi R^2 h = 50,000\ cm^3\ so, h = \frac{50,000}{\pi R^2}$

$SA = 2\pi R^2 + 2\pi Rh = 2\pi R^2 + \frac{2\pi R(50,000)}{\pi R^2} = 2\pi R^2 + \frac{100,000}{R}$

$SA = 2\pi R^2 + \frac{100,000}{R} = SA(R)$

So:

$(SA)' = (2\pi R^2 + \frac{100,000}{R})' = 2\pi(2R) + (100,000R^{-1})' = 4\pi R - 100,000R^{-2} = 0$

$4\pi R - 100,000R^{-2} = 0$

$4\pi R = \frac{100,000}{R^2}$

$4\pi R^3 = 100,000\ so, R^3 = \frac{100,000}{4\pi} = 7961.78$

$R = \sqrt[3]{7961.78} = 19.96\ cm$

$h = \frac{50,000}{\pi R^2} = \frac{50,000}{\pi(19.96)^2} = \frac{50,000}{1250.98} = 39.96\ cm$

9. The profit relation for a company is $P = Revenue - expenses = \frac{1230}{p} - \frac{550}{(p)^2}$ The price per unit (p) for maximum profit P is: $p = \$0.89$

The first derivative of Profit is:

$P'(p) = (\frac{1230}{p})' - (\frac{550}{p^2})' = 1230(p^{-1})' - 550(p^{-2})' = -1230p^{-2} + 1100p^{-3}$

$P'(p) = -1230p^{-2} + 1100p^{-3} = 0$

$1230p^{-2} = 1100p^{-3}$

$1230 = \frac{1100}{p}$

$p = \frac{1100}{1230} = \$0.89$

10. We need to fence two adjacent lots of land. We have 240 m of fence. The dimensions x and y to maximize the area are: x = 3 m and y = 4 m.

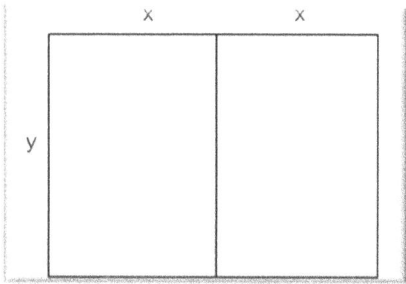

$P = 3y + 4x = 240 \, m$

$4x = 240 - 3y$

$x = \frac{240-3y}{4} = 60 - 0.75y$

$Area \, (A) = 2xy = 2(60 - 0.75y)y = 120y - 1.5y^2 = A(y)$

$A'(y) = (120y)' - (1.5y^2)' = 120 - 3y$

$A'(y) = 0 = 120 - 3y$

$120 = 3y \, so,$

$y = \frac{120}{3} = 40 \, m$

$x = 60 - 0.75y = 60 - 0.75(40) = 60 - 30 = 30 \, m$

CHAPTER 8

Chapter 8. A.   Definition of an integral and notation

1. An integral is a weighted sum of the values of the function times the infinitesimal widths dx.

2. The number of infinitesimal widths has to be as high as it is possible not necessarily 100.

3. The notation of definite integral in the interval [a,b] is: $\int_a^b f(x)dx$.

4. e sign $\int$ represents the sum of products between $f(x)$ and dx.

5. The notation of indefinite integral or antiderivative is: $\int f(x)dx$.

6. The function $f(x)$ under the integral sign it is called integrand.

7. The first documented technique that tried to calculate the integral was used by the ancient Greek astronomer Euxodus around 370 BC.

8. The Monte Carlo method is used in statistics and probabilities to generate random numbers that that follow certain criteria and calculating the average of these numbers.

Chapter 8. B.   Definite and indefinite integrals

1. If $f(x) = 3x^2 - e^x$, then $\int f(x)dx = x^3 - e^x + C$

$\int f(x)dx = \int(3x^2 - e^x)dx = \int(3x^2)dx - \int e^x dx = \frac{3x^{2+1}}{3} - e^x = x^3 - e^x + C$

2. If $f(x) = 2x^4 + \ln(x)$, then $\int f(x)dx = \frac{2}{5}x^5 + x\ln(x) - x + C$

$\int f(x)dx = \int[2x^4 + \ln(x)]dx = \int 2x^4\, dx + \int[\ln(x)]dx = \frac{2}{5}x^5 + x\ln(x) - x + C$

3. If $f(x) = \ln(x) - 2^x$, then $\int f(x)dx = x\ln(x) - x - \frac{2^x}{\ln(2)} + C$

$\int f(x)dx = \int[\ln(x)]dx - \int 2^x dx = x\ln(x) - x - \frac{2^x}{\ln(2)} + C$

4. If $f(x) = 1 + \ln(x)$, then $\int f(x)dx = x\ln(x) + C$

$\int f(x)dx = \int dx + \int[\ln(x)\, dx = x + x\ln(x) - x = x\ln(x) + C$

5. If $f(x) = 2 - \cos(x)$, then $\int f(x)dx = 2x - \sin(x) + C$

$\int f(x)dx = \int 2dx - \int \cos(x)dx = 2x - \sin(x) + C$

6. If $f(x) = x^2 - 2x$, then $\int_0^x f(x)dx = \frac{1}{3}x^3 - x^2$

$\int_0^x f(x)dx = \int_0^x x^2 dx - \int_0^x 2xdx = \frac{1}{3}x^3 - x^2$

7. If $f(x) = 2x^2 + 3x - 4$, then $\int_0^x f(x)dx = -1\frac{5}{6}$

$\int_0^1 f(x) = \int_0^1(2x^2 + 3x - 4)dx = \int_0^1 2x^2 dx + \int_0^1 3xdx - \int_0^1 4dx = \frac{2}{3}(1)^3 + \frac{3}{2}(1)^2 - -4(1) = \frac{2}{3} + \frac{3}{2} - 4 = \frac{4}{6} + \frac{9}{6} - \frac{24}{6} = -\frac{11}{6} = -1\frac{5}{6}$

8. If $f(x) = x^2 - 4\cos(x)$, then $\int_0^6 f(x)dx = 71.58$

$\int_0^6 f(x)dx = \int_0^6[x^2 - 4\cos(x)]dx = \int_0^6 x^2 dx - 4\int_0^6 \cos(x)\, dx = \frac{1}{3}(6)^3 - 4\sin(6) = 72 - 4(0.104) = 71.58$

9. If $f(x) = \cos(x) + 3x$, then $\int_0^4 f(x)dx = 24.069$

$\int_0^4 f(x)dx = \int_0^4 [\cos(x) + 3x]dx = \int_0^4 \cos(x)dx + 3\int_0^4 xdx = \sin(4) + \frac{3}{2}(4)^2 = 0.069 + 24 = 24.069$

10. If $f(x) = 2x + \sqrt{x}$, then $\int_0^3 f(x)dx = 9(1 + \frac{\sqrt{3}}{2})$

$\sqrt{x} = x^{\frac{1}{2}}$ so, $\int x^{\frac{1}{2}}dx = \frac{3x^{\frac{3}{2}}}{2}$

$\int_0^3 f(x)dx = \int_0^3 (2x + \sqrt{x})dx = \int_0^3 2xdx + \int_0^3 \sqrt{x}dx = (3)^2 + \frac{3(3)^{\frac{3}{2}}}{2} = 9 + \frac{9\sqrt{3}}{2} = 9(1 + \frac{\sqrt{3}}{2})$

Chapter 8. C.   Approximations-Riemann sum, rectangle method, trapezoidal method

1. If $f(x) = 2x^2 - 3x$, using Riemann left sum, the integral $\int_{-2}^3 f(x)dx = 20$

The interval $\Delta x = 1$

$\int_{-2}^3 f(x)dx \cong \sum_{-2}^3 f(x)\Delta x = f(-2)\Delta x + f(-1)\Delta x + f(0)\Delta x + f(1)\Delta x + f(2)\Delta x$

$f(-2) = 2(-2)^2 - 3(-2) = 8 + 6 = 14$

$f(-1) = 2(-1)^2 - 3(-1) = 2 + 3 = 5$

$f(0) = 2(0)^2 - 3(0) = 0$

$f(1) = 2(1)^2 - 3(1) = 2 - 3 = -1$

$f(2) = 2(2)^2 - 3(2) = 8 - 6 = 2$

$\sum_{-2}^3 f(x)\Delta x = f(-2)\Delta x + f(-1)\Delta x + f(0)\Delta x + f(1)\Delta x + f(2)\Delta x = 14 + 5 - 1 + 2 = 20$

$\int_{-2}^3 f(x)dx \cong 20$

2. If $f(x) = 4x^3 + 3x^2 - 2x + 1$, using Riemann left sum, the integral $\int_{-3}^4 f(x)dx \cong 91$

The interval $\Delta x = 1$

$\int_{-3}^4 f(x)dx \cong \sum_{-3}^4 f(x)\Delta x = f(-3)\Delta x + f(-2)\Delta x + f(-1)\Delta x + f(0)\Delta x + f(1)\Delta x + f(2)\Delta x + f(3)\Delta x$

$f(-3) = 4(-3)^3 + 3(-3)^2 - 2(-3) + 1 = 4(-27) + 27 + 6 + 1 = -81 + 7 = -74$

$f(-2) = 4(-2)^3 + 3(-2)^2 - 2(-2) + 1 = -32 + 12 + 4 + 1 = -15$

$f(-1) = 4(-1)^3 + 3(-1)^2 - 2(-1) + 1 = -4 + 3 + 2 + 1 = 2$

$f(0) = 4(0)^3 + 3(0)^2 - 2(0) + 1 = 1$

$f(1) = 4(1)^3 + 3(1)^2 - 2(1) + 1 = 4 + 3 - 2 + 1 = 6$

$f(2) = 4(2)^3 + 3(2)^2 - 2(2) + 1 = 32 + 12 - 4 + 1 = 41$

$f(3) = 4(3)^3 + 3(3)^2 - 2(3) + 1 = 4(27) + 27 - 6 + 1 = 130$

$\int_{-3}^4 f(x)dx \cong \sum_{-3}^4 f(x)\Delta x = f(-3)\Delta x + f(-2)\Delta x + f(-2)\Delta x + f(0)\Delta x + f(1)\Delta x + f(2)\Delta x + f(3)\Delta x = -74 - 15 + 2 + 1 + 6 + 41 + 130 = 91$

3. If $f(x) = 3x^3 - 2x^2 + x - 1$, using Rectangle Method, the integral $\int_{-2}^3 f(x)dx \cong 21.875$

The interval $\Delta x = 1$

The mid points are: x=-1.5; x=-0.5; x=0.5; x=1.5; x=2.5

$\int_{-2}^{3} f(x)dx \cong \sum_{-2}^{3} f(x)\Delta x = f(-1.5)\Delta x + f(-0.5)\Delta x + f(0.5)\Delta x + f(1.5)\Delta x + f(2.5)\Delta x$

$f(-1.5) = 3(-1.5)^3 - 2(-1.5)^2 + (-1.5) - 1 = 3(-3.375) - 2(2.25) - 1.5 - 1 = -17.125$

$f(-0.5) = 3(-0.5)^3 - 2(-0.5)^2 + (-0.5) - 1 = 3(-0.125) - 2(0.25) - 0.5 - 1 = -2.375$

$f(0.5) = 3(0.5)^3 - 2(0.5)^2 + (0.5) - 1 = 3(0.125) - 2(0.25) + 0.5 - 1 = -0.625$

$f(1.5) = 3(1.5)^3 - 2(1.5)^2 + (1.5) - 1 = 3(3.375) - 2(2.25) + 1.5 - 1 = 6.125$

$f(2.5) = 3(2.5)^3 - 2(2.5)^2 + (2.5) - 1 = 3(15.625) - 2(6.25) + 2.5 - 1 = 35.875$

$\int_{-2}^{3} f(x)dx = \sum_{-2}^{3} f(x)\Delta x = f(-1.5)\Delta x + f(-0.5)\Delta x + f(0.5)\Delta x + f(1.5)\Delta x + f(2.5)\Delta x =$

$-17.125 - 2.375 - 0.625 + 6.125 + 35.875 = 21.875$

4. If $f(x) = x^2 + 2x + \sqrt[3]{x}$, using Rectangle Method, the integral $\int_{-3}^{3} f(x)dx \cong 17.5$

The interval $\Delta x = 1$

The mid points are: x=-2.5; x=-1.5; x=-0.5; x=0.5; x=1.5; x=2.5

$\int_{-3}^{3} f(x)dx \cong \sum_{-3}^{3} f(x)\Delta x = f(-2.5)\Delta x + f(-1.5)\Delta x + f(-0.5)\Delta x + f(0.5)\Delta x + f(1.5)\Delta x + f(2.5)\Delta x$

$f(-2.5) = (-2.5)^2 + 2(-2.5) + \sqrt[3]{-2.5} = 6.25 - 5 - 1.35 = -0.1$

$f(-1.5) = (-1.5)^2 + 2(-1.5) + \sqrt[3]{-1.5} = 2.25 - 3 - 1.14 = -1.89$

$f(-0.5) = (-0.5)^2 + 2(-0.5) + \sqrt[3]{-0.5} = 0.25 - 1 - 0.79 = -1.54$

$f(0.5) = (0.5)^2 + 2(0.5) + \sqrt[3]{0.5} = 0.25 + 1 + 0.79 = 2.04$

$f(1.5) = (1.5)^2 + 2(1.5) + \sqrt[3]{1.5} = 2.25 + 3 + 1.14 = 6.39$

$f(2.5) = (2.5)^2 + 2(2.5) + \sqrt[3]{2.5} = 6.25 + 5 + 1.35 = 12.6$

$\int_{-3}^{3} f(x)dx \cong \sum_{-3}^{3} f(x)\Delta x = f(-2.5)\Delta x + f(-1.5)\Delta x + f(-0.5)\Delta x + f(0.5)\Delta x + f(1.5)\Delta x + f(2.5)\Delta x = -0.1 - 1.89 - 1.54 + 2.04 + 6.39 + 12.6 = 17.5$

5. If $f(x) = 4x^3 + \sin(x)$, using Riemann left sum, the integral $\int_{-3}^{3} f(x)dx \cong -108.05$

The interval $\Delta x = 1$

$\int_{-3}^{3} f(x)dx \cong \sum_{-3}^{3} f(x)\Delta x = f(-3)\Delta x + f(-2)\Delta x + f(-1)\Delta x + f(0)\Delta x + f(1)\Delta x + f(2)\Delta x$

$f(-3) = 4(-3)^3 + \sin(-3) = 4(-27) - 0.05 = -108 - 0.05 = -108.05$

$f(-2) = 4(-2)^3 + \sin(-2) = -32 - 0.03 = -32.03$

$f(-1) = 4(-1)^3 + \sin(-1) = -4 - 0.01 = -4.01$

$f(0) = 0$

$f(1) = 4(1)^3 + \sin(1) = 4 + 0.01 = 4.01$

$f(2) = 4(2)^3 + \sin(2) = 32 + 0.05 = 32.05$

$\int_{-3}^{3} f(x)dx \cong \sum_{-3}^{3} f(x)\Delta x = f(-3)\Delta x + f(-2)\Delta x + f(-1)\Delta x + f(0)\Delta x + f(1)\Delta x + f(2)\Delta x =$

-108.050

6. If $f(x) = 3x^3 - 2x^2 + 1$, using left side Riemann sum, the integral $\int_{-2}^{2} f(x)dx \cong -32$

The interval $\Delta x = 1$

The points are: x=-2; x=-1; x=0; x=1;

$\int_{-2}^{2} f(x)dx \cong \Sigma_{-2}^{2} f(x)\Delta x = f(-2)\Delta x + f(-1)\Delta x + f(0)\Delta x + f(1)\Delta x$

$f(-2) = 3(-2)^3 - 2(-2)^2 + 1 = -24 - 8 + 1 = -31$

$f(-1) = 3(-1)^3 - 2(-1)^2 + 1 = -3 - 2 + 1 = -4$

$f(0) = 3(0)^3 - 2(0)^2 + 1 = 1$

$f(1) = 3(1)^3 - 2(1)^2 + 1 = 3 - 2 + 1 = 2$

$\int_{-2}^{2} f(x)dx \cong \Sigma_{-2}^{2} f(x)\Delta x = f(-2)\Delta x + f(-1)\Delta x + f(0)\Delta x + f(1)\Delta x = -31 - 4 + 1 + 2 = -32$

7. If $f(x) = 3x^3 - 2x^2 + 1$, using left side Riemann sum, the integral $\int_{-2}^{2} f(x)dx \cong -19$

The interval $\Delta x = 0.5$

The points are: x=-2; x=-1.5; x=-1; x=-0.5; x=0; x=0.5; x=1; x=1.5

$\int_{-2}^{2} f(x)dx \cong \Sigma_{-2}^{2} f(x)\Delta x = f(-2)\Delta x + f(-1.5)\Delta x + f(-1)\Delta x + f(-0.5)\Delta x + f(0)\Delta x +$
$f(0.5)\Delta x + f(1)\Delta x + f(1.5)\Delta x$

$f(-2) = 3(-2)^3 - 2(-2)^2 + 1 = -24 - 4 + 1 = -31$

$f(-1.5) = 3(-1.5)^3 - 2(-1.5)^2 + 1 = 3(-3.375) - 4.5 + 1 = -10.125 - 4.5 + 1 = -13.625$

$f(-1) = 3(-1)^3 - 2(-1)^2 + 1 = -3 - 2 + 1 = -4$

$f(-0.5) = 3(-0.5)^3 - 2(-0.5)^2 + 1 = 3(-0.125) - 0.5 + 1 = 0.125$

$f(0) = 3(0)^3 - 2(0)^2 + 1 = 1$

$f(0.5) = 3(0.5)^3 - 2(0.5)^2 + 1 = 3(0.125) - 0.5 + 1 = 0.875$

$f(1) = 3(1)^3 - 2(1)^2 + 1 = 3 - 2 + 1 = 2$

$f(1.5) = 3(1.5)^3 - 2(1.5)^2 + 1 = 3(3.375) - 4.5 + 1 = 10.125 - 4.5 + 1 = 6.625$

$\int_{-2}^{2} f(x)dx \cong \Sigma_{-2}^{2} f(x)\Delta x = f(-2)(0.5) + f(-1.5)(0.5) + f(-1)(0.5) + f(-0.5)(0.5) +$
$f(0)(0.5) + f(0.5)(0.5) + f(1)(0.5) + f(1.5)(0.5) = (-31)(0.5) + (-13.625)(0.5) + (-4)(0.5) +$
$(0.125)(0.5) + 1 * (0.5) + (0.875)(0.5) + 2 * (0.5) + (6.625)(0.5) = -15.5 - 6.812 - 2 -$
$0.0625 + 0.5 + 0.437 + 1 + 3.312 = -19$

8. If $f(x) = 3x^3 - 2x^2 + 1$, using trapezoidal method, the integral $\int_{-2}^{2} f(x)dx \cong -6$

The interval $\Delta x = 1$

$\int_{-2}^{2} f(x)dx \cong \Sigma_{i=-2}^{i=2} \frac{f(x_i)+f(x_{i+1})}{2}\Delta x_i = \frac{f(-2)+f(-1)}{2} * 1 + \frac{f(-1)+f(0)}{2} * 1 + \frac{f(0)+f(1)}{2} * 1 + \frac{f(1)+f(2)}{2} * 1$

$f(-2) = 3(-2)^3 - 2(-2)^2 + 1 = -24 - 8 + 1 = -31$

$f(-1) = 3(-1)^3 - 2(-1)^2 + 1 = -3 - 2 + 1 = -4$

$f(0) = 3(0)^3 - 2(0)^2 + 1 = 1$

$f(1) = 3(1)^3 - 2(1)^2 + 1 = 3 - 2 + 1 = 2$

$f(2) = 3(2)^3 - 2(2)^2 + 1 = 24 - 8 + 1 = 17$

$\int_{-2}^{2} f(x)dx \cong \sum_{i=-2}^{i=2} \frac{f(x_i)+f(x_{i+1})}{2} \Delta x_i = \frac{-31-4}{2} * 1 + \frac{-4+1}{2} * 1 + \frac{1+2}{2} * 1 + \frac{2+17}{2} * 1 = -17.5 - 1.5 +$

$1.5 + 9.5 = -8$

9. If $f(x) = 3x^3 - 2x^2 + 1$, using trapezoidal method, the integral $\int_{-2}^{2} f(x)dx \cong -7$

The interval $\Delta x = 0.5$

$\int_{-2}^{2} f(x)dx \cong \sum_{i=-2}^{i=2} \frac{f(x_i)+f(x_{i+1})}{2} \Delta x_i = \frac{f(-2)+f(-1.5)}{2} * 0.5 + \frac{f(-1.5)+f(-1)}{2} * 0.5 + \frac{f(-1)+f(-0.5)}{2} * 0.5 +$

$\frac{f(-0.5)+f(0)}{2} * 0.5 + \frac{f(0)+f(0.5)}{2} * 0.5 + \frac{f(0.5)+f(1)}{2} * 0.5 + \frac{f(1)+f(1.5)}{2} * 0.5 + \frac{f(1.5)+f(2)}{2} * 0.5$

$f(-2) = 3(-2)^3 - 2(-2)^2 + 1 = -24 - 8 + 1 = -31$

$f(-1.5) = 3(-1.5)^3 - 2(-1.5)^2 + 1 = 3(-3.375) - 2(2.25) + 1 = -10.125 - 4.5 + 1 = -13.625$

$f(-1) = 3(-1)^3 - 2(-1)^2 + 1 = -3 - 2 + 1 = -4$

$f(-0.5) = 3(-0.5)^3 - 2(-0.5)^2 + 1 = 3(-0.125) - 2(0.25) + 1 = -0.375 - 0.5 + 1 = 0.125$

$f(0) = 3(0)^3 - 2(0)^2 + 1 = 1$

$f(0.5) = 3(0.5)^3 - 2(0.5)^2 + 1 = 3(0.125) - 2(0.25) + 1 = 0.375 - 0.5 + 1 = 0.875$

$f(1) = 3(1)^3 - 2(1)^2 + 1 = 3 - 2 + 1 = 2$

$f(1.5) = 3(1.5)^3 - 2(1.5)^2 + 1 = 3(3.375) - 2(2.25) + 1 = 10.125 - 4.5 + 1 = 6.625$

$f(2) = 3(2)^3 - 2(2)^2 + 1 = 24 - 8 + 1 = 17$

$\int_{-2}^{2} f(x)dx \cong \sum_{i=-2}^{i=2} \frac{f(x_i)+f(x_{i+1})}{2} \Delta x_i = \frac{f(-2)+f(-1.5)}{2} * 0.5 + \frac{f(-1.5)+f(-1)}{2} * 0.5 + \frac{f(-1)+f(-0.5)}{2} * 0.5 +$

$\frac{f(-0.5)+f(0)}{2} * 0.5 + \frac{f(0)+f(0.5)}{2} * 0.5 + \frac{f(0.5)+f(1)}{2} * 0.5 + \frac{f(1)+f(1.5)}{2} * 0.5 + \frac{f(1.5)+f(2)}{2} * 0.5 =$

$\frac{-31-13.625}{2} * 0.5 + \frac{-13.625-4}{2} * 0.5 + \frac{-4-0.125}{2} * 0.5 + \frac{0.125+1}{2} * 0.5 + \frac{1+0.875}{2} * 0.5 + \frac{0.875+2}{2} * 0.5 +$

$\frac{2+6.625}{2} * 0.5 + \frac{6.625+17}{2} * 0.5 = -11.16 - 4.41 - 0.97 + 0.28 + 0.468 + 0.72 + 2.156 + 5.91 = -7$

10. If $f(x) = x^2 - x - 6$, using trapezoidal method, the integral $\int_{1}^{5} f(x)dx \cong 5.50$

The interval $\Delta x = 0.5$

$\int_{1}^{5} f(x)dx = \sum_{i=1}^{i=4} \frac{f(x_i)+f(x_{i+1})}{2} \Delta x_i = \frac{f(1)+f(1.5)}{2} * 0.5 + \frac{f(1.5)+f(2)}{2} * 0.5 + \frac{f(2)+f(2.5)}{2} * 0.5 +$

$\frac{f(2.5)+f(3)}{2} * 0.5 + \frac{f(3)+f(3.5)}{2} * 0.5 + \frac{f(3.5)+f(4)}{2} * 0.5 + \frac{f(4)+f(4.5)}{2} * 0.5 + \frac{f(4.5)+f(5)}{2} * 0.5$

$f(1) = (1)^2 - 1 - 6 = -6$

$f(1.5) = (1.5)^2 - 1.5 - 6 = 2.25 - 1.5 - 6 = -5.25$

$f(2) = (2)^2 - 2 - 6 = 4 - 2 - 6 = -4$

$f(2.5) = (2.5)^2 - 2.5 - 6 = 6.25 - 2.5 - 6 = -2.25$

$f(3) = (3)^2 - 3 - 6 = 9 - 3 - 6 = 0$

$f(3.5) = (3.5)^2 - 3.5 - 6 = 12.25 - 3.5 - 6 = 2.75$

$f(4) = (4)^2 - 4 - 6 = 16 - 4 - 6 = 6$

$f(4.5) = (4.5)^2 - 4.5 - 6 = 20.25 - 4.5 - 6 = 9.75$

$f(5) = (5)^2 - 5 - 6 = 25 - 5 - 6 = 14$

$\int_1^5 f(x)dx = \sum_{i=1}^{i=4} \frac{f(x_i)+f(x_{i+1})}{2} \Delta x_i = \frac{f(1)+f(1.5)}{2} * 0.5 + \frac{f(1.5)+f(2)}{2} * 0.5 + \frac{f(2)+f(2.5)}{2} * 0.5 +$

$\frac{f(2.5)+f(3)}{2} * 0.5 + \frac{f(3)+f(3.5)}{2} * 0.5 + \frac{f(3.5)+f(4)}{2} * 0.5 + \frac{f(4)+f(4.5)}{2} * 0.5 + \frac{f(4.5)+f(5)}{2} * 0.5 = \frac{-6-5.25}{2} *$

$0.5 + \frac{-5.25-4}{2} * 0.5 + \frac{-4-2.25}{2} * 0.5 + \frac{-2.25+0}{2} * 0.5 + \frac{0+2.75}{2} * 0.5 + \frac{2.75+6}{2} * 0.5 + \frac{6+9.75}{2} * 0.5 + \frac{9.75+14}{2} *$

$0.5 = -2.81 - 2.31 - 1.56 - 0.56 + 0.69 + 2.19 + 3.94 + 5.94 = 5.50$

Chapter 8. D.   Fundamental Theorem of Calculus

1. If $f(x) = 3x^2 + 3x - 2$ then, the $\int_{-4}^5 f(x)dx = 184.5$

First, the indefinite integral of $f(x)$ is:

$\int f(x)dx = \int(3x^2 + 3x - 2)dx = \int 3x^2 dx + \int 3x dx - \int 2\,dx = x^3 + \frac{3}{2}x^2 - 2x + C = F(x)$

Applying the Fundamental Theorem of Calculus, we have:

$\int_{-4}^5 f(x)dx = F(5) - F(-4)$

$F(5) = (5)^3 + \frac{3}{2}(5)^2 - 2(5) = 125 + 1.5 * 25 - 10 = 152.5 + C$

$F(-4) = (-4)^3 + \frac{3}{2}(-4)^2 - 2(-4) = -64 + 24 + 8 = -32 + C$

$\int_{-4}^5 f(x)dx = F(5) - F(-4) = 152.5 - (-32) = 184.5$

2. If $f(x) = x^2 + 3\sqrt{x}$ then, the $\int_1^5 f(x)dx = 102.41$

$\int \sqrt{x}dx = \int x^{\frac{1}{2}}dx = \frac{x^{\frac{1}{2}+1}}{\frac{1}{2}+1} + C = \frac{x^{\frac{3}{2}}}{\frac{3}{2}} + C = \frac{2}{3}\sqrt{x^3} + C = \frac{2x\sqrt{x}}{3} + C$

First, the indefinite integral of $f(x)$ is:

$\int f(x)dx = \int(x^2 + 3\sqrt{x})dx = \int x^2 dx + \int 3\sqrt{x}dx = \frac{1}{3}x^3 + 3\frac{2x\sqrt{x}}{3} + C = \frac{1}{3}x^3 + 6\,x\sqrt{x} + C = F(x)$

Applying the Fundamental Theorem of Calculus, we have:

$\int_1^5 f(x)dx = F(5) - F(1)$

$F(5) = \frac{1}{3}x^3 + 6\,x\sqrt{x} = \frac{1}{3}(5)^3 + 6 * 5\sqrt{5} = 41.66 + 30 * 2.23 = 108.74$

$F(1) = \frac{1}{3}x^3 + 6x\sqrt{x} = \frac{1}{3}(1)^3 + 6\,x\sqrt{x} = 0.33 + 6 = 6.83$

$\int_1^5 f(x)dx = F(5) - F(1) = 108.74 - 6.83 = 102.41$

3. If $f(x) = 4x - 3$ then, the $\int_{-4}^5 f(x)dx = -9$

First, the indefinite integral of $f(x)$ is:

$\int f(x)dx = \int(4x - 3)dx = \int 4x dx - \int 3dx = \frac{4x^2}{2} - 3x + C = 2\,x^2 - 3x + C = F(x)$

Applying the Fundamental Theorem of Calculus, we have:

$\int_{-4}^5 f(x)dx = F(5) - F(-4)$

$F(5) = 2(5)^2 - 3(5) = 50 - 15 = 35$

$F(-4) = 2(-4)^2 - 3(-4) = 32 + 12 = 44$

$\int_{-4}^{5} f(x)dx = F(5) - F(-4) = 35 - 44 = -9$

4. If $f(x) = 5x + 6\sqrt{x}$ then, the $\int_{2}^{7} f(x)dx = 175.3$

$\int \sqrt{x}dx = \int x^{\frac{1}{2}}dx = \frac{x^{\frac{1}{2}+1}}{\frac{1}{2}+1} + C = \frac{x^{\frac{3}{2}}}{\frac{3}{2}} + C = \frac{2}{3}\sqrt{x^3} + C = \frac{2x\sqrt{x}}{3} + C$

First, the indefinite integral of $f(x)$ is:

$\int f(x)dx = \int (5x + 6\sqrt{x})dx = \int 5xdx + \int 6\sqrt{x}dx = \frac{5x^2}{2} + 6\frac{2x\sqrt{x}}{3} + C = \frac{5x^2}{2} + 4x\sqrt{x} + C = F(x)$

Applying the Fundamental Theorem of Calculus, we have:

$\int_{2}^{7} f(x)dx = F(7) - F(2)$

$F(7) = \frac{5(7)^2}{2} + 4(7)\sqrt{7} = 122.5 + 28\sqrt{7} = 122.5 + 74.08 = 196.58$

$F(2) = \frac{5(2)^2}{2} + 4(2)\sqrt{2} = 10 + 8(1.41) = 21.28$

$\int_{2}^{7} f(x)dx = F(7) - F(2) = 196.58 - 21.28 = 175.3$

5. If $f(x) = \frac{7}{x} + x - 3$ then, the $\int_{1}^{5} f(x)dx = 11.26$

First, the indefinite integral of $f(x)$ is:

$\int f(x)dx = \int \left(\frac{7}{x} + x - 3\right)dx = \int \frac{7}{x}dx + \int xdx - \int 3dx = 7lnx + \frac{x^2}{2} - 3x + C = F(x)$

Applying the Fundamental Theorem of Calculus, we have:

$\int_{1}^{5} f(x)dx = F(5) - F(1)$

$F(5) = 7ln(5) + \frac{(5)^2}{2} - 3(5) = 7(1.6) + 12.5 - 15 = 11.2 + 12.5 - 15 = 8.76$

$F(1) = 7ln(1) + \frac{(1)^2}{2} - 3(1) = 0 + 0.5 - 3 = -2.5$

$\int_{1}^{5} f(x)dx = F(5) - F(1) = 8.76 - (-2.5) = 11.26$

6. If $f(x) = \frac{4+x^2}{x} - 3x + 7$ then, the $\int_{2}^{8} f(x)dx = -12.49$

$f(x) = \frac{4+x^2}{x} - 3x + 7 = \frac{4}{x} + x - 3x + 7 = \frac{4}{x} - 2x + 7$

First, the indefinite integral of $f(x)$ is:

$\int f(x)dx = \int \left(\frac{4}{x} - 2x + 7\right)dx = \int \frac{4}{x}dx - \int 2xdx + \int 7dx = 4lnx - x^2 + 7x + C = F(x)$

Applying the Fundamental Theorem of Calculus, we have:

$\int_{2}^{8} f(x)dx = F(8) - F(2)$

$F(8) = 4\ln(8) - (8)^2 + 7(8) = 4(2.07) - 64 + 56 = 0.31$

$F(2) = 4\ln(2) - (2)^2 + 7(2) = 4(0.69) - 4 + 14 = 12.77$

$\int_{2}^{8} f(x)dx = F(8) - F(2) = 0.28 - 12.77 = -12.49$

7. If $f(x) = 4x + e^x$ then, the $\int_{-2}^{6} f(x)dx = 467.28$

First, the indefinite integral of $f(x)$ is:

$\int f(x)dx = \int (4x + e^x)dx = \int 4xdx + \int e^x dx = 2x^2 + e^x + C = F(x)$

Applying the Fundamental Theorem of Calculus, we have:

$\int_{-2}^{6} f(x)dx = F(6) - F(-2)$

$F(6) = 2(6)^2 + e^6 = 2(36) + 403.42 = 475.42$

$F(-2) = 2(-2)^2 + e^{-2} = 2(4) + 0.13 = 8.13$

$\int_{-2}^{6} f(x)dx = F(6) - F(-2) = 475.42 - 8.13 = 467.28$

8. If $f(x) = e^x + \sin(x)$ then, the $\int_{1}^{6} f(x)dx = 400.72$

First, the indefinite integral of $f(x)$ is:

$\int f(x)dx = \int [e^x + \sin(x)]dx = \int e^x dx + \int \sin(x)\, dx = e^x - \cos(x) + C = F(x)$

Applying the Fundamental Theorem of Calculus, we have:

$\int_{1}^{6} f(x)dx = F(6) - F(-2)$

$F(6) = e^6 - \cos(6) = 403.42 - 0.994 = 402.46$

$F(1) = e^1 - \cos(1) = 2.71 - 0.999 = 2.17$

$\int_{1}^{6} f(x)dx = F(6) - F(-2) = 402.43 - 1.71 = 400.29$

9. If $f(x) = 2e^x + 3x^2 - 4x + 5$ then, the $\int_{-3}^{8} f(x)dx = 6445.81$

First, the indefinite integral of $f(x)$ is:

$\int f(x)dx = \int (2e^x + 3x^2 - 4x + 5)dx = \int 2e^x dx + \int 3x^2 dx - \int 4xdx + \int 5dx = 2\,e^x + x^3 - 2x^2 + 5x + C = F(x)$

Applying the Fundamental Theorem of Calculus, we have:

$\int_{-3}^{8} f(x)dx = F(8) - F(-3)$

$F(8) = 2e^8 + (8.^3 - 2(8)^2 + 5(8) = 5961.91 + 512 - 128 + 40 = 6385.92$

$F(-3) = 2e^{-3} + (-3)^3 - 2(-3)^2 + 5(-3) = 0.099 - 27 - 18 - 15 = -59.9$

$\int_{-3}^{8} f(x)dx = F(8) - F(-3) = 6385.91 - (-59.9) = 6445.82$

10. If $f(x) = x^2 - 3x + 7$ then, the $\int_{-3}^{3} f(x)dx = 60$

First, the indefinite integral of $f(x)$ is:

$\int f(x)dx = \int (x^2 - 3x + 7)dx = \int x^2 dx - \int 3xdx + \int 7dx = \frac{1}{3}x^3 - \frac{3}{2}x^2 + 7x + C = F(x)$

Applying the Fundamental Theorem of Calculus, we have:

$\int_{-3}^{3} f(x)dx = F(3) - F(-3)$

$F(3) = \frac{1}{3}(3)^3 - \frac{3}{2}(3)^2 + 7(3) = 9 - 13.5 + 21 = 16.5$

$F(-3) = \frac{1}{3}(-3)^3 - \frac{3}{2}(-3)^2 + 7(-3) = -9 - 13.5 - 21 = -43.5$

$\int_{-3}^{3} f(x)dx = F(3) - F(-3) = 16.5 - (-43.5) = 60$

Chapter 8. E. a.   Antiderivatives of functions

1. If $f(x) = \sin(x) + 2x - \frac{3x^2+3x}{x}$

then, the antiderivative of $f(x)$ will be: $F(x) = -\cos(x) - \frac{x^2}{2} - 3x + C$

$f(x) = \sin(x) + 2x - \frac{3x^2+3x}{x} = \sin(x) + 2x - 3x - 3 = \sin(x) - x - 3$

$F(x) = \int f(x)dx = \int (\sin(x) - x - 3)dx = \int [\sin(x)]dx - \int x dx - \int 3 dx = -\cos(x) - \frac{x^2}{2} - 3x + C$

2. If $f(x) = 5x + \frac{4x^3+3x^2-x}{3}$ then,

the antiderivative of $f(x)$ will be: $F(x) = \frac{x^4}{3} + \frac{x^3}{3} + \frac{7x^2}{3} + C$

$f(x) = 5x + \frac{4x^3+3x^2-x}{3} = 5x + \frac{4x^3}{3} + x^2 - \frac{x}{3} = \frac{4x^3}{3} + x^2 + \frac{14x}{3}$

$F(x) = \int f(x)dx = \int (\frac{4x^3}{3} + x^2 + \frac{14x}{3})dx = \int (\frac{4x^3}{3}) dx + \int x^2 dx + \int (\frac{14x}{3}) dx = \frac{4x^4}{3*4} + \frac{x^3}{3} + \frac{14x^2}{3*2} =$

$\frac{x^4}{3} + \frac{x^3}{3} + \frac{7x^2}{3} + C$

3. If $f(x) = 7\tan(x) - \sec^2(x)$ then,

the antiderivative of $f(x)$ will be: $F(x) = -7\ln|\cos(x)| - \tan(x) + C$

$F(x) = \int [f(x)]dx = \int [7\tan(x) - \sec^2(x)]dx = \int [7\tan(x)]dx - \int [\sec^2(x)]dx = -$
$7\ln|\cos(x)| - \tan(x) + C$

4. If $f(x) = e^x - \frac{x^2-3x-10}{x+2} + 7$ then,

the antiderivative of $f(x)$ will be: $F(x) = e^x - \frac{x^2}{2} + 12x + C$

$f(x) = e^x - \frac{x^2-3x-10}{x+2} + 7 = e^x - \frac{(x+2)(x-5)}{x+2} + 7 = e^x - (x-5) + 7 = e^x - x + 12$

$F(x) = \int [f(x)]dx = \int (e^x - x + 12)dx = \int e^x dx - \int x dx + \int 12 dx = e^x - \frac{x^2}{2} + 12x + C$

5. If $f(x) = \cos(x) - \frac{2x^2+3x}{x} + 3$ then,

the antiderivative of $f(x)$ will be: $F(x) = \sin(x) - x^2 + C$

$f(x) = \cos(x) - \frac{2x^2+3x}{x} + 3 = \cos(x) - 2x - 3 + 3 = \cos(x) - 2x$

$F(x) = \int [f(x)]dx = \int [\cos(x) - 2x]dx = \int [\cos(x)]dx - \int 2x dx = \sin(x) - x^2 + C$

6. If $f(x) = \frac{2x^2+3x}{x^2} - 5\sec^2(x)$ then,

the antiderivative of $f(x)$ will be: $F(x) = 2x + 3\ln|x| - 5\tan(x) + C$

$f(x) = \frac{2x^2+3x}{x^2} - 5\sec^2(x) = \frac{2x^2}{x^2} + \frac{3x}{x^2} - 5\sec^2(x) = 2 + \frac{3}{x} - 5\sec^2(x)$

$F(x) = \int [f(x)]dx = \int [2 + \frac{3}{x} - 5\sec^2(x)]dx = \int 2dx + \int \frac{3}{x}dx - \int [5\sec^2(x)]dx = 2x + 3ln|x| -$

$5\tan(x) + C$

7. If $f(x) = \frac{\sin^2(x)+\cos^2(x)}{x} - e^x$ then,

the antiderivative of $f(x)$ will be: $F(x) = ln|x| - e^x + C$

$f(x) = \frac{\sin^2(x)+\cos^2(x)}{x} - e^x = \frac{1}{x} - e^x$

$F(x) = \int [f(x)]dx = \int [\frac{1}{x} - e^x]dx = \int \frac{1}{x}dx - \int edx = ln|x| - e^x + C$

8. If $f(x) = \frac{\sin^2(x)-\cos^2(x)}{\sin(x)+\cos(x)} + 1 + \ln(x)$ then,

the antiderivative of $f(x)$ will be: $F(x) = -\cos(x) - \sin(x) + xln(x) + C$

$f(x) = \frac{\sin^2(x)-\cos^2(x)}{\sin(x)+\cos(x)} + 1 + \ln(x) = \frac{[\sin(x)-\cos(x)][\sin(x)+\cos(x)]}{\sin(x)+\cos(x)} + 1 + \ln(x) = \sin(x) -$

$\cos(x) + 1 + \ln(x)$

$F(x) = \int [f(x)]dx = \int [\sin(x) - \cos(x) + 1 + \ln(x)]dx = \int [\sin(x)]\,dx - \int [\cos(x)]\,dx +$

$\int dx + \int \ln(x)\,dx = -\cos(x) - \sin(x) + x + xln(x) - x = -\cos(x) - \sin(x) + xln(x)$

9. If $f(x) = \sin^2(x) + \cos^2(x) + 1 - \ln(x)$ then,

the antiderivative of $f(x)$ will be: $F(x) = 3x - xln(x) + C$

$f(x) = \sin^2(x) + \cos^2(x) + 1 - \ln(x) = 1 + 1 - \ln(x) = 2 - \ln(x)$

$F(x) = \int [f(x)]dx = \int [2 - \ln(x)]dx = \int 2dx - \int \ln(x)\,dx = 2x - xln(x) + x + C = 3x - xln(x) +$

C

10. If $f(x) = \frac{2}{\sqrt{x}} - \sqrt{x}$ then,

the antiderivative of $f(x)$ will be: $F(x) = 4\sqrt{x} - \frac{3}{2}x\sqrt{x} + C$

$f(x) = 2x^{-\frac{1}{2}} - x^{\frac{1}{2}}$

$F(x) = \int [f(x)]dx = \int [2x^{-\frac{1}{2}} - x^{\frac{1}{2}}]dx = \int (2x^{-\frac{1}{2}})dx - \int (x^{\frac{1}{2}})dx = 2\frac{x^{\frac{1}{2}}}{\frac{1}{2}} - \frac{2}{3}x^{\frac{3}{2}} + C = 4\sqrt{x} - \frac{2}{3}x\sqrt{x} + C$

Chapter 8. E. b.   Methods of Integration - Substitution

1. If $f(x) = 4(2x + 1)$ then, $F(x) = \int f(x)dx = (2x + 1)^2 + C$

$u = 2x + 1$

$u' = \frac{du}{dx} = 2$ so, $dx = \frac{du}{2}$

$\int 4u\frac{du}{2} = \int 2udu = 2\frac{u^2}{2} = u^2$

But $u = 2x + 1$ so $F(x) = \int f(x)dx = (2x + 1)^2 + C$

2. If $f(x) = 30x(5x^2 - 3)^2$ then, $F(x) = \int f(x)dx = (5x - 3)^3 + C$

$u = 5x^2 - 3$

$u' = \frac{du}{dx} = 10x$ so, $dx = \frac{du}{10x}$

$\int 30xu^2 \frac{du}{10x} = \int 3u^2 du = 3\frac{u^3}{3} = u^3$

But $u = 5x^2 - 3$ so $F(x) = \int f(x)dx = (5x^2 - 3)^3 + C$

3. If $f(x) = 2x\cos(x^2)$ then, $F(x) = \int f(x)dx = \sin(x^2) + C$

$u = x^2$

$u' = \frac{du}{dx} = 2x$ so, $dx = \frac{du}{2x}$

$\int 2x\cos(u) \frac{du}{2x} = \int \cos(u)du = \sin(u) = \sin(x^2) + C$

4. If $f(x) = \frac{5}{3x-73}$ then, $F(x) = \int f(x)dx = \frac{5}{3}\ln|3x - 73| + C$

$u = 3x - 73$

$u' = \frac{du}{dx} = 3$ so, $dx = \frac{du}{3}$

$\int \frac{5}{u} \frac{du}{3} = \frac{5}{3}\ln|u| + C$

$F(x) = \int f(x)dx = \frac{5}{3}\ln|3x - 73| + C$

5. If $f(x) = \frac{7\ln(x)}{5x}$ then, $F(x) = \int f(x)dx = \frac{7}{10}[\ln|x|]^2 + C$

$u = \ln(x)$

$u' = \frac{du}{dx} = \frac{1}{x}$ so, $dx = xdu$

$\int f(x)\,dx = \int \frac{7u}{5x}xdu = \int \frac{7u}{5}du = \frac{7u^2}{5*2} = \frac{7}{10}u^2 + C$

$F(x) = \int f(x)dx = \frac{7}{10}[\ln|x|]^2 + C$

6. If $f(x) = 3x^3e^{x^4}$ then, $F(x) = \int f(x)dx = \frac{3}{4}e^{x^4} + C$

$u = x^4$ so, $u' = \frac{du}{dx} = 4x^3$

$dx = \frac{du}{4x^3}$

$\int 3x^3 e^u \frac{du}{4x^3} = \int \frac{3}{4}e^u du = \frac{3}{4}e^u + C$

$F(x) = \int f(x)dx = \frac{3}{4}e^{x^4} + C$

7. If $f(x) = (6x - 4)(3x^2 - 4x + 5)^2$ then, $F(x) = \int f(x)dx = \frac{1}{3}(3x^2 - 4x + 5)^3 + C$

$u = 3x^2 - 4x + 5$ so, $u' = \frac{du}{dx} = 6x - 4$

$dx = \frac{du}{6x-4}$

$\int (6x - 4)u^2 \frac{du}{6x-4} = \int u^2 du = \frac{1}{3}u^3 + C$

$F(x) = \int f(x)dx = \frac{1}{3}(3x^2 - 4x + 5)^3 + C$

8. If $f(x) = \frac{8x^3+3}{\sqrt{2x^4+3x-5}}$ then, $F(x) = \int f(x)dx = 2\sqrt{2x^4+3x-5} + C$

$u = 2x^4 + 3x - 5 \ so, u' = \frac{du}{dx} = 8x^3 + 3$

$dx = \frac{du}{8x^3+3}$

$\int \frac{(8x^3+3)du}{(8x^3+3)\sqrt{u}} = \int \frac{du}{\sqrt{u}} = \int u^{-\frac{1}{2}}du = \frac{u^{-\frac{1}{2}+1}}{\frac{1}{2}} = 2\sqrt{u} + C$

$F(x) = \int f(x)dx = 2\sqrt{2x^4+3x-5} + C$

9. If $f(x) = \frac{9x^2-8x}{\sqrt{3x^3-4x^2+5}}$ then, $\int_1^3 f(x) = F(3) - F(1) = 10.14$

$u = 3x^3 - 4x^2 + 5 \ so, u' = \frac{du}{dx} = 9x^2 - 8x$

$dx = \frac{du}{9x^2-8x}$

$\int \frac{(9x^2-8x)du}{(9x^2-8x)\sqrt{u}} = \int \frac{du}{\sqrt{u}} = 2\sqrt{u} + C$

$\int_1^3 f(x) = F(3) - F(1) = 2\sqrt{3(3)^3 - 4(3)^2 + 5} - 2\sqrt{3(1)^3 - 4(1)^2 + 5} = 2\sqrt{81 - 36 + 5} -$

$2\sqrt{3 - 4 + 5} = 2\sqrt{50} - 2\sqrt{4} = 14.14 - 4 = 10.14$

10. If $f(x) = \frac{1}{\sqrt{x^2+9}} = \frac{1}{\sqrt{x^2+3^2}}$ then, $F(x) = \int f(x)dx = \ln\left|\sec[\tan^{-1}(\frac{x}{3})] + \tan[\tan^{-1}(\frac{x}{3})]\right| + C$

$x = 3\tan(\Theta) \ so, x' = \frac{dx}{d\Theta} = 3\sec^2(\Theta) \ ; \ \tan(\Theta) = \frac{x}{3} \ so, \Theta = \tan^{-1}(\frac{x}{3})$

$dx = 3\sec^2(\Theta) \ d\Theta$

$\tan^2(\Theta) + 1 = \frac{\sin^2(\Theta)}{\cos^2(\Theta)} + 1 = \frac{\sin^2(\Theta)+\cos^2(\Theta)}{\cos^2(\Theta)} = \frac{1}{\cos^2(\Theta)} = \sec^2(\Theta)$

$\int \frac{3\sec^2(\Theta) \ d\Theta}{\sqrt{9\tan^2(\Theta)+9}} = \int \frac{3\sec^2(\Theta) \ d\Theta}{\sqrt{9[\tan^2(\Theta)+1]}} = \int \frac{3\sec^2(\Theta) \ d\Theta}{3\sqrt{[\tan^2(\Theta)+1]}} = \int \frac{\sec^2(\Theta) \ d\Theta}{\sqrt{\sec^2(\Theta)}} = \int \frac{\sec^2(\Theta) \ d\Theta}{\sec(\Theta)} = \int \sec(\Theta) \ d\Theta =$

$\ln|\sec(\Theta) + \tan(\Theta)| + C$

$F(x) = \int f(x)dx = \ln\left|\sec[\tan^{-1}(\frac{x}{3})] + \tan[\tan^{-1}(\frac{x}{3})]\right| + C$

Chapter 8. E. c.   Methods of Integration – By parts

1. If $f(x) = x^4\ln(x)$ then $F(x) = \int f(x)dx = \frac{1}{5}x^5\left[\ln(x) - \frac{1}{5}\right] + C$

$u = \ln(x) \ so \ u' = \frac{du}{dx} = \frac{1}{x}$

$du = \frac{1}{x}dx$

$dv = x^4dx \ so \ v = \int x^4dx = \frac{1}{5}x^5 + C$

By using the integration by parts formula, we have:

$\int u \, dv = uv - \int v \, du$

$\int \ln(x)(x^4 dx) = \ln(x)\frac{1}{5}x^5 - \int \frac{1}{5}x^5\left(\frac{1}{x}\right)dx = \ln(x)\frac{1}{5}x^5 - \frac{1}{5}\int x^4 dx = \ln(x)\frac{1}{5}x^5 - \frac{1}{5}*\frac{x^5}{5} =$

$\ln(x)\frac{1}{5}x^5 - \frac{x^5}{25} = \frac{1}{5}x^5\left[\ln(x) - \frac{1}{5}\right] + C$

2. If $f(x) = x^3 \cos(x)$ then $F(x) = \int f(x)dx = \sin(x)(x^3) + 3x^2 \cos(x) + 6x \sin(x) + 6 \cos(x) + C$

$u = x^3$ so $u' = \frac{du}{dx} = 3x^2$

$du = 3x^2 dx$

$dv = \cos(x)dx$ so $v = \int \cos(x)dx = \sin(x) + C$

By using the integration by parts formula, we have:

$\int u dv = uv - \int v du$

$\int \cos(x)(x^3 dx) = \sin(x)(x^3) - 3\int x^2[\sin(x)\, dx] =$

$3\int x^2[\sin(x)\, dx] =$

$u = x^2, u' = \frac{du}{dx} = 2x$ so, $du = 2x dx$

$dv = \sin(x)\, dx$, so $v = \int \sin(x)\, dx = -\cos(x) + C$

$3\int x^2[\sin(x)\, dx] = 3\{[x^2(-\cos x)] - \int[-\cos(x)]2x dx\} = -3x^2 \cos(x) + 6\int x\cos(x)dx$

So,

$\int \cos(x)(x^3 dx) = \sin(x)(x^3) - [-3x^2 \cos(x) + 6\int x\cos(x)dx]$

$\int x \cos(x)dx =$

$u = x,, u' = \frac{du}{dx} = 1$ so, $du = dx$

$dv = \cos(x), v = \int \cos(x)\, dx = \sin(x)$

$\int u dv = uv - \int v du$

$\int x \cos(x)dx = x\sin(x) - \int \sin(x)\, dx = x\sin(x) - [-\cos(x)] = x\sin(x) + \cos(x)$

$\int \cos(x)(x^3 dx) = \sin(x)(x^3) - [-3x^2 \cos(x) + 6\int x\cos(x)dx] = \sin(x)(x^3) + 3x^2 \cos(x) +$

$6[x\sin(x) + \cos(x)] = \sin(x)(x^3) + 3x^2 \cos(x) + 6x\sin(x) + 6\cos(x)$

3. If $f(x) = \frac{\ln(x)}{x^3}$ then $F(x) = \int f(x)dx = -\frac{1}{2x^2}\left[\ln(x) + \frac{1}{2}\right] + C$

$u = \ln(x)$ so $u' = \frac{du}{dx} = \frac{1}{x}$

$du = \frac{1}{x}dx$

$dv = \frac{1}{x^3}dx$ so $v = \int \frac{1}{x^3}dx = \int x^{-3}dx = -\frac{1}{2}x^{-2} + C$

By using the integration by parts formula, we have:

$\int u dv = uv - \int v du$

$\int \ln(x)\left(\frac{1}{x^3}dx\right) = \ln(x)\left(-\frac{1}{2}x^{-2}\right) - \int\left(-\frac{1}{2}x^{-2}\right)\left(\frac{1}{x}dx\right) = -\frac{\ln(x)}{2x^2} + \int\left(\frac{1}{2}x^{-3}dx\right) = -\frac{\ln(x)}{2x^2} - \frac{1}{4x^2} =$

$-\frac{1}{2x^2}\left[\ln(x) + \frac{1}{2}\right] + C$

4. If $f(x) = x^2 \sin(x)$ then $F(x) = \int f(x)dx = -x^2[\cos(x)] + 2x\sin(x) + 2\cos(x) + C$

$u = x^2$ so $u' = \frac{du}{dx} = 2x$

$du = 2xdx$

$dv = \sin(x)\,dx$ so $v = \int \sin(x)\,dx = -\cos(x) + C$

By using the integration by parts formula, we have:

$\int u\,dv = uv - \int v\,du$

$\int x^2 \sin(x)\,dx = x^2[-\cos(x)] - \int[-\cos(x)]2xdx = -x^2[\cos(x)] + 2\int \cos(x)\,xdx + C$

We use the integration by parts formula again for $\int \cos(x)xdx$;

$u = x$ so, $u' = \frac{du}{dx} = 1$

$du = dx$

$dv = \cos(x)\,dx$ so, $v = \int \cos(x)\,dx = \sin(x) + C$

$\int u\,dv = uv - \int v\,du$

$\int x\cos(x)dx = x\sin(x) - \int \sin(x)\,dx = x\sin(x) + \cos(x) + C$

So, we have

$\int x^2 \sin(x)\,dx = x^2[-\cos(x)] - \int[-\cos(x)]2xdx = -x^2[\cos(x)] + 2\int \cos(x)xdx =$
$-x^2[\cos(x)] + 2[x\sin(x) + \cos(x)] = -x^2[\cos(x)] + 2x\sin(x) + 2\cos(x) + C$

5. If $f(x) = e^x \cos(x)$ then $F(x) = \int f(x)dx = \frac{1}{2}e^x[\sin(x) + \cos(x)] + C$

$u = e^x$ so $u' = \frac{du}{dx} = e^x$

$du = e^x dx$

$dv = \cos(x)\,dx$ so $v = \int \cos(x)\,dx = \sin(x) + C$

By using the integration by parts formula, we have:

$\int u\,dv = uv - \int v\,du$

** $\int e^x \cos(x)\,dx = e^x \sin(x) - \int e^x \sin(x)\,dx + C$

We use the integration by parts formula again for $\int e^x \sin(x)\,dx$:

$u = e^x$ so $u' = \frac{du}{dx} = e^x$

$du = e^x dx$

$dv = \sin(x)\,dx$ so $v = \int \sin(x)\,dx = -\cos(x) + C$

By using the integration by parts formula, we have:

$\int u\,dv = uv - \int v\,du$

$\int e^x \sin(x)\,dx = e^x[-\cos(x)] - \int[-\cos(x)]e^x dx = -e^x[\cos(x)] + \int[\cos(x)]e^x dx + C$

So, the original integral in formula **

$\int[\cos(x)]e^x = e^x \sin(x) - \{-e^x[\cos(x)] + \int[\cos(x)]e^x dx\}$

$\int[\cos(x)]e^x = e^x \sin(x) + e^x \cos(x) - \int[\cos(x)]e^x dx$

$2\int[\cos(x)]e^x = e^x \sin(x) + e^x \cos(x) + C$

$\int[\cos(x)]e^x = \frac{1}{2}e^x[\sin(x) + \cos(x)] + C$

6. If $f(x) = x^2 e^x$ then $F(x) = \int f(x)dx = x^2 e^x - xe^x + e^x + C$

$u = x^2$ so $u' = \frac{du}{dx} = 2x$

$du = 2xdx$

$dv = e^x dx$ so $v = \int e^x dx = e^x + C$

By using the integration by parts formula, we have:

$\int u\,dv = uv - \int v\,du$

$\int x^2 e^x dx = x^2 e^x - \int e^x 2x\,dx = x^2 e^x - 2\int e^x x\,dx$

By using the integration by parts formula again for $\int e^x x\,dx$, we have:

$u = x$ so, so $u' = \frac{du}{dx} = 1$

$du = dx$

$dv = e^x dx$ so $v = \int e^x dx = e^x + C$

$\int u\,dv = uv - \int v\,du$

$\int e^x x\,dx = xe^x - \int e^x dx = xe^x - e^x + C$

So:

$\int x^2 e^x dx = x^2 e^x - 2(xe^x - e^x) = x^2 e^x - 2xe^x + 2e^x + C$

7. If $f(x) = \frac{\ln(x)}{x}$ then $F(x) = \int f(x)dx = \frac{1}{2}\ln^2(x) + C$

$u = \ln(x)$ so, so $u' = \frac{du}{dx} = \frac{1}{x}$

$du = \frac{1}{x}dx$

$dv = \frac{1}{x}dx$ so $v = \int \frac{1}{x}dx = \ln(x) + C$

$\int u\,dv = uv - \int v\,du$

$\int \frac{\ln(x)}{x}dx = \ln(x)\left[\ln(x)\right] - \int \frac{\ln(x)}{x}dx$

$2\int \frac{\ln(x)}{x}dx = \ln^2(x) + C$

$\int \frac{\ln(x)}{x}dx = \frac{1}{2}\ln^2(x) + C$

8. If $f(x) = x\sqrt{1 - sin^2(x)}$ then $F(x) = \int f(x)dx = xsin(x) + cos(x) + C$

$f(x) = x\sqrt{1 - sin^2(x)} = x\sqrt{cos^2(x)} = xcos(x)$

$\int f(x)dx = \int xcos(x)dx$

$u = x$ so, $u' = \frac{du}{dx} = 1$

$du = dx$

$dv = cos(x)dx$, so $v = \int cos(x)\,dx = sin(x) + C$

$\int u\,dv = uv - \int v\,du$

$\int xcos(x)dx = xsin(x) - \int sin(x)\,dx = xsin(x) - [-cos(x)] = xsin(x) + cos(x) + C$

9. If $f(x) = xln(x)$ then $F(x) = \int f(x)dx = \frac{x^2}{2}\left[\ln(x) - \frac{1}{2}\right] + C$

$u = ln(x)$ so, $u' = \frac{du}{dx} = \frac{1}{x}$

$du = \frac{1}{x}dx$

$dv = xdx$ so, $v = \int xdx = \frac{x^2}{2} + C$

$\int udv = uv - \int vdu$

$\int xln(x) = \ln(x)\frac{x^2}{2} - \int \frac{x^2}{2} * \frac{1}{x}dx = \ln(x)\frac{x^2}{2} - \int \frac{x}{2}dx = \ln(x)\frac{x^2}{2} - \frac{x^2}{4} = \frac{x^2}{2}\left[\ln(x) - \frac{1}{2}\right] + C$

10. If $f(x) = e^x\sin(2x)$ then $F(x) = \int f(x)dx = \frac{1}{5}\sin(2x)e^x - \frac{2}{5}e^x\cos(2x) + C$

$u = sin(2x)$ so, $u' = \frac{du}{dx} = 2\cos(2x)$

$du = 2\cos(2x)dx$

$dv = e^xdx$ so, $v = \int e^xdx = e^x + C$

$\int udv = uv - \int vdu$

$\int \sin(2x)e^xdx = sin(2x)e^x - \int e^x 2\cos(2x)\,dx = sin(2x)e^x - 2\int e^x\cos(2x)\,dx$

By using the integration by parts formula again for $\int e^x\cos(2x)dx$, we have:

$u = cos(2x)$ so, $u' = \frac{du}{dx} = -2\sin(2x)$

$du = -2\sin(2x)dx$

$dv = e^xdx$ so, $v = \int e^xdx = e^x + C$

$\int udv = uv - \int vdu$

$\int e^x\cos(2x)\,dx = e^x cos(2x) - \int e^x[-2\sin(2x)dx] = e^x cos(2x) + \int e^x[2\sin(2x)dx] = e^x cos(2x) + 2\int e^x\sin(2x)dx$

$\int \sin(2x)e^xdx = sin(2x)e^x - 2[e^x cos(2x) + 2\int e^x\sin(2x)\,dx] = sin(2x)e^x - 2e^x cos(2x) - 4\int e^x\sin(2x)\,dx$

$5\int e^x\sin(2x)\,dx = sin(2x)e^x - 2e^x cos(2x) + C$

$\int e^x\sin(2x)\,dx = \frac{1}{5}sin(2x)e^x - \frac{2}{5}e^x cos(2x) + C$

Chapter 8. F. a. Aria under a curve, volume of solids, average value of functions

PAGE 1

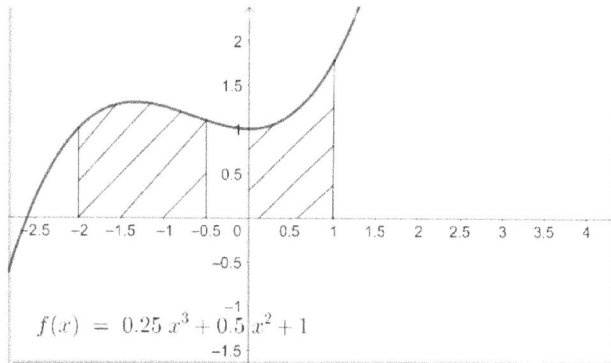

$f(x) = 0.25\,x^3 + 0.5\,x^2 + 1$

1. The area under $f(x) = \frac{1}{4}x^3 + \frac{1}{2}x^2 + 1$ between x=-2 and x=-0.5 is: 1.813

$Area = \int_{-2}^{-0.5} f(x)dx = F(-0.5) - F(-2)$

$F(x) = \int f(x)dx = \int \left[\frac{1}{4}x^3 + \frac{1}{2}x^2 + 1\right]dx = \int \frac{1}{4}x^3dx + \int \frac{1}{2}x^2dx + \int dx = \frac{1}{16}x^4 + \frac{1}{6}x^3 + x + C$

$F(-0.5) = \frac{1}{16}(-0.5)^4 + \frac{1}{6}(-0.5)^3 + (-0.5) = 0.003 - 0.02 - 0.5 = -0.517$

$F(-2) = \frac{1}{16}(-2)^4 + \frac{1}{6}(-2)^3 + (-2) = 1 - 1.33 - 2 = -2.33$

$Area = \int_{-2}^{-0.5} f(x)dx = F(-0.5) - F(-2) = -0.517 + 2.33 = 1.813$

2. The area under $f(x) = \frac{1}{4}x^3 + \frac{1}{2}x^2 + 1$ between x=0 and x=1 is: 1.223

$Area = \int_0^1 f(x)dx = F(1) - F(0)$

$F(x) = \int f(x)dx = \int \left[\frac{1}{4}x^3 + \frac{1}{2}x^2 + 1\right]dx = \int \frac{1}{4}x^3 dx + \int \frac{1}{2}x^2 dx + \int dx = \frac{1}{16}x^4 + \frac{1}{6}x^3 + x + C$

$F(1) = \frac{1}{16}(1)^4 + \frac{1}{6}(1)^3 + (1) = 0.0625 + 0.16 + 1 = 1.223$

$F(0) = \frac{1}{16}(0)^4 + \frac{1}{6}(0)^3 + (0) = 0$

$Area = \int_0^1 f(x)dx = F(1) - F(0) = 1.223$

3. The area under $f(x) = 0.18x^4 - 0.6x^3 + 0.5x^2 + x + 2$ between x=0.1 and x=3 is:11.39

$Area = \int_{0.1}^3 f(x)dx = F(3) - F(0.1)$

$F(x) = \int f(x) = \int[0.18x^4 - 0.6x^3 + 0.5x^2 + x + 2]dx = \int 0.18x^4 dx - \int 0.6x^3 dx +$

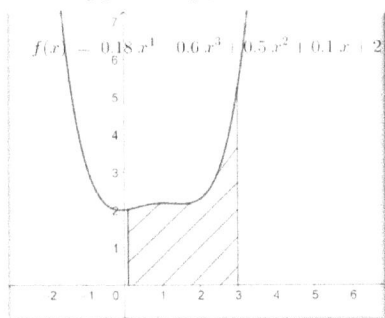

$\int 0.5x^2 + \int xdx + \int 2dx = \frac{0.18}{5}x^5 - \frac{0.6}{4}x^4 + \frac{0.5}{3}x^3 + \frac{1}{2}x^2 + 2x + C$

$F(3) = \frac{0.18}{5}(3)^5 - \frac{0.6}{4}(3)^4 + \frac{0.5}{3}(3)^3 + \frac{1}{2}(3)^2 + 2(3) = 8.74 - 12.15 + 4.5 + 4.5 + 6 = 11.59$

$F(0.1) = \frac{0.18}{5}(0.1)^5 - \frac{0.6}{4}(0.1)^4 + \frac{0.5}{3}(0.1)^3 + \frac{1}{2}(0.1)^2 + 2(0.1) = 0 - 0.00001 + 0.0001 + 0.005 + 0.2 = 0.2$

$Area = \int_{0.1}^3 f(x)dx = F(3) - F(0.1) = 11.59 - 0.2 = 11.39$

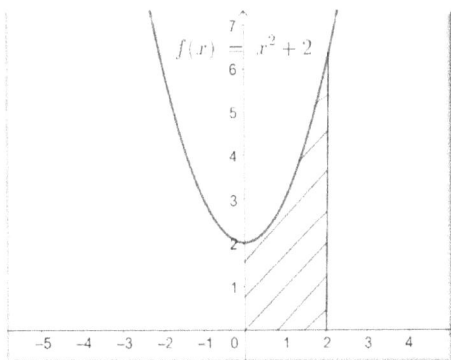

4. The volume of the object that is made by the function $f(x) = x^2 + 2$ that rotate around x axis between x=0 and x=2 is: V=25.06 π

$V = \int_0^2 \pi[f(x)]^2 dx = \pi \int_0^2 (x^2 + 2)^2 dx = \pi \int_0^2 (x^4 + 4x^2 + 4)dx = \pi[F(2) - F(0)]$

$F(x) = \int(x^4 + 4x^2 + 4)dx = \int x^4 dx + \int 4x^2 dx + \int 4dx = \frac{1}{5}x^5 + \frac{4}{3}x^3 + 4x + C$

$F(2) = \frac{1}{5}(2)^5 + \frac{4}{3}(2)^3 + 4(2) = 6.4 + 10.6 + 8 = 25.06$

$F(0) = \frac{1}{5}(0)^5 + \frac{4}{3}(0)^3 + 4(0) = 0$

$V = \pi[F(2) - F(0)] = 25.06\pi$

5. The volume of the object that is made by the function $f(x) = -0.3x^2 + 4$ that rotate around x axis between x=0.5 and x=3 is: V= 22.87π

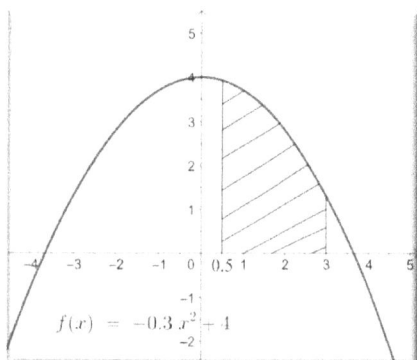

$f(x) = -0.3x^2 + 4$

$V = \int_{0.5}^{3} \pi[f(x)]^2 dx = \pi \int_{0.5}^{3} (-0.3x^2 + 4)^2 dx =$

$\pi \int_{0.5}^{3} (0.09x^4 - 2.4x^2 + 16)dx = \pi[F(3) - F(0.5)]$

$F(x) = \int (0.09x^4 - 2.4x^2 + 16)dx = \int 0.09x^4 dx -$

$\int 2.4x^2 dx + \int 16 dx = \frac{0.09}{5}x^5 - \frac{2.4}{3}x^3 + 16x + C$

$F(3) = \frac{0.09}{5}(3)^5 - \frac{2.4}{3}(3)^3 + 16(3) = 4.37 - 21.6 + 48 = 30.77$

$F(0.5) = \frac{0.09}{5}(0.5)^5 - \frac{2.4}{3}(0.5)^3 + 16(0.5) = 0.00056 - 0.1 + 8 = 7.9$

$V = \pi[F(3) - F(0.5)] = \pi(30.77 - 7.9) = 22.87\pi$

6. The volume of the sphere that is made by the function $4 = x^2 + y^2$ that rotate around x axis between x=-2 and x=2 is: V= 10.66π

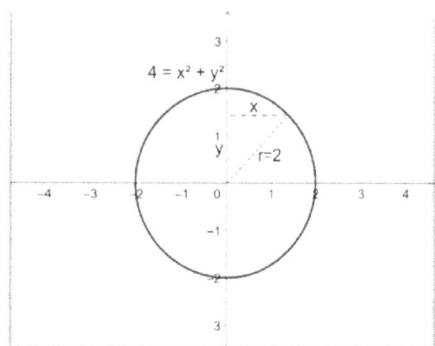

We have:

$x^2 = 4 - y^2$

Area of the circle with the radius x is:

$A = \pi x^2 = \pi(4 - y^2)$

The volume is:

$V(y) = \int A(y)dy = \int \pi(4 - y^2)dy = 4\pi \int dy -$

$\pi \int y^2 dy = 4\pi y - \frac{\pi}{3}y^3 + C$

If we consider the limits as -2 and 2 we will have:

$V = \int_{-2}^{2} (\pi(4 - y^2)dy = F(2) - F(-2)$

$F(2) = 4\pi(2) - \frac{\pi}{3}(2)^3 = 8\pi - 2.66\pi = 5.33\pi$

$F(-2) = 4\pi(-2) - \frac{\pi}{3}(-2)^3 = -8\pi + 2.66\pi = -5.33\pi$

$V = \int_{-2}^{2} (\pi(4 - y^2)dy = F(2) - F(-2) = 5.33\pi - (-5.33)\pi = 10.66\pi$

7. The volume of the object that is made by the function $f(x) = x^2$ that rotate around x axis between x=0 and x=3 is: 48.6π

$V = \int_{0}^{3} \pi[f(x)]^2 dx = \pi \int_{0}^{3} (x^2)^2 dx = \pi \int_{0}^{3} (x^4)dx = \pi[F(3) - F(0)]$

$F(x) = \int x^4 dx = \frac{1}{5}x^5 + C$

$F(3) = \frac{1}{5}(3)^5 = 48.6$

$F(0) = \frac{1}{5}(0)^5 = 0$

$V = \pi[F(3) - F(0)] = \pi(48.6 - 0) = 48.6\pi$

8. The volume of the object that is made by the function $f(x) = x$ that rotate around x axis between x=1 and x=5 is: 41.33π

$V = \int_1^5 \pi [f(x)]^2 dx = \pi \int_1^5 (x)^2 dx = \pi \int_1^5 (x^2) dx = \pi [F(5) - F(1)]$

$F(x) = \int x^2 dx = \frac{1}{3}x^3 + C$

$F(5) = \frac{1}{3}(5)^3 = 41.66$

$F(1) = \frac{1}{3}(1)^3 = 0.33$

$V = \pi [F(5) - F(1)] = \pi (41.66 - 0.33) = 41.33\pi$

9. The average value of the function $f(x) = 2x + \sin(x)$ on the interval [1,3] is: 4

$\int_1^3 f(x) dx = F(3) - F(1)$

$F(x) = \int f(x) dx = \int [2x + \sin(x)] dx = \int 2x dx + \int \sin(x) dx = x^2 - \cos(x) + C$

$F(3) = (3)^2 - \cos(3) = 9 - 0.998 = 8.002$

$F(1) = (1)^2 - \cos(1) = 1 - 0.999 = 0$

$\int_1^3 f(x) dx = F(3) - F(1) = 8 - 0 = 8$

$Average\ value = \frac{1}{3-1} \int_1^3 f(x) dx = \frac{8}{2} = 4$

10. The average value of the function $f(x) = x^2 + \frac{1}{x}$ on the interval [1,5] is: 10.73

$\int_1^5 f(x) dx = F(5) - F(1)$

$F(x) = \int f(x) dx = \int (x^2 + \frac{1}{x}) dx = \int x^2 dx + \int \frac{1}{x} dx = \frac{1}{3}x^3 + \ln(x) + C$

$F(5) = \frac{1}{3}(5)^3 + \ln(5) = 41.66 + 1.6 = 43.26$

$F(1) = \frac{1}{3}(1)^3 + \ln(1) = 0.33 + 0 = 0.33$

$\int_1^5 f(x) dx = F(5) - F(1) = 43.26 - 0.33 = 42.93$

$Average\ value = \frac{1}{5-1} \int_1^5 f(x) dx = \frac{42.93}{4} = 10.73$

Chapter 8. F. b.   Differential equations, Initial value problems, Slope fields

The trajectory of the Space Shuttle in the first minutes is represented by:

$h(t) = 2008 - 0.047t^3 + 18.3t^2 - 345t$

1. The velocity of the Space Shuttle at 20 seconds is: 330.6 m/s.

$h'(t) = -(0.047t^3)' + (18.3t^2)' - (345t)' = -0.047(3)t^2 + 18.3(2)t - 345$

$h'(20) = -0.047(3)(20)^2 + 18.3(2)(20) - 345 = -56.4 + 732 - 345 = 330.6 m/s$

2. The acceleration of the Space Shuttle at 20 seconds is $30.9\ m/s^2$.

$h''(t) = (-0.141t^2)' + (36.6t)' = -0.141(2)t + 36.6$

$h''(20) = -0.282(20) + 36.6 = -5.64 + 36.6 = 30.9 \, m/s^2$

A formula one speed car has the acceleration formula: a(t)=t+4 m/s^2.

3. The velocity of the car after 10 seconds is: 90 m/s

$v(t) = \int a(t)dt = \int(t+4)dt = \frac{1}{2}t^2 + 4t + C$

$v(10) = \frac{1}{2}(10)^2 + 4(10) = 50 + 40 = 90 \, m/s.$

4. The distance traveled by the car after 10 seconds is: 366.6 m.

$s(t) = \int v(t)dt = \int(\frac{1}{2}t^2 + 4t)dt = \frac{1}{6}t^3 + \frac{4}{2}t^2 + C$

$s(10) = \frac{1}{6}(10)^3 + 2(10)^2 = 166.6 + 200 = 366.6 \, m.$

5. If the differential equation is $y' = \frac{dy}{dx} = 15 + x$, the initial condition is y(0)=3, then:

$y = 15x + \frac{1}{2}x^2 + 3$

$dy = (15 + x)dx$

We integrate both sides;

$\int dy = \int(15 + x)dx = \int 15dx + \int xdx = 15x + \frac{1}{2}x^2 + C$

$y = 15x + \frac{1}{2}x^2 + C$

We have that y (0) =3

$3 = 15(0) + \frac{1}{2}(0)^2 + C$

$3 = C$

$y = 15x + \frac{1}{2}x^2 + 3$

6. If the differential equation is $y' = \frac{dy}{dx} = 2x^2 - 3x$, the initial condition is y(1)=2, then:

$y = \frac{2}{3}x^3 - \frac{3}{2}x^2 + 2.83$

$dy = (2x^2 - 3x)dx$

We integrate both sides:

$\int dy = \int(2x^2 - 3x)dx = \int 2x^2 dx - \int 3xdx = \frac{2}{3}x^3 - \frac{3}{2}x^2 + C$

$y = \frac{2}{3}x^3 - \frac{3}{2}x^2 + C$ so, $2 = \frac{2}{3}(1)^3 - \frac{3}{2}(1)^2 + C$

$2 - \frac{2}{3} + \frac{3}{2} = C$

$\frac{12}{6} - \frac{4}{6} + \frac{9}{6} = \frac{17}{6} = 2.83 = C$

$y = \frac{2}{3}x^3 - \frac{3}{2}x^2 + 2.83$

7. If the differential equation is $y' = \frac{dy}{dx} = e^x + 4x$, the initial condition is y(0)=5, then:

$y = e^x + 2x^2 + 4$

$dy = (e^x + 4x)dx$

We integrate both sides:

$\int dy = \int(e^x + 4x)dx = \int e^x dx + \int 4x dx = e^x + 2x^2 + C$

$y = e^x + 2x^2 + C$

$5 = e^0 + 2(0)^2 + C$

$5 = 1 + C$

$4 = C$

$y = e^x + 2x^2 + 4$

8. One of the functions whose derivative is $f(x) = 4x$ is $F(x) = 2x^2 + 37$

$F(x) = \int 4x dx = 2x^2 + C$

So, one of the functions is $F(x) = 2x^2 + 37$

9. One of the functions whose derivative is $f(x) = 3x^{-1}$ is $F(x) = 3\ln(x) + 3$

$F(x) = \int 3x^{-1}dx = 3\ln(x) + C$

So, one of the functions is $F(x) = 3\ln(x) + 3$

10. One of the functions whose derivative is $f(x) = e^x$ is $F(x) = e^x + 3$

$F(x) = \int e^x dx = e^x + C$

So, one of the functions is $F(x) = e^x + 3$

FORMULAS

Cartesian System Formulas

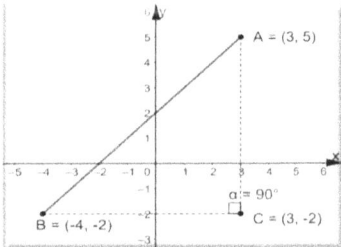

$$AB = \sqrt{BC^2 + AC^2} = \sqrt{(x_c - x_b)^2 + (y_c - y_b)^2}$$

$$x = \frac{x_2 + x_1}{2}$$

$$y = \frac{y_2 + y_1}{2}$$

$$Slope = \frac{Rise}{Run} = \frac{Vertical\ distance}{Horizontal\ distance} = \frac{y_2 - y_1}{x_2 - x_1} = m \qquad\qquad y = mx + b \qquad \underline{general\ form}$$

$y - y_1 = m(x - x_1)$ is called <u>point slope equation</u>.

lines are perpendicular if $m_1 \times m_2 = -1, or\ m_1 = \frac{-1}{m_2}$

lines are parallel if $m_1 = m_2$

Cartesian System Formulas

Area of rectangle A = L x h $\qquad\qquad\qquad\qquad Area\ of\ triangle = \frac{base \times heigth}{2}$

$$Area = \frac{(big\ base + small\ base) \times heigth}{2} \qquad\qquad Area = \pi R^2$$

$$Volume\ prism = L \times W \times h = Area\ of\ base \times heigth$$

$$Volume = Area\ of\ the\ base \times height = \pi R^2 h$$

$$Volume\ of\ a\ cone = \frac{Area\ of\ the\ base \times heigth}{3} = \frac{\pi R^2 h}{3} \qquad\qquad rad =$$

$length\ of\ the\ subtended\ arc$
$\overline{\quad radius\ of\ the\ circle\quad}$

Trigonometry Formulas

$$\sin ∢\phi = \frac{opposite}{hypotenuse} \qquad \cos ∢\phi = \frac{adjacent}{hypotenuse} \qquad \tan ∢\phi = \frac{opposite}{adjacent} \qquad \cot ∢\phi =$$

$adjacent$
$\overline{opposite}$

	30^0	60^0	45^0
Sin(Φ)	$\frac{1}{2}$	$\frac{\sqrt{3}}{2}$	$\frac{\sqrt{2}}{2}$
Cos(Φ)	$\frac{\sqrt{3}}{2}$	$\frac{1}{2}$	$\frac{\sqrt{2}}{2}$
Tan(Φ)	$\frac{\sqrt{3}}{3}$	$\sqrt{3}$	1

Power rules

We consider a, b as non-zero real numbers, and m, n integers.

1. The product of powers with the same base

$$a^n \times a^m = a^{n+m}$$

2. The division of powers with the same base

$$a^n \div a^m = a^{n-m}$$

3. The product of two numbers at an exponent.

$$(a \times b)^n = a^n \times b^n$$

4. The quotient of two different numbers at an exponent.

$$\left(\frac{a}{b}\right)^n = \frac{a^n}{b^n} \quad b \neq 0$$

5. The negative exponent.

$$a^{-n} = \frac{1}{a^n} \quad a \neq$$

7. Exponent zero

$$a^0 = 1$$

6. Exponent at an exponent

$$(a^m)^n = a^{m \times n}$$

8. Fractional exponents

$$a^{\frac{m}{n}} = \sqrt[n]{a^m}$$

Rules of radicals

a. $\sqrt[n]{a} \times \sqrt[n]{b} = \sqrt[n]{a \times b}$

b. $\sqrt[n]{\frac{a}{b}} = \frac{\sqrt[n]{a}}{\sqrt[n]{b}} = \sqrt[n]{a} \div \sqrt[n]{b}$

c. $\sqrt[3]{-27} = -3$

The odd root from negative numbers is a negative number.

The even root from negative numbers is an imaginary number.

$$\sqrt{-1} = i \ or \ imaginary$$

$\frac{a}{b} = \frac{c}{d}$, b and d are non-zero numbers

Then,

$$a \times d = c \times b$$

Rules of logarithms

1. Multiplication into addition

$$\log_a(b \times c) = \log_a b + \log_a c$$

3. The argument at an exponent

$$\log_a b^w = w \times \log_a b \quad b > 0$$

2. Division into subtraction

$$\log_a(b \div c) = \log_a b - \log_a c$$

4. Changing the base

$$\log_a b = \frac{\log_c b}{\log_c a} \quad a, b > 0$$

Special FOIL Formulas

$(x + a)^2 = x^2 + 2xa + a^2, a = constant$

$(x - a)^2 = x^2 - 2xa + a^2, a = constant$

$x^2 - c^2 = (x - c)(x + c)$

Quadratic Formula

$$x_{1,2} = \frac{-b \pm \sqrt{b^2 - 4ac}}{2a}$$

Differentiation Formulas

$x' = 0$

$[f(x) * g(x)]' = f'(x) * g(x) + f(x) * g'(x)$

$[f(x) \mp g(x)]' = f'(x) \mp g'(x)$

$[c * f(x)]' = c * f'(x)$

$\left[\frac{f(x)}{g(x)}\right]' = \frac{[f(x)]'g(x) - f(x)[g(x)]'}{[g(x)]^2}$

$\{f[g(x)]\}' = f'[g(x)] * g'(x)$

$[x^n]' = n * x^{n-1}$

$[\sin(x)]' = \cos(x)$

$[\cos(x)]' = -\sin(x)$

$[\tan(x)]' = \sec^2(x)$

$[\cot(x)]' = -\csc^2(x)$

$[\sec(x)]' = \sec(x) * \tan(x)$

$[\csc(x)]' = -\csc(x) * \cot(x)$

$[e^x]' = e^x$

$[b^x]' = b^x * \ln(b)$

$[\ln(x)]' = \frac{1}{x}$

$[\sin^{-1}(x)]' = \frac{1}{\sqrt{1-x^2}}$

$[\cos^{-1}(x)]' = \frac{-1}{\sqrt{1-x^2}}$

$[\tan^{-1}(x)]' = \frac{1}{x^2+1}$

$[\cot^{-1}(x)]' = \frac{-1}{x^2+1}$

$[\sec^{-1}(x)]' = \frac{1}{|x|\sqrt{x^2-1}}$

$[\csc^{-1}(x)]' = \frac{-1}{|x|\sqrt{x^2-1}}$

Integration Formulas

$\int dx = x + C$

$\int x^n dx = \frac{x^{n+1}}{n+1} + C$

$\int x^3 dx = \frac{x^4}{4} + C$

$\int a^x dx = \frac{a^x}{\ln(a)} + C$

$\int \frac{dx}{x} = \ln|x| + C$

$\int \cos(x)\, dx = \sin(x) + C$

$\int e^x dx = e^x + C$

$\sqrt{x} = x^{\frac{1}{2}}$ so,

$\int \ln(x)\, dx = x\ln(x) - x + C$

$\int \sqrt{x}\, dx = \int x^{\frac{1}{2}}\, dx = \frac{2x^{\frac{3}{2}}}{3} + C$

$\int \sin(x)\, dx = -\cos(x) + C$

$\int \tan(x)\, dx = -\ln|\cos(x)| + C$

$\int \cot(x)\, dx = \ln|\sin(x)| + C$

$\int \sec(x)\, dx = \ln|\sec(x) + \tan(x)| + C$

$\int \csc(x)\, dx = -\ln|\csc(x) + \cot(x)| + C$

$\int \sec^2(x)\, dx = \tan(x) + C$

$\int \csc^2(x)\, dx = -\cot(x) + C$

$\int \sec(x)\tan(x)\, dx = \sec(x) + C$

$\int \csc(x)\cot(x)\, dx = -\csc(x) + C$

$\int \frac{dx}{\sqrt{b^2-x^2}} = \sin^{-1}\left(\frac{x}{b}\right) + C$

$\int \frac{dx}{b^2+x^2} = \frac{1}{b}\tan^{-1}\left(\frac{x}{b}\right) + C$

$\int \frac{dx}{x\sqrt{x^2-b^2}} = \frac{1}{b}\sec^{-1}\left(\frac{|x|}{b}\right) + C$

Some Used Trigonometric Formulas

$Sec(x) = \frac{1}{\cos(x)}$

$\csc(x) = \frac{1}{\sin(x)}$

$sin(x)^2 + cos(x)^2 = 1$

$\sin(x) = \pm\sqrt{1 - \cos(x)^2}$

$\cos(x) = \pm\sqrt{1 - \sin(x)^2}$

$\tan(x)^2 = sec(x)^2 - 1$

APPENDIX

<u>The slope of a tangent line to a curve</u>

Remember the geometric illustration we showed at the beginning of chapter two. Let's suppose we have a circle.

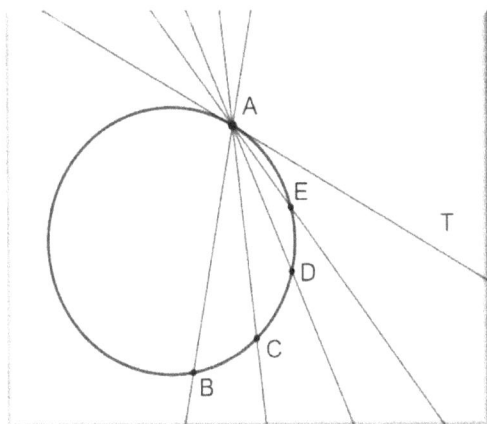

The <u>chord</u> AB is the longest chord compared with the others in this example. As we are approaching the point A going on the circle from point B towards point A, through points B, C, D, E, the length of the chord is becoming smaller and smaller.

When we are at a point which is extremely close to point A, the length of the chord is extremely small going towards zero. The moment we are in point A the chord becomes a tangent to the circle, line AT.

The word <u>tangent</u> comes from the Latin word **tangens** which means "touching". Any tangents to any curve, touch that particular curve in one and only one point.

As we can see below, we have a line that goes through the points S (point 1) and M (point 2). On these points this line intersects the graph of the function:

$f(x) = 4x^3 + 0.7(x-1)^2 + x - 0.5$

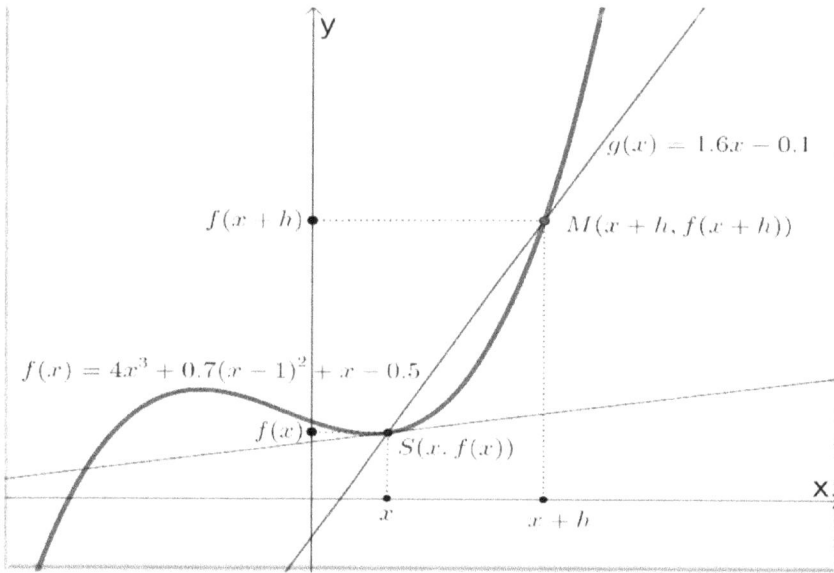

The slope of the line SM can be calculated with the formula:

slope of SM =
$$\frac{y_2 - y_1}{x_2 - x_1} =$$
$$\frac{f(x+h) - f(x)}{x+h-x} =$$
$$\frac{f(x+h) - f(x)}{h}$$

Now, we calculate the slope of the tangent that touches the graph of the function f(x) in the point S.

To be able to calculate this tangent we have to consider that the difference between the x coordinates of the points M and S x and x+h respectively, is super small, going towards zero. We need to use the limit when the difference of the coordinates x and x+h goes towards zero.

So,
$$x_2 - x_1 = x + h - x = h$$
The slope the tangent will be calculated using the limit of the slope formula:
$$\lim_{h \to 0} \frac{f(x+h) - f(x)}{h}$$

Bibliography

1) A Brief History of Calculus –

https://www.wyzant.com/resources/lessons/math/calculus/introduction/history_of_calculus

About the Author

Dr. Marcel Sincraian has been working with numbers whole his life, for more than 30 years, as an Engineer, Accountant and Math teacher. While an Engineer, he got his Ph.D. in Civil Engineering. He participated in European Union funded engineering research projects in soil dynamics. As a Math teacher, he noticed that a lot of his students struggle with math and especially with Calculus. Because of his math passion, he decided to help the students with a book in Introduction to Calculus that is easy to use and challenges the students through testing themselves. He lives with his family in Burnaby, BC, Canada.

Author, Marcel Sincraian, Ph.D.

www.ingramcontent.com/pod-product-compliance
Lightning Source LLC
Chambersburg PA
CBHW061616210326
41520CB00041B/7458